Innovative Catalytic Materials for Environmental Remediation and Energy Applications

Innovative Catalytic Materials for Environmental Remediation and Energy Applications

Guest Editors

Georgios Bampos
Athanasia Petala
Zacharias Frontistis

MDPI

Basel • Beijing • Wuhan • Barcelona • Belgrade • Novi Sad • Cluj • Manchester

Guest Editors
Georgios Bampos
Department of Chemical Engineering
University of Patras
Patras
Greece

Athanasia Petala
Department of Environment
Ionian University
Zakynthos
Greece

Zacharias Frontistis
Department of Chemical Engineering
University of Western Macedonia
Kozani
Greece

Editorial Office
MDPI AG
Grosspeteranlage 5
4052 Basel, Switzerland

This is a reprint of the Special Issue, published open access by the journal *Catalysts* (ISSN 2073-4344), freely accessible at: https://www.mdpi.com/journal/catalysts/special_issues/catalytic_materials_for_environmental_remediation_and_energy_applications.

For citation purposes, cite each article independently as indicated on the article page online and as indicated below:

Lastname, A.A.; Lastname, B.B. Article Title. *Journal Name* **Year**, *Volume Number*, Page Range.

ISBN 978-3-7258-3944-5 (Hbk)
ISBN 978-3-7258-3943-8 (PDF)
https://doi.org/10.3390/books978-3-7258-3943-8

Contents

catalysts

MDPI

Editorial

Innovative Catalytic Materials for Environmental Remediation and Energy Applications

Georgios Bampos [1,*], Athanasia Petala [2,*] and Zacharias Frontistis [3,*]

1 Department of Chemical Engineering, University of Patras, GR-26504 Patras, Greece
2 Department of Environment, Ionian University, GR-29100 Zakynthos, Greece
3 Department of Chemical Engineering, University of Western Macedonia, GR-50132 Kozani, Greece
* Correspondence: geoba@chemeng.upatras.gr (G.B.); apetala@ionio.gr (A.P.); zfronistis@uowm.gr (Z.F.)

Citation: Bampos, G.; Petala, A.; Frontistis, Z. Innovative Catalytic Materials for Environmental Remediation and Energy Applications. *Catalysts* **2024**, *14*, 102. https://doi.org/10.3390/catal14020102

Received: 17 January 2024
Accepted: 24 January 2024
Published: 25 January 2024

1. Introduction

The need for low-cost and environmentally friendly energy is greater than ever nowadays due to the global population growth as well as the modern lifestyle. Considering this, research has focused on the development of novel cost-effective, eco-friendly technologies with high performance for power production and energy storage in addition to conventional technology. The reforming processes for hydrogen or syn-gas production [1,2], fuel cells producing electrical power [3–5] or chemical energy (H_2, CH_4, etc.) [6,7], various types of batteries [8,9] and supercapacitors [10–12] are listed among them. Environmental remediation technologies have also experienced tremendous growth over the last few decades in order to effectively remove organic contaminants from wastewater [13–15]. The activation of persulfate [16,17] and a Fenton-like process [18,19] are some indicatively advanced oxidation processes (AOPs) typically applied for the degradation of organic contaminants. The common ground for most of the abovementioned technologies, considering both energy production and storage and environmental management, is that they are based on catalytic processes. Therefore, the design and synthesis of innovative and economically affordable catalytic materials for various catalytic processes are of particular importance for their potential application. The present Special Issue includes ten research articles and one review article highlighting new perspectives on the design and implementation of innovative catalytic materials for a series of processes. The materials studied in the presented works are related to both catalytic and electrocatalytic processes, thus applying to a wide range of researchers.

2. Overview of Published Articles

Investigating the development of electrocatalytic materials, Zhu and Kamali (contribution 1) introduced the synthesis of the α-MoO_3 nanoribbon structure, which is incorporated into MoS_2 and graphene nanosheets derived from natural graphite and MoS_2 minerals. They implement a rapid preparation method in order to harvest its storage properties as potential electrode material for Li-ion batteries. The proposed preparation method resulted in the formation of a molybdenum oxycarbide layer in the electrode/electrolyte interface, which can facilitate the charge and ion transfer and significantly reduces the electrochemical resistance of the electrode, thus favoring the Li-ion storage capacity (773.5 mAh g^{-1}) compared to that obtained from the bare MoS_2 (176.8 mAh g^{-1}). Next, the unique significance of the hydrogen production from the acetic acid (AA) process is underscored by Alshammari et al. (contribution 2) since it provides clean hydrogen fuel compared to reforming processes while also contributing to agricultural waste management. For this purpose, they studied the hydrogen evolution reaction (HER) over glassy carbon decorated with Au (Au/GCE) using a neutral KCl electrolyte solution in the presence of AA in a three-compartment cell, implementing the rotating disk electrode (RDE) technique. The proton reduction was observed at a lower overpotential (−0.2 V vs. RHE), exhibiting higher current values in employed Au/GCE electrodes compared to the bare GCE electrode.

Kokkinou et al. investigated the electrochemical promotion of CO_2 hydrogenation over Rh-based catalysts deposited on an O^{2-}-conducting Y_2O_3-stabilized ZrO_2 (YSZ) solid electrolyte (contribution 3). This exceptionally detailed study is among the few works related to the CO_2 hydrogenation reaction for CH_4 production over Rh/YSZ at the laboratory scale in such narrow CO_2/H_2 ratios. It was found that applying a positive or negative potential or current to the electrode acted beneficially on the electrocatalytic performance towards the CH_4 production rate, thus confirming the electrochemical promotion of the catalysis (EPOC) theory.

The dry reforming of methane (DRM) towards syngas (CO and H_2) was studied over $Nd_{2-x}Sr_xNiO_{4\pm\delta}$ ($0 \leq x \leq 1.4$) solid solutions with a K_2NiF_4 structure synthesized via a freeze-drying method by Shlyakhtin et al. (contribution 4). Enhanced DRM activity was exhibited by samples with x = 0 and x > 1. These two peaks observed in the DRM performance were attributed to the contribution of Nd_2O_3 and $SrCO_3$, respectively. Hydrogen production was also investigated through the decomposition of H_2S, employing Co_3O_4/CeO_2 catalytic systems (contribution 5). Kraia et al. studied the hydrogen disulfide decomposition under water in an effort to simulate the concentration of H_2S in deep water layers of the Black Sea. The significance of this work is two-fold since it combines the production of chemical energy with the environmental management of the topical ecosystem, producing multiple benefits for coastal countries. The 30 wt.% Co_3O_4/CeO_2 catalyst exhibited the highest performance, which was attributed to the in situ sulfidation of cobalt species forming stable phases (Co_3S_4, CoS).

The purification of the H_2-rich gas stream obtained from the various reforming processes (mostly steam reforming of hydrocarbons) was conducted via a series of processes, including water gas shift (WGS) reaction, methanation reaction, CO oxidation, etc.. Kouroumlidis et al. investigated the WGS reaction at high (HT) and low temperatures (LT) over a series of modified particulates and structure Pt/TiO_2 catalytic systems (contribution 6). The modification of the TiO_2 support was performed with alkali- or alkaline earth metals (Na, Cs, Ca, Sr), whereas, in another series, the metal phase of Pt (0.5 wt.%) was enriched with another metal (namely Ru, Cr, Fe, Cu). Among all the tested samples, 0.5 wt.% Pt/TiO_2(Ca) exhibited the best performance and was tested further in the form of pellets and monoliths. Its high WGS activity was attributed to metal support interaction effects.

Innovative $La_{0.8}Sr_{0.2}Co_xFe_{1-x}O_{3-\delta}$ perovskite oxides, with a varying Co to Fe ratio (x = 0, 0.1, 0.2, 0.4, 0.6, 0.8, 0.9, 1), were synthesized in the form of powders via the in situ combustion synthesis method and were tested via CO and propane combustion for CO_2 production by Safakas et al. (contribution 7). Increasing the Co/Fe ratio resulted in an increase in the catalytic performance of both reactions, which was attributed to the occurrence of the interfacial redox mechanism. The latter was due to the enhancement of the number of catalytically active sites, strongly related to oxygen-vacant sites neighboring the Co sites, as well as to the lower redox stability of Co cations compared to Fe ones. CO oxidation was also studied over a series of transition metal oxides—this was performed on CuO-based catalytic systems, which were supported on an α-MnO_2 nanowire structure employing the precipitation synthesis method (contribution 8). Specifically, in this detailed work, Zhang et al. examined the effect on the catalytic performance of the type of transition metal oxide used (Fe_2O_3, Co_3O_4, NiO), the loading wt.% of the oxide and the calcination temperature. The highest performance in terms of CO conversion was found for the 3 wt.% Co_3O_4—10 wt.% CuO/α-MnO_2 catalytic system which was calcined at 400 °C. This enhanced catalytic activity was related to its high specific surface area, which resulted in an increase in oxygen vacancies, thus favoring the CO oxidation on the basis of the Mars–van Krevelen mechanism.

The beneficial effect of the combination of Ni-based catalytic systems with molten salt on the olive kernel–CO_2 gasification was thoroughly studied by Lampropoulos et al. (contribution 9). The gasification of biomass feedstock is a process of high significance since it enhances the total amount of renewable energy, whereas the exploitation of olive kernel as feedstock is a step towards the energy independence of Mediterranean countries. The presence of molten salt improved the mass and heat transport properties, thus favoring catalytic activity.

The production of α,β-unsaturated compounds via condensation between an aldehyde or a ketone with an active hydrogen compound (Knoevenagel reaction) employing a basic catalytic system is a very important process towards the synthesis of various chemical compounds. Chowdhury et al. investigated Knoevenagel condensation over a porous magnesium–aluminum mixed-metal phosphate, which was prepared via a hydrothermal method (contribution 10). The basic sites of the synthesized catalyst were found to enhance its efficiency towards the condensation reaction, whereas its remarkable recyclability highlighted its potential application as a catalytic system for similar reactions.

Syngas conversion reactions are very important for the production of various chemicals. The recent trends of the unique role of magnetic nanomaterials as catalytic systems for syngas production and conversion were discussed in the review article authored by Chidhambaram et al. (contribution 11). Various synthesis methods and experimental parameters are discussed in detail in this very informative review paper, highlighting the role of the applied magnetic field and the beneficial effect of the addition of various modifiers.

3. Conclusions

Considering the published works in the current Special Issue, it is concluded that various technologies based on catalytic processes were investigated. New insights concerning both novel synthesis methods and innovative materials are revealed, paving the way for further developing power-producing and storage technologies as well as purification processes.

Funding: This research received no external funding.

Acknowledgments: We are thankful to all the authors for submitting their impressive work in this Special Issue and the reviewers for their time and effort in reviewing the manuscripts.

Conflicts of Interest: The authors declare no conflicts of interest.

List of Contributions

1. Zhu, W.; Kamali, A.R. Molten Salt-Assisted Catalytic Preparation of MoS_2/α-MoO_3/Graphene as High-Performance Anode of Li-Ion Battery. *Catalysts* **2023**, *13*, 499.
2. Alshammari, B.H.; Begum, H.; Ibrahim, F.A.; Hamdy, M.S.; Oyshi, T.A.; Khatun, N.; Hasnat, M.A. Electrocatalytic Hydrogen Evolution Reaction from Acetic Acid over Gold Immobilized Glassy Carbon Surface. *Catalysts* **2023**, *13*, 744.
3. Kokkinou, N.; Xydas, F.; Brosda, S.; Kyriakou, G.; Katsaounis, A. Electrochemical Promotion of CO_2 Hydrogenation Using Rh Catalysts Supported on O^{2-} Conducting Solid Electrolyte. *Catalysts* **2023**, *13*, 1014.
4. Shlyakhtin, O.A.; Timofeev, G.M.; Malyshev, S.A.; Loktev, A.S.; Mazo, G.N.; Shatalova, T.; Arkhipova, V.; Roslyakov, I.V.; Dedov, A.G. $Nd_{2-x}Sr_xNiO_4$ Solid Solutions: Synthesis, Structure and Enhanced Catalytic Properties of Their Reduction Products in the Dry Reforming of Methane. *Catalysts* **2023**, *13*, 966.
5. Kraia, T.; Varvoutis, G.; Marnellos, G.E.; Konsolakis, M. Unveiling the Role of In Situ Sulfidation and H_2O Excess on H_2S Decomposition to Carbon-Free H_2 over Cobalt/Ceria Catalysts. *Catalysts* **2023**, *13*, 504.
6. Kouroumlidis, A.; Bampos, G.; Panagiotopoulou, P.; Kondarides, D.I. Performance of Particulate and Structured Pt/TiO_2-Based Catalysts for the WGS Reaction under Realistic High- and Low-Temperature Shift Conditions. *Catalysts* **2023**, *13*, 372.
7. Safakas, A.; Kournoutis, V.C.; Bampos, G.; Bebelis, S. CO and Propane Combustion on $La_{0.8}Sr_{0.2}Co_xFe_{1-x}O_{3-\delta}$ Perovskites: Effect of Fe-to-Co Ratio on Catalytic Activity. *Catalysts* **2023**, *13*, 1342.
8. Zhang, H.; Zhang, Y.; Song, H.; Cui, Y.; Xue, Y.; Wu, C.; Pan, C., Xu, J.; Qiu, J.; Xu, L.; et al. Transition Metal (Fe_2O_3, Co_3O_4 and NiO)-Promoted CuO-Based α-MnO_2 Nanowire Catalysts for Low-Temperature CO Oxidation. *Catalysts* **2023**, *13*, 588.

9. Lampropoulos, A.; Karakoulia, S.A.; Varvoutis, G.; Spyridakos, S.; Binas, V.; Zouridi, L.; Stefa, S.; Konsolakis, M.; Marnellos, G.E. The Combined Impact of Ni-Based Catalysts and a Binary Carbonate Salts Mixture on the CO_2 Gasification Performance of Olive Kernel Biomass Fuel. *Catalysts* **2023**, *13*, 596.
10. Chowdhury, A.; Bhattacharjee, S.; Chongdar, S.; Malakar, B.; Maity, A.; Bhaumik, A. A New Mixed-Metal Phosphate as an Efficient Heterogeneous Catalyst for Knoevenagel Condensation Reaction. *Catalysts* **2023**, *13*, 1053.
11. Chidhambaram, N.; Kay, S.J.; Priyadharshini, S.; Meenakshi, R.; Sakthivel, P.; Dhanbalan, S.; Shanavas, S.; Kamaraj, S.-K.; Thirumurugan, A. Magnetic Nanomaterials as Catalysts for Syngas Production and Conversion. *Catalysts* **2023**, *13*, 440.

References

1. Acar, C.; Dincer, I. Review and evaluation of hydrogen production options for better environment. *J. Clean. Prod.* **2019**, *218*, 835–849. [CrossRef]
2. Ogo, S.; Sekine, Y. Recent progress in ethanol steam reforming using non-noble transition metal catalysts: A review. *Fuel Process. Technol.* **2020**, *199*, 106238. [CrossRef]
3. Jiao, K.; Xuan, J.; Du, Q.; Bao, Z.; Xie, B.; Wang, B.; Zhao, Y.; Fan, L.; Wang, H.; Hou, Z.; et al. Designing the next generation of proton-exchange membrane fuel cells. *Nature* **2021**, *595*, 361–369. [CrossRef]
4. Hua, Z.; Zheng, Z.; Pahon, E.; Péra, M.-C.; Gao, F. A review on lifetime prediction of proton exchange membrane fuel cells system. *J. Power Sources* **2022**, *529*, 231256. [CrossRef]
5. Yaqoob, A.A.; Mohamad Ibrahim, M.N.; Rafatullah, M.; Chua, Y.S.; Ahmad, A.; Umar, K. Recent Advances in Anodes for Microbial Fuel Cells: An Overview. *Materials* **2020**, *13*, 2078. [CrossRef]
6. Ferreira, A.P.R.A.; Oliveira, R.C.P.; Mateus, M.M.; Santos, D.M.F. A Review of the Use of Electrolytic Cells for Energy and Environmental Applications. *Energies* **2023**, *16*, 1593. [CrossRef]
7. Seelajaroen, H.; Spiess, S.; Haberbauer, M.; Hassel, M.M.; Aljabour, A.; Thallner, S.; Guebitz, G.M.; Sariciftci, N.S. Enhanced methane producing microbial electrolysis cells for wastewater treatment using poly(neutral red) and chitosan modified electrodes. *Sustain. Energy Fuels* **2020**, *4*, 4238–4248. [CrossRef]
8. Nitta, N.; Wu, F.; Lee, J.T.; Yushin, G. Li-ion battery materials: Present and future. *Mater. Today* **2015**, *18*, 252–264. [CrossRef]
9. Eftekhari, A.; Fang, B. Electrochemical hydrogen storage: Opportunities for fuel storage, batteries, fuel cells, and supercapacitors. *Int. J. Hydrogen Energy* **2017**, *42*, 25143–25165. [CrossRef]
10. Wang, Y.; Song, Y.; Xia, Y. Electrochemical capacitors: Mechanism, materials, systems, characterization and applications. *Chem. Soc. Rev.* **2016**, *45*, 5925–5950. [CrossRef] [PubMed]
11. Chakraborty, S.; Mary, N.L. Review—An Overview on Supercapacitors and Its Applications. *J. Electrochem. Soc.* **2022**, *169*, 020552. [CrossRef]
12. Poonam; Sharma, K.; Arora, A.; Tripathi, S.K. Review of supercapacitors: Materials and devices. *J. Energy Storage* **2019**, *21*, 801–825. [CrossRef]
13. Taoufik, N.; Boumya, W.; Achak, M.; Sillanpää, M.; Barka, N. Comparative overview of advanced oxidation processes and biological approaches for the removal pharmaceuticals. *J. Environ. Manag.* **2021**, *288*, 112404. [CrossRef] [PubMed]
14. Giannakis, S.; Lin, K.-Y.A.; Ghanbari, F. A review of the recent advances on the treatment of industrial wastewaters by Sulfate Radical-based Advanced Oxidation Processes (SR-AOPs). *Chem. Eng. J.* **2021**, *406*, 127083. [CrossRef]
15. Trojanowicz, M.; Bojanowska-Czajka, A.; Bartosiewicz, I.; Kulisa, K. Advanced Oxidation/Reduction Processes treatment for aqueous perfluorooctanoate (PFOA) and perfluorooctanesulfonate (PFOS)–A review of recent advances. *Chem. Eng. J.* **2018**, *336*, 170–199. [CrossRef]
16. Wang, J.; Wang, S. Activation of persulfate (PS) and peroxymonosulfate (PMS) and application for the degradation of emerging contaminants. *Chem. Eng. J.* **2018**, *334*, 1502–1517. [CrossRef]
17. Matzek, L.W.; Carter, K.E. Activated persulfate for organic chemical degradation: A review. *Chemosphere* **2016**, *151*, 178–188. [CrossRef] [PubMed]
18. Fedorov, K.; Dinesh, K.; Sun, X.; Darvishi Cheshmeh Soltani, R.; Wang, Z.; Sonawane, S.; Boczkaj, G. Synergistic effects of hybrid advanced oxidation processes (AOPs) based on hydrodynamic cavitation phenomenon–A review. *Chem. Eng. J.* **2022**, *432*, 134191. [CrossRef]
19. Wu, K.; Si, X.; Jiang, J.; Si, Y.; Sun, K.; Yousaf, A. Enhanced degradation of sulfamethoxazole by Fe–Mn binary oxide synergetic mediated radical reactions. *Environ. Sci. Pollut. Res.* **2019**, *26*, 14350–14361. [CrossRef] [PubMed]

catalysts

MDPI

Article

CO and Propane Combustion on $La_{0.8}Sr_{0.2}Co_xFe_{1-x}O_{3-\delta}$ Perovskites: Effect of Fe-to-Co Ratio on Catalytic Activity

Alexandros Safakas, Vasileios Ch. Kournoutis, Georgios Bampos and Symeon Bebelis

Department of Chemical Engineering, University of Patras, Caratheodory 1, University Campus, GR-26504 Patras, Greece; asafakas@chemeng.upatras.gr (A.S.); vkournoutis@chemeng.upatras.gr (V.C.K.); geoba@chemeng.upatras.gr (G.B.)

* Correspondence: simeon@chemeng.upatras.gr; Tel.: +30-2610-969511

Abstract: Perovskites are promising alternative catalysts for oxidation reactions due to their lower cost compared to noble metals, and their greater thermal stability. The catalytic oxidation of CO is essential in order to control CO emissions in a series of applications whereas the catalytic combustion of propane is considered an economical and environmentally acceptable solution for energy production and gaseous pollutant management, since propane is among the organic compounds involved in photochemical reactions. This work concerns the effect of the Co/Fe ratio in the B-sites of a series of eight $La_{0.8}Sr_{0.2}Co_xFe_{1-x}O_{3-\delta}$ perovskites, with x ranging from 0 to 1, on the catalytic activity towards CO and C_3H_8 oxidation. The perovskite oxides were synthesized using the combustion synthesis method and characterized with respect to their specific surface areas, structures, and reduction properties. Increasing the Co/Fe ratio resulted in an increase in CO and propane conversion under both oxidative and stoichiometric conditions. The increase in Co content is considered to facilitate the formation of oxygen vacancies due to the lower redox stability of the cobalt cations compared to iron cations, favoring oxygen ion mobility and oxygen exchange between the gas phase and the oxide surface, thus enhancing the catalytic performance.

Keywords: CO combustion; propane combustion; CO oxidation; propane oxidation; perovskites; $La_{0.8}Sr_{0.2}Co_xFe_{1-x}O_{3-\delta}$; lanthanum strontium cobaltite ferrite

Citation: Safakas, A.; Kournoutis, V.C.; Bampos, G.; Bebelis, S. CO and Propane Combustion on $La_{0.8}Sr_{0.2}Co_xFe_{1-x}O_{3-\delta}$ Perovskites: Effect of Fe-to-Co Ratio on Catalytic Activity. *Catalysts* **2023**, *13*, 1342. https://doi.org/10.3390/catal13101342

Academic Editor: Patrick Da Costa

Received: 30 June 2023
Revised: 19 September 2023
Accepted: 28 September 2023
Published: 4 October 2023

1. Introduction

Increasing energy demands, mostly attributed to rising global population, result in an increase in the use of internal combustion engine vehicles that are considered as the main source of emissions of CO as well as other toxic compounds, such as NO_x, due to incomplete fossil fuel combustion [1]. These exhaust emissions are strongly related to a series of health issues and environmental problems [2,3]. The exhaust gas harmful components are reduced via oxidation in catalytic converters, using, typically, expensive noble metals as catalysts [4]. The oxidation of CO is described by the following equation [5]:

$$CO + \frac{1}{2}O_2 \rightarrow CO_2 \qquad \Delta H^o_{298\,K} = -283\ \text{kJ mol}^{-1} \tag{1}$$

So far, the use of hybrid technology vehicles combining internal combustion engines with electric motors powered by batteries, and of electrical vehicles, powered exclusively by batteries or fuel cells with zero pollutant emissions [6], has not managed to reduce CO emissions to the desired levels, since the vast majority of human population does not have access to these advanced technologies. Nonetheless, the use of the aforementioned environmentally friendly technologies includes processes where CO production is inevitable. An indicative example is the production of significant amounts of CO in the hydrocarbons reforming processes for H_2 production to feed H_2-fuelled proton exchange membrane fuel cells (PEMFCs) [7,8]. The reduction in CO concentration to the required level (<50 ppm) to avoid poisoning of the Pt-based electrocatalysts typically used in PEFMCs is achieved via the water–gas shift reaction, followed by preferential oxidation of CO in the presence of H_2 [9].

Considering the above, the catalytic CO oxidation is essential in both conventional and alternative power applications. Catalysts typically used for CO oxidation are noble metal-based materials [10], which are characterized by high cost and low resistance to high CO concentration [11]. Towards addressing these issues, the development of alternative catalytic materials with enhanced resistance to CO poisoning, low cost, and similar activity to that of noble metal-based catalysts has attracted significant research interest [11,12].

Perovskite oxides of the ABO_3 type, where A-sites are occupied by alkaline earth or alkali metal cations and B-sites by transition metal cations, are considered as alternative catalytic materials for CO oxidation, exhibiting good activity and resistance to poisoning [12–14]. The chemical composition and structure of perovskite oxides play a crucial role in their catalytic performance [12–16], as they affect their redox [17,18] and/or acid–base [19] properties. The catalytic activity of a perovskite depends primarily on the transition metals occupying the B-sites, whereas A-sites are considered to play a secondary role in catalytic performance [12,13,20]. Perovskitic oxides with La or other rare earth metals in the A-sites and Co, Fe, Ni, or Mn in the B-sites have been reported as the most active for CO oxidation [12,16]. In addition to parameters that typically affect catalytic activity, such as the nature of the metal cations and the catalytic surface area [20], their enhanced catalytic performance has been attributed to structural defects and rapid surface oxygen exchange [12].

CO oxidation, as well as other oxidation reactions, on perovskite oxides can proceed via two different catalytic mechanisms, namely, the "suprafacial" and "intrafacial" mechanisms [12,14–16], which can also operate concurrently [21]. According to the suprafacial mechanism, CO or an organic molecule is oxidized by oxygen adsorbed on the catalytic surface from the gas phase, whereas the perovskite surface acts as a fixed template providing atomic orbitals with suitable energy and symmetry to facilitate the adsorption and surface reaction of the reactants. A Langmuir–Hinshelwood or an Eley–Rideal kinetic model can be used in this case, depending on whether the oxidized reactant is adsorbed or reacts directly from the gas phase. According to the intrafacial mechanism, which resembles the Mars–van Krevelen surface redox model [22], the lattice oxygen of the perovskite oxide participates in the CO or organic molecule oxidation step, and the reaction is carried out via a redox cycle where the consumed lattice oxygen is afterwards replenished by incorporation in the resultant surface oxygen vacancies of oxygen from the gas phase in the rate determining step. Provided that they have sufficient mobility, oxygen anions migrating from the bulk can also fill in the surface oxygen vacancies, thus virtually participating in the oxidation process, contrary to the suprafacial mechanism where only surface oxygen species participate in the reaction [14,16]. In general, the suprafacial mechanism is expected for low-temperature processes in which removal/incorporation of oxygen from/into the perovskite lattice is much slower than adsorption/desorption and reactions between adsorbed surface species [14–16]. The intrafacial mechanism, on the other hand, is preferably followed at relatively higher temperatures, where the mobility of lattice oxygen and the surface oxygen exchange are favored [14–16].

Partial substitution of the metal cations at the A- and B-sites with metal cations of different oxidation numbers results in alteration of the oxygen sub-stoichiometry with parallel formation of vacant lattice sites or/and change in the oxidation state of the transition metal ions at the B-sites or/and formation of positively charged holes in order to maintain overall electro-neutrality, while the perovskitic structure remains stable [5,14]. This modification can enhance the catalytic performance of the perovskite oxide. For example, in $La_{1-x}Sr_xMnO_3$ perovskites, the La^{3+} cations are partially substituted by Sr^{+2} cations at the A-sites, causing a deviation from the electro-neutrality, which is counterbalanced by partial transformation of Mn^{3+} to Mn^{4+} at the B-sites [5]. On the other hand, in $La_{1-x}Sr_xFeO_3$ perovskites, partial substitution of lanthanum by strontium at the A sites is followed by the formation of oxygen vacancies as the main mechanism to maintain overall electro-neutrality [23,24]. In general, partial substitution of both A- and B-sites in a ABO_3 perovskite by an alkaline earth metal A′ and another transition metal B′, respectively, to form a AA′BB′O_3 type perovskite, results both in change in the oxidation state of the metal ions at the B-sites and in the formation of lattice oxygen vacancies, facilitating oxygen dissociation on the surface, increasing lattice oxygen anion mobility, and improving the ionic conductivity of the parent material [14,25].

Combining two catalytically active transition metals at the B-sites of a perovskitic oxide can result in enhanced catalytic activity owing to a synergistic effect between the two metals, which alters the adsorptive properties of the perovskite [5]. Co, Mn, and Fe cations are typically found at the B-sites of the perovskite catalysts [5]. It is generally accepted that the co-presence of cations of two different transition metals at the B-sites facilitates electron transfer in the redox catalytic cycle, which enhances the reducibility of the perovskite oxide and improves its catalytic performance [5]. Zhang et al. [26] used oxygen temperature-programmed desorption (O_2-TPD) to study the effect of partial substitution of La by Sr and Co by Fe at the A- and B-sites, respectively, on the desorption properties of the perovskite $La_{1-x}Sr_xCo_{1-y}Fe_yO_3$. They found that coexistence of Co and Fe was beneficial for the oxygen adsorption/desorption properties of the perovskite, thus enhancing its catalytic activity towards n-butane complete oxidation. Oxygen adsorption/desorption at temperatures lower than 300 °C was favored whereas partial substitution of La by Sr at the A-sites significantly enhanced the activity of the perovskite. Scott et al. [27] studied the effect of temperature, chemical composition, and oxygen partial pressure on oxygen sub-stoichiometry δ of $La_{1-x}Sr_xCo_{1-y}Fe_yO_{3-\delta}$. They found that δ increased with increasing temperature, decreasing oxygen partial pressure, and increasing Sr and Co content at the A- and B-sites, respectively. Lankhorst and ten Elshof [28] investigated the effect of partial substitution of Co by Fe at the B-sites of $La_{0.2}Sr_{0.8}Co_{1-y}Fe_yO_{3-\delta}$ (y = 0–0.6) on oxygen sub-stoichiometry δ and found that the δ value decreased with increasing Fe/Co ratio. This was attributed to an increase in the binding energy of O^{2-} in the perovskite lattice, which favors the formation of holes over oxygen vacancies [28].

Levasseur and Kaliaguine [29] studied the effect of the Co/Fe ratio at the B-sites of $La_{1-x}Ce_xCo_{1-y}Fe_yO_3$ (x = 1, 0.9 and y = 0, 0.2, 0.4, 1) on catalytic activity for methanol, CO, and CH_4 oxidation. Partial substitution of La by Ce at the A-sites caused an enhancement of the catalytic performance. Increasing Fe content resulted in a decrease in the catalytic activity and in a decrease in the β-oxygen peak in the O_2-TPD curves, accompanied by a shift in the desorption peaks of all oxygen species to higher temperatures. Moreover, decreasing the Co/Fe ratio resulted in an increase in the reduction temperature of Co^{3+} and Co^{2+} species to Co^{2+} and Co^0 species, respectively, as manifested by H_2 temperature-programmed reduction measurements. Therefore, it was concluded that an increase in the Fe content reduces the redox activity of the B-sites and decreases the lattice oxygen mobility. Similar results were reported by Tanaka et al. [30] concerning the catalytic activity towards the oxidation of a mixture of hydrocarbons, NO, and CO over $La_{0.9}Ce_{0.1}Co_{1-y}Fe_yO_3$ (y = 0, 0.2, 0.4, 0.6, 0.8, 1). Isupova et al. [31] observed a maximum CO oxidation activity of $La_{0.7}Sr_{0.3}Co_{1-y}Fe_yO_3$ (y = 0, 0.05, 0.1, 0.7) for a Co/Fe atomic ratio equal to 0.95/0.05. This behavior was attributed to an increase in the number of structural defects which in turn resulted in an increase in the density of surface defects, thus favoring CO oxidation reaction on the basis of its mechanism.

Catalytic combustion of propane, described by Equation (2), is also a reaction of significant research interest, both for energy production, especially in areas where there is no access to natural gas, and for management of air pollutants since propane belongs to volatile organic compounds (VOCs) [32].

$$C_3H_8 + 5O_2 \rightarrow 3CO_2 + 4H_2O(g) \quad \Delta H^o_{298\ K} = \quad 2044\ kJ\ mol^{-1} \quad (2)$$

Noble metal based catalysts are considered highly efficient towards propane combustion and have been extensively studied [33–38]. However, the high cost of noble metals has motivated the development of alternative noble metal-free catalytic systems. Ma et al. [39]

investigated propane oxidation on Co_3O_4 nanorods, as alternative to Pt/Al_2O_3. They highlighted the differences in propane combustion mechanism over the two catalysts, reporting a Langmuir–Hinshelwood mechanism for Co_3O_4 nanorods and an Eley–Rideal mechanism for Pt/Al_2O_3. Perovskites have also been used as catalysts for propane combustion, although the corresponding studies are significantly fewer than those concerning CO oxidation [14]. Klvana et al. [40] investigated the effect of the synthesis method of $La_{0.66}Sr_{0.34}Co_{0.2}Fe_{0.8}O_3$ and specific surface area on its activity for propane combustion. Kinetic experiments over a wide range of gas phase compositions indicated a Mars–van Krevelen mechanism [22,40,41]. Alifanti et al. [42] compared methane and propane oxidation on $La_{1-x}Sr_xM_{1-y}M'_yO_{3-\delta}$ (M: Mn, Cu, Ni, Co and M′: Ni, Fe), focusing on the relative role of specific surface area and oxygen mobility. The effect of partial substitution of La by Sr, Ce, and Ca at the A-sites has been studied in propane oxidation on $La_{1-x}Sr_xCoO_3$ [43], on $La_{1-x}Sr_xFeO_3$, $La_{1-x}Ce_xFeO_3$ and $La_{1-x}Ce_xCoO_3$ [44], and on $La_{1-x}Ca_xCoO_3$ [25]. Merino et al. [45] studied the effect of the partial substitution of Co by Fe at the B-sites in propane combustion on $LaCo_{1-y}Fe_yO_{3\pm\lambda}$ (y = 0, 0.1, 0.3, 0.5). Partial substitution of Co by Fe by 10% (y = 0.1) was found to favor the catalytic performance, whereas a further increase in the Fe content resulted in a gradual reduction in the catalytic activity [45]. Song et al. [46] investigated the kinetics of propane oxidation on $La_{0.66}Sr_{0.34}Ni_{0.3}Co_{0.7}O_3$ and proposed an extended Mars–van Krevelen kinetic model, taking also into account the inhibiting effect of adsorbed CO_2 and H_2O products.

Considering the abovementioned studies of CO and propane combustion on perovskite oxides, it can be concluded that relevant research interest has been mainly focused on the investigation of the catalytic effect of the transition metals occupying the B-sites, with particular interest in Co and Fe. Along this direction, eight (8) perovskite oxides $La_{0.8}Sr_{0.2}Co_xFe_{1-x}O_{3-\delta}$, with a fixed La/Sr ratio at the A-sites and a varying Co and Fe content at the B-sites (x = 0, 0.1, 0.2, 0.4, 0.6, 0.8, 0.9, 1) were synthesized via the combustion synthesis method and tested for their catalytic activity towards CO and propane combustion. The synthesized oxides were characterized as it concerns their specific surface area, structure, and reducibility in H_2 atmosphere. The observed differences in catalytic activity among the tested perovskites were mainly correlated with differences in oxygen vacancies, surface oxygen exchange, and oxygen ion mobility induced by changing the relative content of Co and Fe at the B-sites. The novelty of the present work lies on the fact that, to the best of our knowledge, it is the first systematic study of the effect of the Co/Fe ratio on the catalytic performance of La-Sr-Co-Fe perovskites for both CO and propane total oxidation, two reactions of significant environmental and technological interest, allowing comparison of the activity trends for the two reactions.

2. Results

2.1. Physicochemical and Structural Characterization

The composition, the notation, and the measured via N_2 physical adsorption (BET method) SSA values of the synthesized perovskites are shown in Table 1. No significant differences were observed between their nominal composition and their stoichiometry determined via inductively coupled plasma optical emission spectroscopy (ICP-OES). The SSA varied from 4.1 $m^2\ g^{-1}$ for LSCF_8291 to 7.9 $m^2\ g^{-1}$ for LSCF_8219. The low SSAs are attributed to the elevated calcination temperature (900 °C) used to obtain the perovskite structure [47].

The XRD spectra of the eight synthesized perovskites are presented in Figure 1. Six of them (LSC, LSCF_8282, LSCF_8264, LSCF_8246, LSCF_8228, and LSF) have been already presented in a previous work of our group [47]. After identification of the formed phases using ICDD data files, the formation of the perovskite phase was confirmed (main peak at 2θ between 32° and 34°) without the presence of secondary phases at detectable levels. A gradual shift in all XRD peaks towards lower 2θ values was observed with increasing partial substitution of Co by Fe (Figure 1b). This shift was accompanied by a change in the main peak from doublet ($0.4 \leq x \leq 1$) to single (x = 0, 0.1, 0.2), indicating a change in the $La_{0.8}Sr_{0.2}Co_xFe_{1-x}O_{3-\delta}$ perovskite structure from rhombohedral to orthorhombic [25,48,49].

This is in accordance with the results of Tai et al. [49] for $La_{0.8}Sr_{0.2}Co_{1-y}Fe_yO_3$ ($0 \leq y \leq 1$) sintered in air at 1250 to 1350 °C, who reported that partial substitution of Co by Fe to an iron percentage of 80% or higher (y = 0.8, 0.9, 1) resulted in a change in the perovskite unit cell from rhombohedral to orthorhombic. Similarly, Natile et al. [48] reported a change from rhombohedral to orthorhombic structure for $La_{0.6}Sr_{0.4}Co_{1-y}Fe_yO_{3-\delta}$ perovskites (y = 0.2, 0.5, 0.8) calcined at 800 °C, but for a lower Fe content ($y \geq 0.5$), which implies that the La/Sr ratio at the A-sites and, possibly, the thermal treatment conditions can affect the Co/Fe ratio for which the aforementioned change in unit cell geometry occurs. The observed gradual shift in the XRD peaks towards higher 2θ values with increasing partial substitution of Fe by Co (Figure 1) indicates a contraction of the perovskite unit cell, which can be attributed to the substitution of Fe^{3+} by Co^{3+}, the latter having a smaller ionic radius [45,48,50]. Moreover, this shift becomes relatively smaller with increasing Co content (Figure 1b), which implies a smaller reduction in the unit cell size [47]. This behavior has also been reported for $LaCo_xFe_{1-x}O_{3-\delta}$, associated with changes in the relative number of Co^{2+} to Co^{3+} and Fe^{4+} to Fe^{3+} accompanying the substitution of Fe^{3+} by Co^{3+} [50].

Table 1. Composition, notation, specific surface area (SSA), and average crystallite size of the synthesized perovskites $La_{0.8}Sr_{0.2}Co_xFe_{1-x}O_{3-\delta}$.

Perovskite	Notation	SSA ($m^2\ g^{-1}$)	Crystallite Size (nm)
$La_{0.8}Sr_{0.2}FeO_{3-\delta}$	LSF	7.2 [1]	29.1
$La_{0.8}Sr_{0.2}Co_{0.1}Fe_{0.9}O_{3-\delta}$	LSCF_8219	7.9	19.7
$La_{0.8}Sr_{0.2}Co_{0.2}Fe_{0.8}O_{3-\delta}$	LSCF_8228	5.6 [1]	16.9
$La_{0.8}Sr_{0.2}Co_{0.4}Fe_{0.6}O_{3-\delta}$	LSCF_8246	4.7 [1]	50.2
$La_{0.8}Sr_{0.2}Co_{0.6}Fe_{0.4}O_{3-\delta}$	LSCF_8264	6.8 [1]	32.5
$La_{0.8}Sr_{0.2}Co_{0.8}Fe_{0.2}O_{3-\delta}$	LSCF_8282	5.5 [1]	32.5
$La_{0.8}Sr_{0.2}Co_{0.9}Fe_{0.1}O_{3-\delta}$	LSCF_8291	4.1	60.2
$La_{0.8}Sr_{0.2}CoO_{3-\delta}$	LSC	5.7 [1]	34.2

[1] ref. [47].

The average crystallite size, estimated via application of the Scherrer equation [51,52] for the XRD peak located at 2θ between 46° and 48°, revealed a significant difference among the synthesized materials (Table 1). The largest crystallite size, equal to 60.2 nm, was calculated for LSCF_8291, whereas the smallest, equal to 16.9 nm, was calculated for LSCF_8228. Although no clear trend was observed, it could be argued that increasing the Fe content results in an increase in the SSA and a decrease in the average crystallite size, as also reported for $La_{0.6}Sr_{0.4}Co_{1-y}Fe_yO_{3-\delta}$ by Natile et al. [48]. The observed changes in the crystallite size were accompanied by changes in the lattice strain, ε (%), which was calculated using the Williamson–Hall method [52,53]. Specifically, the calculated lattice strain values for the synthesized perovskites follow the increasing order, LSCF_8291 (0.14%) < LSCF_8246 (0.15%) < LSC (0.24%) < LSCF_8282 (0.25%) < LSCF_8264 (0.26%) < LSF (0.28%) < LSCF_8219 (0.38%) < LSCF_8228 (0.4%), which corresponds to a decreasing order of crystallite size (Table 1).

2.2. *Investigation of the Perovskite Reducibility*

The synthesized perovskites were characterized with respect to their reducibility using H_2 temperature-programmed reduction (H_2-TPR). The recorded mass spectrometer signal (ms signal) for H_2 ($m/z = 2$) as a function of temperature (TPR profile) is shown in Figure 2 for all tested samples. The main feature of the TPR curves is a broad inverse peak that starts to develop at temperatures between ca. 200 °C and 250 °C depending on the perovskite composition and which seems to consist of more than one peak, clearly for Co/Fe ratio in the range 6/4 to 1/9. For Co/Fe ratios larger than 6/4, additional reduction features appear at temperatures higher than ca. 595 °C that were not completed due to limitations of the experimental setup, with the exception of a peak located between ca. 595 °C and 685 °C in the TPR profile of the LSC oxide.

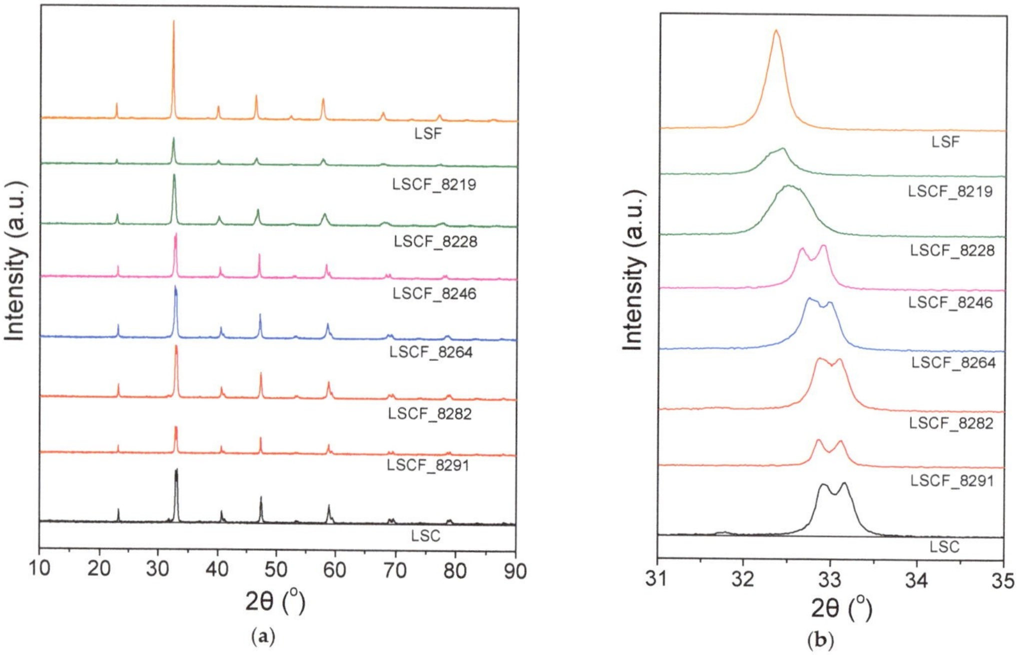

Figure 1. (**a**) XRD patterns of the synthesized perovskite oxides. (**b**) Detailed view of the XRD patterns in the 2θ range from 31° to 35°.

Interpretation of the H_2-TPR features of LSCF perovskite oxides depends on the temperature range of their appearance. The reduction of Fe is typically performed in two consecutive steps, i.e., reduction of Fe^{4+} to Fe^{3+} at 350 to 450 °C and subsequent reduction of Fe^{3+} to Fe^{2+} and possibly further reduction to metallic Fe^0, at temperatures higher than 500 °C [54]. The reduction of cobalt cations is also carried out in two steps, Co^{3+} being reduced to Co^{2+} and Co^{2+} to Co^0 at temperatures lower and higher than 500 °C, respectively [54]. However, according to García-López et al. [55], the reduction of Fe^{3+} to Fe^{2+} can take place at temperatures lower than 400 °C or in the temperature range where reduction of Co^{3+} occurs, resulting in overlapping of the corresponding TPR peaks. Merino et al. [45] also observed, in TPR profiles of $LaCo_{1-y}Fe_yO_3$ oxides, the appearance of a peak in the temperature range 355–385 °C, which they attributed to the reduction of the Fe^{4+} species.

In the present work, broad reduction peaks were recorded for all tested perovskites that could be explained by the potential participation of the bulk lattice oxygen in the reduction reactions, since the surface oxygen is consumed rapidly with increasing temperature. As shown in Figure 2, the increase in the Co/Fe ratio (from LSF to LSC) was accompanied by a shift in the onset reduction temperature to lower values. This can be attributed to the facilitation of the surface lattice oxygen release, associated with the lower redox stability of cobalt cations in LSCF compared to iron cations [56,57]. Moreover, the parallel increase in the number of oxygen ion vacancies and oxygen ion mobility [12,14,58] facilitates the bulk lattice oxygen diffusion to the oxide surface, thus favoring the reaction with hydrogen. The appearance of a doublet peak for LSCF_8219 to LSCF_8264 can be associated with the onset of the reduction of Co^{3+} to Co^{2+}, occurring parallel to the reduction of Fe^{4+} to Fe^{3+} [45]. For LSCF_8282 and LSCF_8291, the peak associated with the reduction of Fe^{4+} to Fe^{3+} is not discernible, as it is presumably overlapped by the dominant peak associated with the reduction of Co^{3+} to Co^{2+} [45].

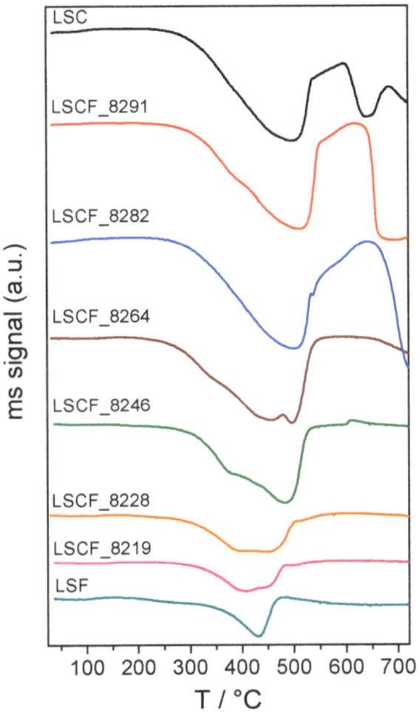

Figure 2. H_2 temperature-programmed reduction profiles of the synthesized perovskites.

2.3. Catalytic Activity for CO Combustion

In Figure 3a, the CO conversion (X_{CO}) vs. temperature curves obtained under oxidative conditions (feed composition: 1% CO/5% O_2/balance He) for the synthesized perovskites are presented. As shown in the figure, complete CO conversion was achieved for all oxides in the temperature range between 265 °C and 460 °C, whereas at each temperature conversion increased with increasing Co/Fe ratio at the B-sites. In Figure 3b, the light-off temperatures T_{50} for the tested materials, i.e., the temperatures corresponding to a CO conversion equal to 50%, are compared. Increasing the Co/Fe ratio resulted in a gradual decrease in T_{50}, which was more pronounced for Co/Fe ratios equal or smaller than 6/4. As shown in the figure, the T_{50} value was practically the same for the LSCF_8264, LSCF_8282, and LSCF_8291 perovskites, which is also evident by comparing the corresponding CO conversion curves (Figure 3a).

Figure 3c shows, in the form of Arrhenius plots, the temperature dependence of the CO_2 production rate r_{CO_2} for the tested perovskites, under oxidative conditions. The CO_2 production rate was measured under practically differential conditions and was normalized to the SSA of each sample to take into account the existing differences in the specific surface area of the tested catalysts (Table 1) and compare their intrinsic activity, assuming that the change in the number of active sites follows the change in SSA. By increasing the Co/Fe ratio, the same r_{CO_2} was achieved at lower temperatures, which implies enhancement of the intrinsic catalytic activity. In order to highlight the differences in the intrinsic catalytic activity of the tested perovskites, the temperatures corresponding to r_{CO_2} equal to 10^{-7} mol s^{-1} m^{-2} (Figure 3c) are compared in Figure 3d. As shown in the figure, this temperature decreased from 283 °C for LSF to 116 °C for LSC, indicating that among the tested perovskites (Table 1) LSC and LSF were the most active and the less active, respectively, for CO oxidation in terms of intrinsic activity, in agreement with the conclusions drawn by comparing the T_{50} values (Figure 3b).

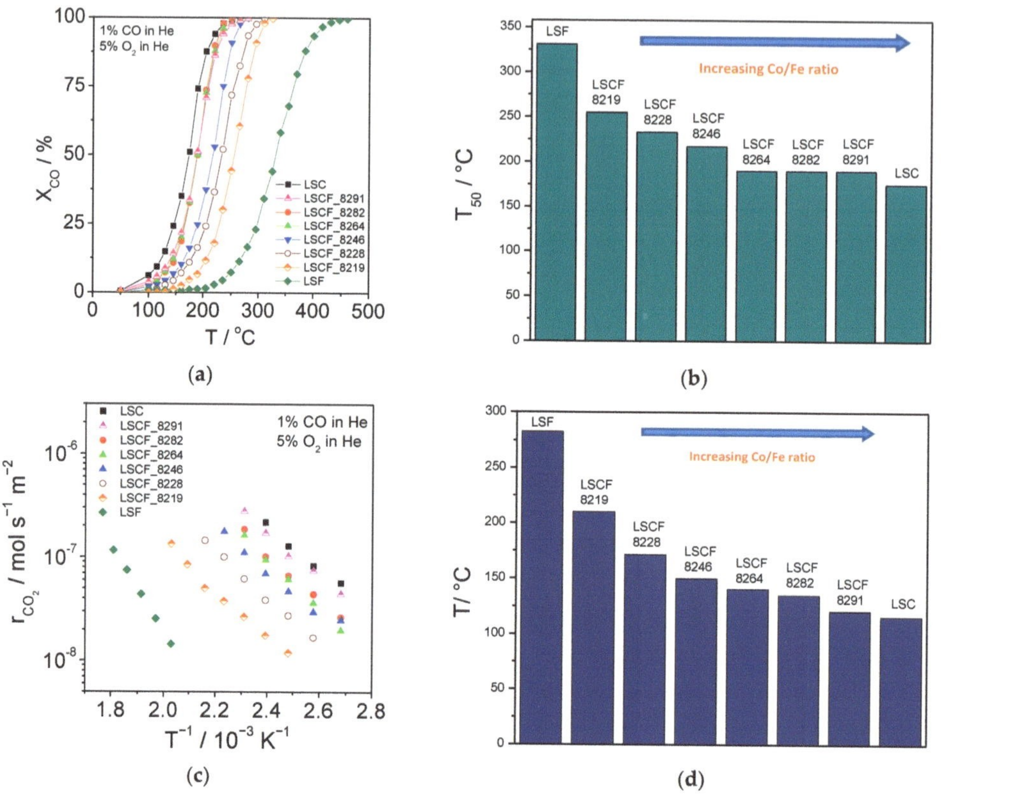

Figure 3. CO combustion on $La_{0.8}Sr_{0.2}Co_xFe_{1-x}O_{3-\delta}$ perovskites for oxidative feed conditions (1%CO/5%O_2/balance He): (**a**) CO conversion, X_{CO}, as a function of temperature, and (**b**) comparison of the corresponding light-off temperatures, T_{50}. (**c**) Temperature dependence of the CO_2 production rate, r_{CO_2} (Arrhenius plots), measured under practically differential conditions, and (**d**) comparison of the temperatures corresponding to r_{CO_2} equal to 10^{-7} mol s^{-1} m^{-2}.

In Figure 4, the catalytic performance of the synthesized $La_{0.8}Sr_{0.2}Co_xFe_{1-x}O_{3-\delta}$ perovskites for CO combustion under stoichiometric feed conditions (1% CO/0.5% O_2/balance He) is compared. Figure 4a shows the dependence of CO conversion (X_{CO}) on temperature. As shown in the figure, complete conversion of CO is reached for temperatures ranging between ca. 430 °C and 640 °C, which are considerably higher than the corresponding ones for oxidative feed conditions (Figure 3a). This difference implies a positive reaction order with respect to oxygen, considering the much lower percentage of O_2 in the feed under stoichiometric conditions (0.5%) compared to that under oxidative conditions (5%). As in the latter case, decreasing the Fe content at the B-sites of the LSCF perovskites resulted in an increase in the CO conversion for the same temperature for Co/Fe $\leq 6/4$, whereas for higher Co/Fe ratios, the differences in CO conversion were small (Figure 4a). Similarly, as shown in Figure 4b, T_{50} decreased from ca. 325 °C for the LSF to ca. 200 °C for the LSCF_8264 and remained practically constant by further decreasing the Fe content (with parallel increase in the Co content) to zero (LSC sample).

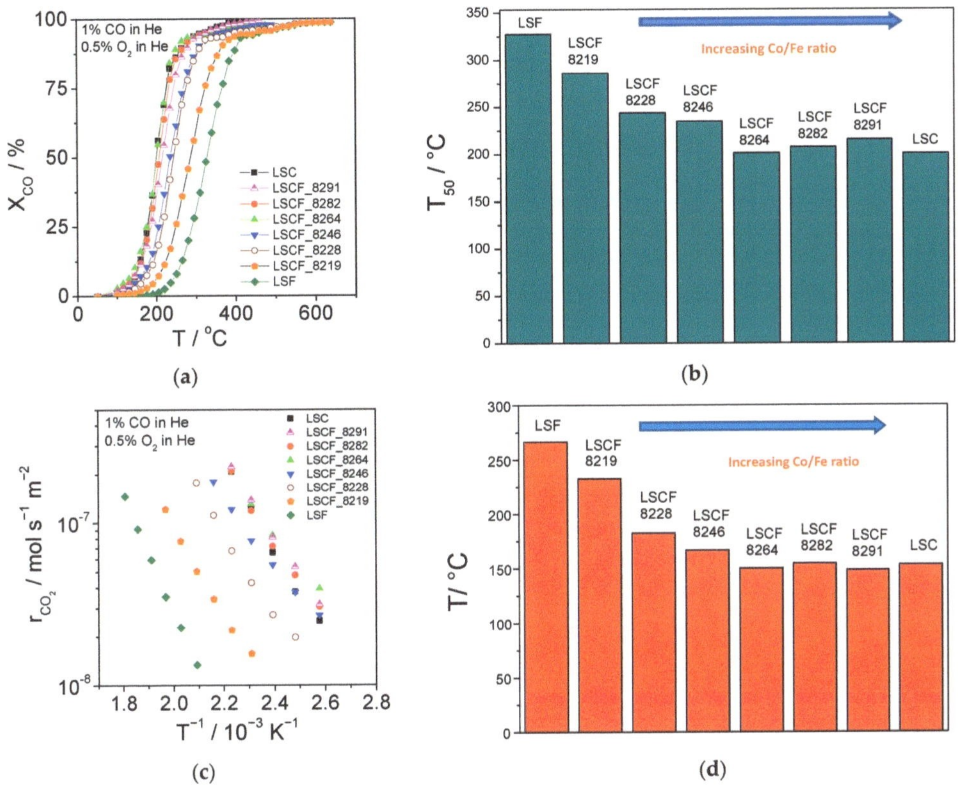

Figure 4. CO combustion on $La_{0.8}Sr_{0.2}Co_xFe_{1-x}O_{3-\delta}$ perovskites for stoichiometric feed conditions (1% CO/0.5%O_2/balance He): (**a**) CO conversion, X_{CO}, as a function of temperature, and (**b**) comparison of the corresponding light-off temperatures, T_{50}. (**c**) Temperature dependence of the CO_2 production rate, r_{CO_2} (Arrhenius plots), measured under practically differential conditions, and (**d**) comparison of the temperatures corresponding to r_{CO_2} equal to 10^{-7} mol s^{-1} m^{-2}.

Figure 4c shows, in the form of Arrhenius plots, the temperature dependence of r_{CO_2} for the tested perovskites, as measured under practically differential conditions and for a stoichiometric CO/O_2 ratio in the feed (1% CO/0.5% O_2/balance He). By increasing the Co/Fe ratio up to 6/4, the same r_{CO_2} was obtained at lower temperatures, indicating an increase in the intrinsic catalytic activity, whereas for higher Co/Fe ratios, the effect was not significant (Figure 4c). This is clearly shown in Figure 4d, where the temperatures corresponding to r_{CO_2} equal to 10^{-7} mol s^{-1} m^{-2} are presented. As shown in the figure, this rate value was obtained at 267 °C for LSF and at 150 °C for LSCF_8264, whereas for the perovskite oxides with higher Co/Fe ratios, the corresponding temperature remained practically constant, varying between 150 and 154 °C (Figure 4d).

In Figure 5, the catalytic performance for CO combustion of the synthesized $La_{0.8}Sr_{0.2}Co_xFe_{1-x}O_{3-\delta}$ perovskites under oxidative and stoichiometric conditions is compared on the basis of the temperature required to obtain a r_{CO_2} equal to 10^{-7} mol s^{-1} m^{-2} (Figure 5a) and on the basis of the light-off temperature, T_{50} (Figure 5b). As shown in Figure 5a, the temperature corresponding to a CO_2 production rate of 10^{-7} mol s^{-1} m^{-2} decreased with increasing Co/Fe ratio under oxidative feed conditions over the entire range of Co/Fe ratios (LSF to LSC), whereas under stoichiometric feed conditions it decreased up to a Co/Fe ratio equal to 6/4, remaining practically the same for higher Co content. The same trend was observed concerning the T_{50} values (Figure 5b). Moreover, with the exception of the less active LSF oxide, both the temperature corresponding to r_{CO_2} equal to 10^{-7} mol s^{-1} m^{-2} and T_{50} for each of the tested perovskites were lower under oxidative conditions compared to stoichiometric conditions, implying a higher catalytic performance in the former case. Considering the above observations, it can be concluded that the catalytic activity for CO oxidation, under both oxidative and stoichiometric feed conditions, depends strongly on the relative content of Co and Fe at the B-sites of the $La_{0.8}Sr_{0.2}Co_xFe_{1-x}O_{3-\delta}$ perovskite oxides. Increasing the Co/Fe ratio results in an increase in the catalytic activity for CO combustion, more pronounced under oxidative reaction conditions. The lowest and highest activities among the tested materials were exhibited by the LSF and the LSC perovskites, respectively.

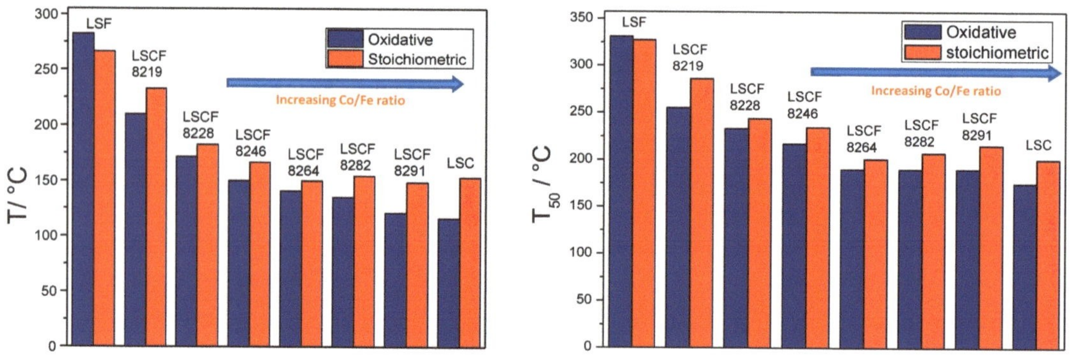

Figure 5. CO combustion on $La_{0.8}Sr_{0.2}Co_xFe_{1-x}O_{3-\delta}$ perovskites under oxidative (1% CO/5% O_2/balance He) and stoichiometric (1% CO/0.5%O_2/balance He) feed conditions: (**a**) Temperatures corresponding to a CO_2 production rate of 10^{-7} mol s^{-1} m^{-2}, measured under practically differential conditions and (**b**) light-off temperatures, T_{50}.

Kinetic measurements with varying O_2 partial pressures were conducted for the most active LSC catalyst, under practically differential conditions. Figure 6a shows, in the form of Arrhenius plot, the temperature dependence of the CO_2 production rate (r_{CO_2}), normalized to the SSA, for four different O_2 partial pressures P_{O_2} (0.5, 1, 2 and 5 kPa) and a constant CO partial pressure, P_{CO}, equal to 1 kPa. As expected, for all oxygen partial pressures, the CO_2 production rate increased exponentially with increasing temperature, whereas the apparent activation energy that was determined from the slopes of the Arrhenius plots ranged between 9.8 kcal mol^{-1} for P_{O_2}= 0.5 kPa and 13.1 kcal mol^{-1} for P_{O_2} = 5 kPa. Moreover, at each temperature, r_{CO_2} was found to increase with increasing O_2 partial pressure, which implies a positive order of the reaction with respect to O_2, in agreement with the higher activity observed under oxidative conditions compared to stoichiometric conditions (Figure 5). The latter behavior was observed for all tested perovskites; thus, it can be assumed that CO combustion is positive order with respect to O_2 on all of them.

The positive order of CO combustion with respect to O_2 is more clearly shown in Figure 6b, where is presented in the form of logarithmic plot the dependence of the CO_2 production rate on O_2 partial pressure at three different temperatures (115, 130, and 145 °C), for constant CO partial pressure equal to 1 kPa. The slopes of the plots for the three temperatures do not differ significantly, corresponding to an apparent reaction order in oxygen of ca. 0.5.

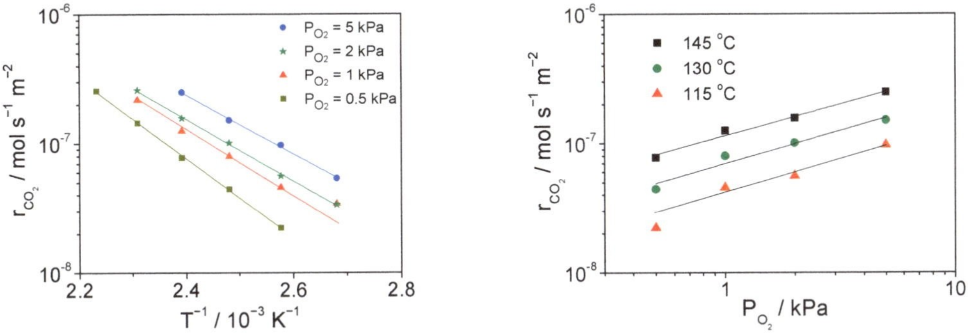

(a) (b)

Figure 6. CO combustion on LSC. (**a**) Temperature dependence of the CO_2 production rate (Arrhenius plots) for four different O_2 partial pressures (P_{O_2}: 0.5, 1, 2 and 5 kPa) and P_{CO} = 1 kPa. (**b**) Effect of P_{O_2} on the CO_2 production rate, at constant P_{CO} = 1 kPa, for three different temperatures (115, 130, and 145 °C).

2.4. Catalytic Activity for Propane Combustion

The synthesized $La_{0.8}Sr_{0.2}Co_xFe_{1-x}O_{3-\delta}$ perovskites were also tested as catalysts for complete oxidation of propane under oxidative and stoichiometric conditions. Figure 7a shows, for all synthesized perovskites, the dependence of propane conversion ($X_{C_3H_8}$) on temperature, under oxidative feed conditions (0.5% C_3H_8/5% O_2/balance He). For all tested perovskite oxides, complete conversion of propane was achieved at temperatures ranging between 520 °C and 670 °C, which are higher than the corresponding ones for CO oxidation (Figure 3a). For the same temperature, increasing the Co/Fe ratio was accompanied by an increase in propane conversion, more pronounced for Co/Fe ratios up to 6/4 (Figure 7a). Equivalently, the light-off temperature (T_{50}) decreased with increasing Co/Fe ratio. Considering the above, it is concluded that the dependence of catalytic activity of $La_{0.8}Sr_{0.2}Co_xFe_{1-x}O_{3-\delta}$ perovskites on the Co/Fe ratio is similar for both CO and propane combustion, i.e., the activity is reduced with increasing Fe and decreasing Co content at the B-sites. The higher temperatures required for complete conversion of propane compared to CO (Figures 3a and 7a), under oxidative feed conditions, are expected since propane is a more complex molecule compared to CO.

Figure 7b,c present for all tested perovskites, in the form of Arrhenius plots, the temperature dependence of the CO_2 production rate (r_{CO_2}) in propane combustion under oxidative (0.5% C_3H_8/5% O_2/balance He) and stoichiometric (0.5% C_3H_8/2.5% O_2/balance He) feed conditions, respectively, with the rate being determined under practically differential conditions and normalized to SSA. As shown in Figure 7b, for oxidative conditions, with increasing Co content the same r_{CO_2} was obtained at lower temperature, more clearly for Co/Fe $\leq$ 6/4, which implies an increase in the catalytic activity for propane combustion. The temperature corresponding to a CO_2 production rate of 10^{-7} mol s^{-1} m^{-2} was equal to 345 °C and 244 °C for the less active (LSF) and the most active (LSC) perovskite catalysts, respectively. As shown in Figure 7c, practically the same behavior was observed for propane combustion under stoichiometric conditions, i.e., the same r_{CO_2} value was achieved at lower temperatures with increasing Co/Fe ratio, the effect being clearer for Co/Fe ratios up to 6/4. In this case, the temperature corresponding to a CO_2 production rate of 10^{-7} mol s^{-1} m^{-2} was 355 °C for the LSF and 255 °C for the LSC perovskites (Figure 7c). Also, the comparison of the data shown in Figure 7b,c reveals that the temperatures corresponding to a specific r_{CO_2} value, for example, 10^{-7} mol s^{-1} m^{-2}, were lower under oxidative conditions compared to stoichiometric conditions for all the tested perovskites, which, similar to CO combustion, indicates a higher catalytic activity under oxidative conditions.

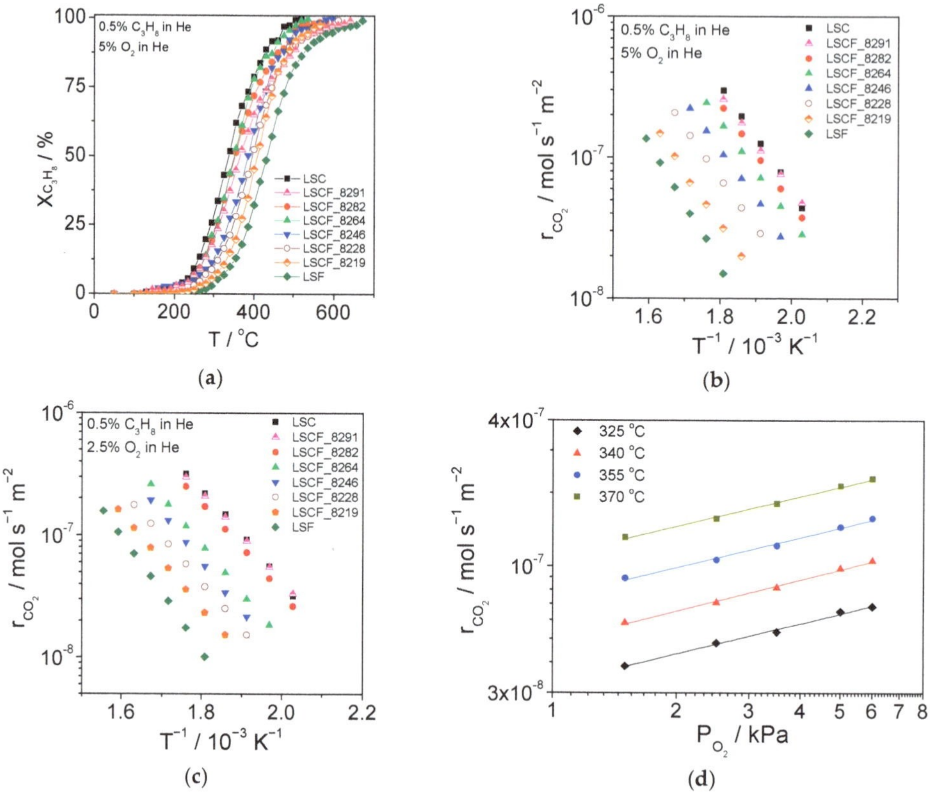

Figure 7. Propane combustion on $La_{0.8}Sr_{0.2}Co_xFe_{1-x}O_{3-\delta}$ perovskites: (**a**) C_3H_8 conversion as a function of temperature under oxidative feed conditions (0.5% C_3H_8/5%O_2/He), (**b**) temperature dependence of the CO_2 production rate under oxidative feed conditions (0.5% C_3H_8/5% O_2/balance He), (**c**) temperature dependence of the CO_2 production rate under stoichiometric feed conditions (0.5% C_3H_8/2.5% O_2/balance He), and (**d**) effect of oxygen partial pressure (P_{O_2}) on the CO_2 production rate, for the LSC catalyst, at constant propane partial pressure $P_{C_3H_8}$ = 0.5 kPa and for four different temperatures (325 °C, 340 °C, 355 °C, and 370 °C).

Kinetic measurements were performed for propane combustion on the most active LSC perovskite. Figure 7d shows, in the form of logarithmic plots, the dependence of the CO_2 production rate on O_2 partial pressure at four different temperatures (325, 340, 355, and 370 °C), for constant propane partial pressure equal to 0.5 kPa. A positive order with respect to O_2 is observed, practically the same for all tested temperatures. This can explain the higher activity for propane combustion observed for LSC under oxidative conditions compared to stoichiometric conditions. As the same behavior was observed for the other tested perovskites, it can be assumed that also for them propane combustion is positive order with respect to O_2 (close to 0.5) in agreement with the observed higher activity under oxidative conditions.

3. Discussion

Oxygen adsorption on perovskites is a complex process, considered to occur mainly at the vicinity of B-sites of perovskite catalysts [13,59,60]. Adler et al. [61] assumed that two neighboring vacant surface lattice sites, similar to those in the bulk of the oxide, are involved in the dissociative adsorption of oxygen and surface oxygen exchange. Royer et al. in their review work [16] highlight the significant role of the oxidation state of the B-site

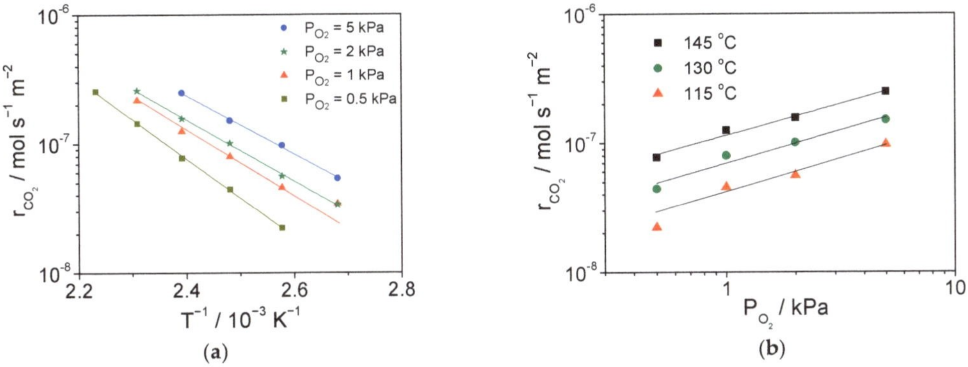

Figure 6. CO combustion on LSC. (**a**) Temperature dependence of the CO_2 production rate (Arrhenius plots) for four different O_2 partial pressures (P_{O_2}: 0.5, 1, 2 and 5 kPa) and P_{CO} = 1 kPa. (**b**) Effect of P_{O_2} on the CO_2 production rate, at constant P_{CO} = 1 kPa, for three different temperatures (115, 130, and 145 °C).

2.4. Catalytic Activity for Propane Combustion

The synthesized $La_{0.8}Sr_{0.2}Co_xFe_{1-x}O_{3-\delta}$ perovskites were also tested as catalysts for complete oxidation of propane under oxidative and stoichiometric conditions. Figure 7a shows, for all synthesized perovskites, the dependence of propane conversion ($X_{C_3H_8}$) on temperature, under oxidative feed conditions (0.5% C_3H_8/5% O_2/balance He). For all tested perovskite oxides, complete conversion of propane was achieved at temperatures ranging between 520 °C and 670 °C, which are higher than the corresponding ones for CO oxidation (Figure 3a). For the same temperature, increasing the Co/Fe ratio was accompanied by an increase in propane conversion, more pronounced for Co/Fe ratios up to 6/4 (Figure 7a). Equivalently, the light-off temperature (T_{50}) decreased with increasing Co/Fe ratio. Considering the above, it is concluded that the dependence of catalytic activity of $La_{0.8}Sr_{0.2}Co_xFe_{1-x}O_{3-\delta}$ perovskites on the Co/Fe ratio is similar for both CO and propane combustion, i.e., the activity is reduced with increasing Fe and decreasing Co content at the B-sites. The higher temperatures required for complete conversion of propane compared to CO (Figures 3a and 7a), under oxidative feed conditions, are expected since propane is a more complex molecule compared to CO.

Figure 7b,c present for all tested perovskites, in the form of Arrhenius plots, the temperature dependence of the CO_2 production rate (r_{CO_2}) in propane combustion under oxidative (0.5% C_3H_8/5% O_2/balance He) and stoichiometric (0.5% C_3H_8/2.5% O_2/balance He) feed conditions, respectively, with the rate being determined under practically differential conditions and normalized to SSA. As shown in Figure 7b, for oxidative conditions, with increasing Co content the same r_{CO_2} was obtained at lower temperature, more clearly for Co/Fe $\leq$ 6/4, which implies an increase in the catalytic activity for propane combustion. The temperature corresponding to a CO_2 production rate of 10^{-7} mol s^{-1} m^{-2} was equal to 345 °C and 244 °C for the less active (LSF) and the most active (LSC) perovskite catalysts, respectively. As shown in Figure 7c, practically the same behavior was observed for propane combustion under stoichiometric conditions, i.e., the same r_{CO_2} value was achieved at lower temperatures with increasing Co/Fe ratio, the effect being clearer for Co/Fe ratios up to 6/4. In this case, the temperature corresponding to a CO_2 production rate of 10^{-7} mol s^{-1} m^{-2} was 355 °C for the LSF and 255 °C for the LSC perovskites (Figure 7c). Also, the comparison of the data shown in Figure 7b,c reveals that the temperatures corresponding to a specific r_{CO_2} value, for example, 10^{-7} mol s^{-1} m^{-2}, were lower under oxidative conditions compared to stoichiometric conditions for all the tested perovskites, which, similar to CO combustion, indicates a higher catalytic activity under oxidative conditions.

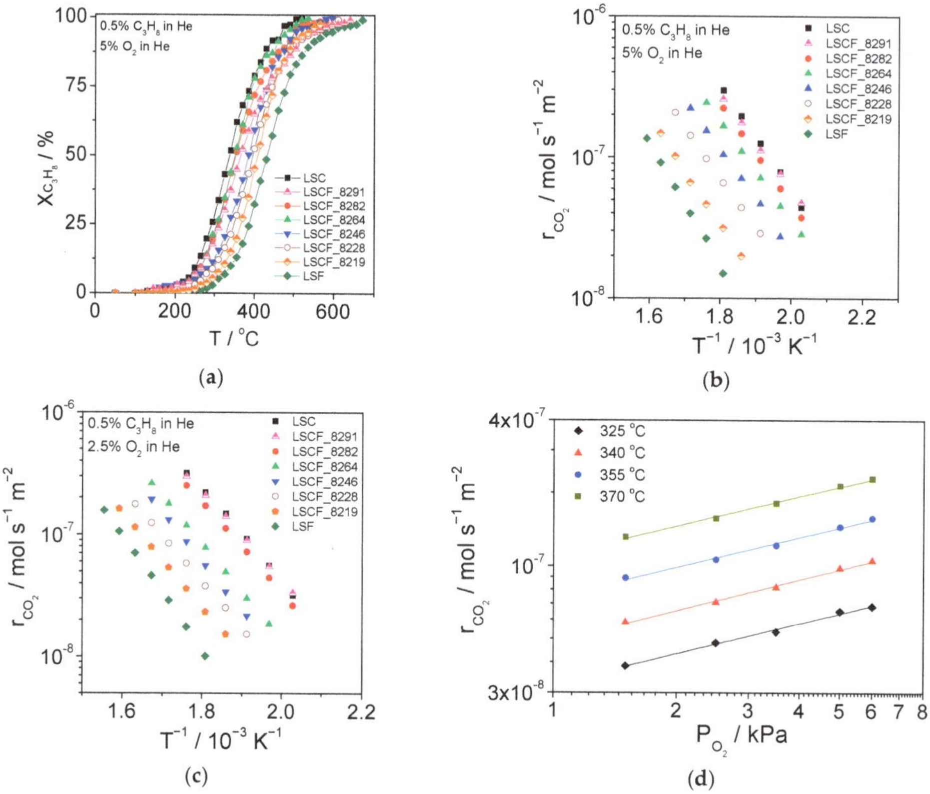

Figure 7. Propane combustion on $La_{0.8}Sr_{0.2}Co_xFe_{1-x}O_{3-\delta}$ perovskites: (**a**) C_3H_8 conversion as a function of temperature under oxidative feed conditions (0.5% C_3H_8/5%O_2/He), (**b**) temperature dependence of the CO_2 production rate under oxidative feed conditions (0.5% C_3H_8/5% O_2/balance He), (**c**) temperature dependence of the CO_2 production rate under stoichiometric feed conditions (0.5% C_3H_8/2.5% O_2/balance He), and (**d**) effect of oxygen partial pressure (P_{O_2}) on the CO_2 production rate, for the LSC catalyst, at constant propane partial pressure $P_{C_3H_8}$ = 0.5 kPa and for four different temperatures (325 °C, 340 °C, 355 °C, and 370 °C).

Kinetic measurements were performed for propane combustion on the most active LSC perovskite. Figure 7d shows, in the form of logarithmic plots, the dependence of the CO_2 production rate on O_2 partial pressure at four different temperatures (325, 340, 355, and 370 °C), for constant propane partial pressure equal to 0.5 kPa. A positive order with respect to O_2 is observed, practically the same for all tested temperatures. This can explain the higher activity for propane combustion observed for LSC under oxidative conditions compared to stoichiometric conditions. As the same behavior was observed for the other tested perovskites, it can be assumed that also for them propane combustion is positive order with respect to O_2 (close to 0.5) in agreement with the observed higher activity under oxidative conditions.

3. Discussion

Oxygen adsorption on perovskites is a complex process, considered to occur mainly at the vicinity of B-sites of perovskite catalysts [13,59,60]. Adler et al. [61] assumed that two neighboring vacant surface lattice sites, similar to those in the bulk of the oxide, are involved in the dissociative adsorption of oxygen and surface oxygen exchange. Royer et al. in their review work [16] highlight the significant role of the oxidation state of the B-site cations and of oxygen vacancies (created by partial substitution of the cations at the A- and/or B-sites) on the perovskite surface oxygen exchange and transport properties, which are also affected by the material microstructure. CO is chemisorbed molecularly on many transition metal oxides, via donation of electron density towards the metal cations, whereas the chemisorption of alkanes on transition metal oxides is generally dissociative [62]. Rhee and Lee [60] refer to two different adsorption states of CO on the surface of $LaCoO_3$ depending on temperature, namely, CO adsorbed in the form of carbonyl species on Co^{3+} cations at 167 °C and CO in the form of bidentate complexes with the lattice oxygen or/and with the surface adsorbed oxygen at 360 °C. It has also been reported that the main CO adsorption site on $LaCoO_3$ perovskite surface is provided by surface lattice oxygen, with formation of surface carbonate species [59,63]. Propane has been reported to adsorb dissociatively on unsubstituted and P-doped $LaCoO_3$ with the surface Co^{3+} sites acting as active sites for initial cleavage of a C-H bond and formation of an adsorbed alkyl radical [64]. Promotion of propane adsorption and dissociation has been also reported for $LaCO_3$ perovskites with increased defect content, synthesized via a defect engineering route, being attributed to an increase in the number of Co^{3+} on the surface combined with an increase in lattice oxygen mobility and surface oxygen capacity, as also corroborated using DFT calculations [65]. Such calculations have been also used to study propane adsorption on macroporous $La_{0.8}Sr_{0.2}CoO_3$ perovskite oxides [66].

As mentioned in the Introduction section, two oxidation mechanisms are considered as dominant in AA'BB'O_3 perovskites, namely, the suprafacial mechanism, which involves only surface oxygen species, and the intrafacial mechanism, corresponding to a redox cycle involving surface lattice oxygen with possible participation of oxygen species originating from the bulk of the perovskite oxide, the prevailing reaction mechanism depending strongly on the perovskite composition and the reaction [12,14–16]. CO oxidation catalyzed by substituted AA'BO_3 and AA'BB'O_3 perovskites, with Co and Fe at the B and B′-sites, is considered to proceed via the intrafacial mechanism due to their higher ionic conductivity compared to other perovskites, associated mainly with their high lattice oxygen mobility [14]. This agrees with the observed high surface exchange in oxygen between the gas phase and the lattice structure of $La_{1-x}Sr_xCoO_3$ (x: 0 – 0.6) at low temperatures (T < 300 °C) [67] at which ABO_3 type perovskites tend to follow the suprafacial mechanism [14,15].

In the present study, a similar trend in the catalytic activity of $La_{0.8}Sr_{0.2}Co_xFe_{1-x}O_{3-\delta}$ for both CO and propane combustion was observed as x varied from 0 (LSF) to 1 (LSC). This indicates that the intrafacial mechanism is probably prevailing, as in this case the rate determining step would be the surface lattice oxygen replenishment for both reactions. The prevalence of the intrafacial mechanism is also corroborated by the fact that, as mentioned above, the tested perovskites exhibit a higher number of oxygen ion vacancies, an improved reducibility of the B-site ions, and a higher surface oxygen exchange as well as a higher mobility of oxygen ions compared to $LaFeO_3$ and $LaCoO_3$ [5,12,14,44,58,67] for which a suprafacial mechanism for CO oxidation at temperatures below 350 °C has been suggested [5,12,14,15]. In this respect, the observed increase in catalytic activity with increasing substitution of iron by cobalt at the B-sites can be mainly attributed to the induced change in the number of lattice oxygen vacancies associated with the easier alteration of the redox state of the Co cations compared to Fe cations [56–58], which increases lattice oxygen anion mobility and enhances surface oxygen exchange [14,54,58], thus favoring the aforementioned redox cycle corresponding to the intrafacial mechanism. The easier release of oxygen from the perovskite oxides with increasing Co/Fe ratio (from LSF to LSC) is corroborated by the observed shift in the onset reduction temperature to lower values in the H_2-TRP profiles (Figure 2). The preferential release of oxygen from oxygen sites in the vicinity of Co rather than Fe sites, as a result of the easier reduction of Co cations compared to Fe cations, has also been observed by Itoh et al. [68] for $La_{0.6}Sr_{0.4}Co_{0.2}Fe_{0.8}O_{3-\delta}$ using in situ X-ray absorption spectroscopy. The increase in oxygen non-stoichiometry, or equivalently in the number of oxygen vacancies, induced by substitution of Fe by Co at the

B-sites of $La_{1-x}Sr_xCo_{1-y}Fe_yO_{3-\delta}$ perovskites has been demonstrated by Scott et al. [27] using solid electrolyte coulometry and CO-TPR measurements. Similarly, Lankhorst and ten Elshof [28] using coulometric titration at 650–950 °C reported a decrease in the oxygen non-stoichiometry in $La_{0.6}Sr_{0.4}Co_{1-y}Fe_yO_{3-\delta}$ with increasing Fe content at the B-sites (changing y from 0 to 0.6), which they associated with an increase in the O^{2-} binding energy in the lattice and with the preferential formation of holes over oxygen vacancies.

The results of the present study are in agreement with the reported by Levasseur and Kaliaguine [29] decrease in the catalytic activity of $La_{1-y}Ce_yCo_{1-x}Fe_xO_3$ for CO combustion with partial substitution of Co by Fe at the B-sites, which they attributed to a decrease in the reducibility of the B-site cations and in the mobility of oxygen in the perovskite bulk. These results also agree with the observed by Tanaka et al. [30] increase in the catalytic activity for propane combustion of $La_{0.9}Ce_{0.1}Co_{1-x}Fe_xO_3$ (x: 0 to 1) with increasing Co content and with the reported by Merino et al. [45] increase in the catalytic activity of $LaCo_{1-y}Fe_yO_{3\pm\lambda}$ (y = 0.1, 0.5, 1) for the same reaction with decreasing y (partial substitution of Fe by Co).

It is noted that the observed variation in catalytic activity of the tested perovskites for CO and propane combustion with changing Fe/Co ratio could be also attributed, to a lesser extent, to the observed changes in their structural characteristics (Section 2.1), which are associated with the surface electronic structure and, thus, may affect the chemisorptive bond strengths of the reactants [16,31,48,50].

4. Materials and Methods

4.1. Synthesis of the Perovskite Oxides

The synthesis method of the perovskite oxides affects significantly their physical properties, including particle size, porosity, and specific surface area (SSA). Synthesis of a perovskite material with high SSA and porosity is expected to enhance its catalytic or electrocatalytic performance. Therefore, the perovskite synthesis methods reported in the literature focus on fine-tuning of the aforementioned physical properties for specific applications [69]. In the present work, the in situ combustion method, with citric acid as fuel, was applied for synthesis of the tested $La_{0.8}Sr_{0.2}Co_xFe_{1-x}O_{3-\delta}$ perovskites, as described in detail in a previous study of our group [47].

The following metal nitrates, in the form of solutions in triple-distilled water, were used as metal precursors: $La(NO_3)_3{\cdot}6H_2O$ (99.9% REO, Alfa Aesar, Karlsruhe, Germany), $Sr(NO_3)_2$ (99+%, ACS reag., Sigma-Aldrich Chemie, Steinheim, Germany), $Fe(NO_3)_3{\cdot}9H_2O$ (≥99.0%, Merck, Darmstadt, Germany), and $Co(NO_3)_2{\cdot}6H_2O$) (≥99.0% KT, ACS reag., Sigma-Aldrich Chemie, Steinheim, Germany). Citric acid monohydrate ($C_6H_8O_7 \cdot H_2O$, 99.5–100.5% assay, Merck, Darmstadt, Germany) and ammonium nitrate (NH_4NO_3, ≥99.0%, Sigma-Aldrich, St. Louis, MO, USA) were used as fuel and extra oxidant, respectively [47]. Citric acid was diluted in triple-distilled water to form a 1.7 M solution, and an appropriate amount of this solution was added under continuous stirring to the metal precursors solution to obtain a citric acid-to-metal ions molar ratio equal to 2:1 [47,70]. An amount of NH_4NO_3 was also added to obtain a molar ratio of NH_4NO_3-to-metal ions equal to 1:1 [47]. The presence of ammonium nitrate, which does not affect the chemical composition of the synthesized perovskite, both accelerates the combustion, reducing sintering time after the formation of the perovskite, and yields in the production of extra amount of gasses that have a dilatative effect on the microstructure of the material, which results in an increase in the SSA of the oxide [71]. An appropriate amount of ammonia solution (30 wt.%, Carlo Erba) was then added dropwise until reaching a pH value of ca. 9 [47,70] in order to neutralize the excess citric acid and form a colloidal solution (sol) [72]. This solution was heated under magnetic stirring on a hot plate until evaporation of H_2O. The resulting slurry was then heated up to ca. 400 °C, using a heat gun, in order to initiate ignition and form a thin crust. The powder, which resulted from shattering the crust, was calcined at 900 °C for 5 h in stagnant air to form the crystalline perovskite phase, and, finally, was finely grounded in a mortar [47].

4.2. Physicochemical Characterization of the Perovskite Oxides (BET, XRD, and ICP-OES)

The specific surface areas (SSAs) of the synthesized perovskite powders were determined via nitrogen physisorption at liquid N_2 temperature (BET method), using a TriStar 3000 surface area and porosimetry analyzer (Micromeritics, Norcross, GA, USA) preceded by degassing of the perovskite samples under vacuum at 300 °C for 1 h [47].

X-ray diffraction (XRD) measurements were performed in order to determine the structural characteristics of the synthesized perovskite materials. XRD spectra were obtained for 2θ values ranging from 10° to 90° at a rate of 0.04 ° s^{-1}, using a Bruker AXS D8 Advance diffractometer, which was equipped with a Cu-Kα lamp (λ = 1.54062 Å) and a Ni filter [47]. The identification and analysis of the observed phases were performed using the EVA software (Bruker AXS, Karlsruhe, Germany).

After previous acid digestion of the perovskite powders, their exact stoichiometry was determined with inductively coupled plasma optical emission spectroscopy (ICP-OES) in an Optima 7000 DV ICP-OES system (Perkin Elmer, Waltham, MA, USA) [47].

4.3. H_2 Temperature-Programmed Reduction Characterization

H_2 temperature-programmed reduction measurements were performed in order to assess and compare the reducibility of the synthesized perovskites. The apparatus used for the experiments consisted of a gas flow system, a quartz reactor positioned in an open-ended vertical electric furnace, and an Omnistar™ GSD301 O1 quadrupole mass spectrometer (Pfeiffer Vacuum, Asslar, Germany) interfaced to a personal computer for on-line monitoring of the TPR effluent gas, using the Quadstar 32-bit (version 7.03) software (Pfeiffer Vacuum, Asslar, Germany). The perovskite mass used in each experiment was 150 mg. The sample was placed in the reactor in the form of a fixed bed deposited on quartz wool and was initially heat treated under He flow at 500 °C for 15 min to remove any adsorbed species [73]. Heat treatment continued under flow of 21% O_2/He mixture at the same temperature for 30 min, followed by free cooling to room temperature (ca. 25 °C). Then, the feed was sequentially switched to He for 15 min and to 3.9% H_2/He mixture for 10 min. Afterwards, the TPR experiment was started by linearly increasing the temperature of the sample up to 720 °C at a rate of 30 °C min^{-1}, using a Eurotherm 815 controller/programmer, and simultaneously monitoring hydrogen in the reactor effluent by recording the transient mass spectrometer signal at mass-to-charge ratio m/z = 2. The temperature of the sample was measured by means of a type-K thermocouple. The gas feed flow rate was equal to 30 cm^3 min^{-1} in all the above steps.

4.4. Catalytic Oxidation Measurements

The experimental apparatus used for the CO and propane combustion measurements consisted of a fixed-bed quartz tube reactor positioned in an open-ended vertical electric furnace, a gas flow system, and an analysis system. The composition and flow rate of the gas mixture fed to the reactor in down-flow mode were adjusted using three electronic mass flow controllers (Aera FC-7700C connected to a ROD-4 operating unit, Advanced Energy Industries, Inc., Fort Collins, CO, USA). Reaction gases were high purity O_2/He, CO or C_3H_8/He mixtures, and He. The perovskite powder (120 mg) was placed on quartz wool in the middle of the quartz tube reactor (8 mm internal diameter). The temperature of the perovskite catalyst was measured in the middle of the catalyst bed using a K-type thermocouple enclosed in a quartz tube entering the upper part of the reactor through a tee connection and was controlled using a Eurotherm 2216e temperature controller. The analysis system consisted of a non-dispersive infrared CO–CO_2 analyzer (BINOS® 100, Rosemount Analytical/Emerson Process Management, Hasselroth, Germany), for continuous monitoring of CO and CO_2 concentration, and a gas chromatograph (TRACE™ CG Ultra, Thermo Fischer Scientific, Inc., Waltham, MA, USA) equipped with a thermal conductivity detector (TCD) and a 6-port sampling valve (VICI Valco). A Carbosieve SII 80/100 (1.8"O.D. × 8 ft) stainless steel packed column operating at 120 °C was used for analysis of the reaction mixture in CO combustion, whereas a cascade of a 23% SP®-1700 on 80/100 Chromosorb P AW (1.8″O.D. × 8 ft) and a 10% TCEP on 100/120 Chromosorb P AW (1.8″O.D. × 8 ft) stainless steel packed columns, operating at 40 °C, were used for analysis of the reaction mixture in propane combustion, using He as carrier gas in both cases. Analysis of the reactor feed was possible by bypassing the reactor using a 4-port valve (VICI Valco). The experiments were carried out under practically atmospheric pressure in the temperature range 50–670 °C, using a total feed flow rate of 100 cm^3 min^{-1}. For this flow rate, the pressure drop across the catalyst fixed-bed reactor, as measured using a manometer, was negligible (less than 0.01 atm). The gas hourly space velocity (GHSV) was equal to 27,500 h^{-1}, as calculated using a measured bulk catalyst bed density value of 0.55 g cm^{-3}. Oxidative (1% CO–5% O_2 and 0.5% C_3H_8–5% O_2) and stoichiometric (1% CO–0.5% O_2 and 0.5% C_3H_8–2.5% O_2) feed compositions were used, with He as balance gas. All measurements were performed after previous establishment of steady-state conditions. No indication of catalyst deactivation due to the produced H_2O(g) was observed during the catalytic measurements concerning propane combustion.

5. Conclusions

In the present work, a series of eight $La_{0.8}Sr_{0.2}Co_xFe_{1-x}O_{3-\delta}$ perovskites, with x = 0, 0.1, 0.2, 0.4, 0.6, 0.8, 0.9, and 1, were synthesized via the in situ combustion synthesis method, and their catalytic activities for CO and C_3H_8 combustion were compared. The perovskite powders were characterized with respect to their specific surface area (SSA), structure, and reducibility via N_2 physisorption (BET method), XRD, and H_2-TPR, respectively, while their chemical composition was determined via ICP-OES.

Their SSA values ranged between 4.1 ($La_{0.8}Sr_{0.2}Co_{0.9}Fe_{0.1}O_{3-\delta}$) and 7.9 m^2 g^{-1} ($La_{0.8}Sr_{0.2}Co_{0.1}Fe_{0.9}O_{3-\delta}$). XRD confirmed the formation of the perovskite structure for all synthesized materials and the absence of secondary phases. A shift in the XRD peaks to higher 2θ was observed with increasing partial substitution of Fe by Co at the B-sites, indicating a contraction of the perovskite unit cell. This shift was accompanied by a gradual change in the main peak (located at 2θ between 32 and 33°) from single to doublet, implying a change in the perovskite symmetry from orthorhombic to rhombohedral.

The H_2-TPR profiles of all synthesized perovskites were characterized by the presence of a broad main reduction peak that can be attributed to the reaction of hydrogen with surface oxygen partly replenished by lattice oxygen migrating from the oxide bulk as the temperature increased. The onset temperature of the reduction process was shifted towards lower values with increasing Co/Fe ratio accompanied by an increase in the main reduction peak, clearly for Co/Fe $\leq$ 6/4. This indicates an easier release of the surface lattice oxygen species and an easier transport of oxygen from the oxide bulk with increasing Co/Fe ratio, which can be mainly associated with the more facile change in the oxidation state of the cobalt cations compared to iron cations.

The catalytic performance of the synthesized $La_{0.8}Sr_{0.2}Co_xFe_{1-x}O_{3-\delta}$ perovskites for CO and propane combustion was investigated in a fixed bed reactor under oxidative and stoichiometric feed conditions. Increase in the Co/Fe ratio, from $La_{0.8}Sr_{0.2}FeO_{3-\delta}$ (LSF) to $La_{0.8}Sr_{0.2}CoO_{3-\delta}$ (LSC), resulted in an enhancement of the catalytic activity for both CO and propane combustion, as concluded both from conversion vs. temperature curves and from CO_2 production rate measurements under practically differential conditions. For CO combustion, the effect of changing the relative content of Co and Fe at the B-sites on catalytic activity was more pronounced under oxidative feed conditions (1% CO/5% O_2/balance He), whereas under stoichiometric feed conditions (1% CO/0.5% O_2/balance He) it was more significant for Co/Fe $\leq$ 6/4. A higher activity of the tested catalytic materials was observed under oxidative reaction conditions compared to stoichiometric conditions, which can be attributed to positive order kinetics with respect to oxygen. The latter was confirmed with kinetic measurements performed for both CO and propane oxidation using $La_{0.8}Sr_{0.2}CoO_{3-\delta}$ as catalyst.

The observed enhancement of the catalytic activity of $La_{0.8}Sr_{0.2}Co_xFe_{1-x}O_{3-\delta}$ perovskites for CO and propane combustion with increasing Co/Fe ratio can be mainly explained on the basis of a prevailing redox catalytic cycle (intrafacial mechanism), considering that the increase in the Co content at the expense of iron induces an increase in the number of catalytically active sites, as they are related to oxygen vacancies at the vicinity of the Co B-sites, and enhances oxygen surface exchange kinetics, due to the lower redox stability of cobalt cations compared iron cations, as also corroborated by the results of the H_2-TPR experiments.

Author Contributions: Conceptualization, A.S., V.C.K. and S.B.; methodology, A.S., V.C.K. and S.B.; investigation, A.S., V.C.K., G.B. and S.B.; visualization; A.S., V.C.K. and G.B.; writing—original draft preparation, A.S., G.B. and S.B.; writing—review and editing, G.B. and S.B.; resources, S.B.; supervision, S.B.; project administration; S.B. All authors have read and agreed to the published version of the manuscript.

Funding: This research received no external funding.

Data Availability Statement: Any data not presented in the manuscript are available upon reasonable written request from the corresponding author.

Acknowledgments: The assistance of Vassilios Dracopoulos, FORTH/ICE-HT, with the XRD characterization and Dimitra Kanellopoulou with the ICP-OES characterization of the perovskite powders is gratefully acknowledged. We also appreciate the permission of D. I. Kondarides to use the TPR apparatus in the Laboratory of Heterogeneous Catalysis of our Department.

Conflicts of Interest: The authors declare no conflict of interest.

References

1. Dey, S.; Sun, S.; Mehta, N.S. Carbon monoxide catalytic oxidation over various iron-based nanoparticles at ambient conditions: A Review. *Carbon Capture Sci. Technol.* **2021**, *1*, 100013. [CrossRef]
2. Yuan, C.; Wu, H.B.; Xie, Y.; Lou, X.W. Mixed transition-metal oxides: Design, synthesis, and energy-related applications. *Angew. Chem. Int. Ed.* **2014**, *53*, 1488–1504. [CrossRef]
3. Huang, W. Oxide Nanocrystal Model Catalysts. *Acc. Chem. Res.* **2016**, *49*, 520–527. [CrossRef] [PubMed]
4. Yakoumis, I.; Panou, M.; Moschovi, A.M.; Panias, D. Recovery of platinum group metals from spent automotive catalysts: A review. *Clean. Eng. Technol.* **2021**, *3*, 100112. [CrossRef]
5. Laguna, O.H.; Bobadilla, L.F.; Hernández, W.Y.; Centeno, M.A. Low-Temperature CO Oxidation. In *Perovskites and Related Mixed Oxides: Concepts and Applications*; Granger, P., Parvulescu, V.I., Kaliaguine, S., Prellier, W., Eds.; Wiley-VCH Verlag GmbH & Co. KGaA: Weinheim, Germany, 2016; pp. 451–473. [CrossRef]
6. Pramuanjaroenkij, A.; Kakaç, S. The fuel cell electric vehicles: The highlight review. *Int. J. Hydrogen Energy* **2023**, *48*, 9401–9425. [CrossRef]
7. Jiao, K.; Xuan, J.; Du, Q.; Bao, Z.; Xie, B.; Wang, B.; Zhao, Y.; Fan, L.; Wang, H.; Hou, Z.; et al. Designing the next generation of proton-exchange membrane fuel cells. *Nature* **2021**, *595*, 361–369. [CrossRef]
8. Bampos, G.; Bebelis, S. Performance of a Pd-Zn Cathode Electrode in a H_2 Fueled Single PEM Fuel Cell. *Electronics* **2022**, *11*, 2776. [CrossRef]
9. Liu, H.; Li, D.; Guo, J.; Li, Y.; Liu, A.; Bai, Y.; He, D. Recent advances on catalysts for preferential oxidation of CO. *Nano Res.* **2022**, *16*, 4399–4410. [CrossRef]
10. Feng, C.; Liu, X.; Zhu, T.; Tian, M. Catalytic oxidation of CO on noble metal-based catalysts. *Environ. Sci. Pollut. Res. Int.* **2021**, *28*, 24847–24871. [CrossRef]
11. Prasad, R.; Singh, P. A Review on CO Oxidation Over Copper Chromite Catalyst. *Catal. Rev.* **2012**, *54*, 224–279. [CrossRef]
12. Royer, S.; Duprez, D. Catalytic Oxidation of Carbon Monoxide over Transition Metal Oxides. *ChemCatChem* **2011**, *3*, 24–65. [CrossRef]
13. Peña, M.A.; Fierro, J.L.G. Chemical Structures and Performance of Perovskite Oxides. *Chem. Rev.* **2001**, *101*, 1981–2018. [CrossRef] [PubMed]
14. Yamazoe, N.; Teraoka, Y. Oxidation catalysis of perovskites—relationships to bulk structure and composition (valency, defect, etc.). *Catal. Today* **1990**, *8*, 175–199. [CrossRef]
15. Voorhoeve, R.J.H. Perovskite-Related Oxides as Oxidation—Reduction Catalysts. In *Advanced Materials in Catalysis*; Burton, J.J., Garten, R.L., Eds.; Academic Press, Inc.: New York, NY, USA, 1977; pp. 129–180. [CrossRef]
16. Royer, S.; Duprez, D.; Can, F.; Courtois, X.; Batiot-Dupeyrat, C.; Laassiri, S.; Alamdari, H. Perovskites as Substitutes of Noble Metals for Heterogeneous Catalysis: Dream or Reality. *Chem. Rev.* **2014**, *114*, 10292–10368. [CrossRef]

17. Giordano, L.; Akkiraju, K.; Jacobs, R.; Vivona, D.; Morgan, D.; Shao-Horn, Y. Electronic Structure-Based Descriptors for Oxide Properties and Functions. *Acc. Chem. Res.* **2022**, *55*, 298–308. [CrossRef]
18. Yang, C.; Grimaud, A. Factors Controlling the Redox Activity of Oxygen in Perovskites: From Theory to Application for Catalytic Reactions. *Catalysts* **2017**, *7*, 149. [CrossRef]
19. Polo-Garzon, F.; Wu, Z. Acid–base catalysis over perovskites: A review. *J. Mater. Chem. A* **2018**, *6*, 2877–2894. [CrossRef]
20. Zhu, H.; Zhang, P.; Dai, S. Recent Advances of Lanthanum-Based Perovskite Oxides for Catalysis. *ACS Catal.* **2015**, *5*, 6370–6385. [CrossRef]
21. Wang, T.; Zhang, C.; Wang, J.; Li, H.; Duan, Y.; Liu, Z.; Lee, J.Y.; Hu, X.; Xi, S.; Du, Y.; et al. The interplay between the suprafacial and intrafacial mechanisms for complete methane oxidation on substituted $LaCoO_3$ perovskite oxides. *J. Catal.* **2020**, *390*, 1–11. [CrossRef]
22. Mars, P.; van Krevelen, D.W. Oxidations carried out by means of vanadium oxide catalysts. *Chem. Eng. Sci.* **1954**, *3*, 41–59. [CrossRef]
23. Pecchi, G.; Jiliberto, M.G.; Buljan, A.; Delgado, E.J. Relation between defects and catalytic activity of calcium doped $LaFeO_3$ perovskite. *Solid State Ion.* **2011**, *187*, 27–32. [CrossRef]
24. Kizaki, H.; Kusakabe, K. *Ab-initio* study of Sr-doping effects on nitric oxide adsorption on the LaO (001) surface of $LaFeO_3$. *Surf. Sci.* **2012**, *606*, 1783–1789. [CrossRef]
25. Merino, N.A.; Barbero, B.P.; Grange, P.; Cadús, L.E. $La_{1-x}Ca_xCoO_3$ perovskite-type oxides: Preparation, characterisation, stability, and catalytic potentiality for the total oxidation of propane. *J. Catal.* **2005**, *231*, 232–244. [CrossRef]
26. Zhang, H.M.; Shimizu, Y.; Teraoka, Y.; Miura, N.; Yamazoe, N. Oxygen sorption and catalytic properties of $La_{1-x}Sr_xCo_{1-y}Fe_yO_3$ perovskite-type oxides. *J. Catal.* **1990**, *121*, 432–440. [CrossRef]
27. Scott, S.P.; Mantzavinos, D.; Hartley, A.; Sahibzada, M.; Metcalfe, I.S. Reactivity of LSCF perovskites. *Solid State Ion.* **2002**, *152–153*, 777–781. [CrossRef]
28. Lankhorst, M.H.R.; ten Elshof, J.E. Thermodynamic Quantities and Defect Structure of $La_{0.6}Sr_{0.4}Co_{1-y}Fe_yO_{3-\delta}$ (y=0–0.6) from High-Temperature Coulometric Titration Experiments. *J. Solid State Chem.* **1997**, *130*, 302–310. [CrossRef]
29. Levasseur, B.; Kaliaguine, S. Effects of iron and cerium in $La_{1-y}Ce_yCo_{1-x}Fe_xO_3$ perovskites as catalysts for VOC oxidation. *Appl. Catal. B Environ.* **2009**, *88*, 305–314. [CrossRef]
30. Tanaka, H.; Mizuno, N.; Misono, M. Catalytic activity and structural stability of $La_{0.9}Ce_{0.1}Co_{1-x}Fe_xO_3$ perovskite catalysts for automotive emissions control. *Appl. Catal. A Gen.* **2003**, *244*, 371–382. [CrossRef]
31. Isupova, L.A.; Sadykov, V.A.; Tsybulya, S.V.; Kryukova, G.N.; Ivanov, V.P.; Petrov, A.N.; Kononchuk, O.F. Effect of structural disorder on the catalytic activity of mixed La−Sr−Co−Fe−O perovskites. *React. Kinet. Catal. Lett.* **1997**, *62*, 129–135. [CrossRef]
32. David, E.; Niculescu, V.-C. Volatile Organic Compounds (VOCs) as Environmental Pollutants: Occurrence and Mitigation Using Nanomaterials. *Int. J. Environ. Res. Public Health* **2021**, *18*, 13147. [CrossRef]
33. Avila, M.S.; Vignatti, C.I.; Apesteguía, C.R.; Garetto, T.F. Effect of support on the deep oxidation of propane and propylene on Pt-based catalysts. *Chem. Eng. J.* **2014**, *241*, 52–59. [CrossRef]
34. Luo, Y.; Zuo, J.; Feng, X.; Qian, Q.; Zheng, Y.; Lin, D.; Huang, B.; Chen, Q. Good interaction between well dispersed Pt and $LaCoO_3$ nanorods achieved rapid Co^{3+}/Co^{2+} redox cycle for total propane oxidation. *Chem. Eng. J.* **2019**, *357*, 395–403. [CrossRef]
35. O'Brien, C.P.; Jenness, G.R.; Dong, H.; Vlachos, D.G.; Lee, I.C. Deactivation of Pt/Al_2O_3 during propane oxidation at low temperatures: Kinetic regimes and platinum oxide formation. *J. Catal.* **2016**, *337*, 122–132. [CrossRef]
36. Hu, Z.; Wang, Z.; Guo, Y.; Wang, L.; Guo, Y.; Zhang, J.; Zhan, W. Total Oxidation of Propane over a Ru/CeO_2 Catalyst at Low Temperature. *Environ. Sci. Technol.* **2018**, *52*, 9531–9541. [CrossRef] [PubMed]
37. Hu, Z.; Liu, X.; Meng, D.; Guo, Y.; Guo, Y.; Lu, G. Effect of Ceria Crystal Plane on the Physicochemical and Catalytic Properties of Pd/Ceria for CO and Propane Oxidation. *ACS Catal.* **2016**, *6*, 2265–2279. [CrossRef]
38. Wang, W.; Li, D.; Yu, H.; Liu, C.; Tang, C.; Chen, J.; Lu, J.; Luo, M. Insights into Different Reaction Behaviors of Propane and CO Oxidation over Pt/CeO_2 and Pt/Nb_2O_5: The Crucial Roles of Support Properties. *J. Phys. Chem. C* **2021**, *125*, 19301–19310. [CrossRef]
39. Ma, L.; Geng, Y.; Chen, X.; Yan, N.; Li, J.; Schwank, J.W. Reaction mechanism of propane oxidation over Co_3O_4 nanorods as rivals of platinum catalysts. *Chem. Eng. J.* **2020**, *402*, 125911. [CrossRef]
40. Klvana, D.; Song, K.S.; Kirchnerova, J. Catalytic performance of $La_{0.66}Sr_{0.34}Co_{0.2}Fe_{0.8}O_3$ perovskite in propane combustion: Effect of preparation and specific surface area. *Korean J. Chem. Eng.* **2002**, *19*, 932–939. [CrossRef]
41. Vannice, M.A. An analysis of the Mars–van Krevelen rate expression. *Catal. Today* **2007**, *123*, 18–22. [CrossRef]
42. Alifanti, M.; Kirchnerova, J.; Delmon, B.; Klvana, D. Methane and propane combustion over lanthanum transition-metal perovskites: Role of oxygen mobility. *Appl. Catal. A Gen.* **2004**, *262*, 167–176. [CrossRef]
43. Nakamura, T.; Misono, M.; Yoneda, Y. Reduction-oxidation and catalytic properties of $La_{1-x}Sr_xCoO_3$. *J. Catal.* **1983**, *83*, 151–159. [CrossRef]
44. Nitadori, T.; Misono, M. Catalytic properties of $La_{1-x}A'_xFeO_3$ (A′ = Sr,Ce) and $La_{1-x}Ce_xCoO_3$. *J. Catal.* **1985**, *93*, 459–466. [CrossRef]

45. Merino, N.A.; Barbero, B.P.; Ruiz, P.; Cadús, L.E. Synthesis, characterisation, catalytic activity and structural stability of $LaCo_{1-y}Fe_yO_{3\pm\lambda}$ perovskite catalysts for combustion of ethanol and propane. *J. Catal.* **2006**, *240*, 245–257. [CrossRef]
46. Song, K.S.; Klvana, D.; Kirchnerova, J. Kinetics of propane combustion over $La_{0.66}Sr_{0.34}Ni_{0.3}Co_{0.7}O_3$ perovskite. *Appl. Catal. A Gen.* **2001**, *213*, 113–121. [CrossRef]
47. Safakas, A.; Bampos, G.; Bebelis, S. Oxygen reduction reaction on $La_{0.8}Sr_{0.2}Co_xFe_{1-x}O_{3-\delta}$ perovskite/carbon black electrocatalysts in alkaline medium. *Appl. Catal. B Environ.* **2019**, *244*, 225–232. [CrossRef]
48. Natile, M.M.; Poletto, F.; Galenda, A.; Glisenti, A.; Montini, T.; De Rogatis, L.; Fornasiero, P. $La_{0.6}Sr_{0.4}Co_{1-y}Fe_yO_{3-\delta}$ Perovskites: Influence of the Co/Fe Atomic Ratio on Properties and Catalytic Activity toward Alcohol Steam-Reforming. *Chem. Mater.* **2008**, *20*, 2314–2327. [CrossRef]
49. Tai, L.-W.; Nasrallah, M.M.; Anderson, H.U.; Sparlin, D.M.; Sehlin, S.R. Structure and electrical properties of $La_{1-x}Sr_xCo_{1-y}Fe_yO_3$. Part 1. The system $La_{0.8}Sr_{0.2}Co_{1-y}Fe_yO_3$. *Solid State Ion.* **1995**, *76*, 259–271. [CrossRef]
50. Gholizadeh, A. The effects of A/B-site substitution on structural, redox and catalytic properties of lanthanum ferrite nanoparticles. *J. Mater. Res. Technol.* **2019**, *8*, 457–466. [CrossRef]
51. Cullity, B.D.; Stock, S.R. *Elements of X-ray Diffraction*, 3rd ed.; Pearson Education, Ltd.: Harlow, UK, 2001; p. 174.
52. Mote, V.D.; Purushotham, Y.; Dole, B.N. Williamson-Hall analysis in estimation of lattice strain in nanometer-sized ZnO particles. *J. Theor. Appl. Phys.* **2012**, *6*, 6. [CrossRef]
53. Bampos, G.; Sygellou, L.; Bebelis, S. Oxygen reduction reaction activity of Pd-based bimetallic electrocatalysts in alkaline medium. *Catal. Today* **2020**, *355*, 685–697. [CrossRef]
54. Deng, J.; Dai, H.; Jiang, H.; Zhang, L.; Wang, G.; He, H.; Au, C.T. Hydrothermal Fabrication and Catalytic Properties of $La_{1-x}Sr_xM_{1-y}Fe_yO_3$ (M = Mn, Co) That Are Highly Active for the Removal of Toluene. *Environ. Sci. Technol.* **2010**, *44*, 2618–2623. [CrossRef] [PubMed]
55. García-López, E.; Marcì, G.; Puleo, F.; La Parola, V.; Liotta, L.F. $La_{1-x}Sr_xCo_{1-y}Fe_yO_{3-\delta}$ perovskites: Preparation, characterization and solar photocatalytic activity. *Appl. Catal. B Environ.* **2015**, *178*, 218–225. [CrossRef]
56. Kuhn, J.N.; Ozkan, U.S. Effect of Co Content Upon the Bulk Structure of Sr- and Co-doped $LaFeO_3$. *Catal. Lett.* **2008**, *121*, 179–188. [CrossRef]
57. Siebert, E.; Roux, C.; Boréave, A.; Gaillard, F.; Vernoux, P. Oxido-reduction properties of $La_{0.7S}Sr_{0.3}Co_{0.8}Fe_{0.2}O_{3-\delta}$ perovskite oxide catalyst. *Solid State Ion.* **2011**, *183*, 40–47. [CrossRef]
58. Chang, H.; Bjørgum, E.; Mihai, O.; Yang, J.; Lein, H.L.; Grande, T.; Raaen, S.; Zhu, Y.-A.; Holmen, A.; Chen, D. Effects of Oxygen Mobility in La–Fe-Based Perovskites on the Catalytic Activity and Selectivity of Methane Oxidation. *ACS Catal.* **2020**, *10*, 3707–3719. [CrossRef]
59. Tascón, J.M.D.; García Fierro, J.L.; González Tejuca, L. Kinetics and Mechanism of CO Oxidation on $LaCoO_3$. *Z. Phys. Chem.* **1981**, *124*, 249–257. [CrossRef]
60. Rhee, C.K.; Lee, H.-I. CO oxidation on $LaCoO_3$ perovskite. *Korean J. Chem. Eng.* **1994**, *11*, 48–54. [CrossRef]
61. Adler, S.B.; Chen, X.Y.; Wilson, J.R. Mechanisms and rate laws for oxygen exchange on mixed-conducting oxide surfaces. *J. Catal.* **2007**, *245*, 91–109. [CrossRef]
62. Campbell, I.M. *Catalysis at Surfaces*; Chapman and Hall: London, UK, 1988; p. 145.
63. Tascón, J.M.D.; González Tejuca, L. Adsorption of CO on the Perovskite-Type Oxide $LaCoO_3$. *Z. Phys. Chem.* **1980**, *121*, 63–78. [CrossRef]
64. Luo, Y.; Zheng, Y.; Feng, X.; Lin, D.; Qian, Q.; Wang, X.; Zhang, Y.; Chen, Q.; Zhang, X. Controllable P Doping of the $LaCoO_3$ Catalyst for Efficient Propane Oxidation: Optimized Surface Co Distribution and Enhanced Oxygen Vacancies. *ACS Appl. Mater. Interfaces* **2020**, *12*, 23789–23799. [CrossRef]
65. Feng, C.; Gao, Q.; Xiong, G.; Chen, Y.; Pan, Y.; Fei, Z.; Li, Y.; Lu, Y.; Liu, C.; Liu, Y. Defect engineering technique for the fabrication of $LaCoO_3$ perovskite catalyst via urea treatment for total oxidation of propane. *Appl. Catal. B Environ.* **2022**, *304*, 121005. [CrossRef]
66. Yang, J.; Shi, L.; Li, L.; Fang, Y.; Pan, C.; Zhu, Y.; Liang, Z.; Hoang, S.; Li, Z.; Guo, Y. Surface modification of macroporous $La_{0.8}Sr_{0.2}CoO_3$ perovskite oxides integrated monolithic catalysts for improved propane oxidation. *Catal. Today* **2021**, *376*, 168–176. [CrossRef]
67. Nakamura, T.; Misono, M.; Yoneda, Y. Catalytic Properties of Perovskite-type Mixed Oxides, $La_{1-x}Sr_xCoO_3$. *Bull. Chem. Soc. Jpn.* **1982**, *55*, 394–399. [CrossRef]
68. Itoh, T.; Shirasaki, S.; Ofuchi, H.; Hirayama, S.; Honma, T.; Nakayama, M. Oxygen partial pressure dependence of *in situ* X-ray absorption spectroscopy at the Co and Fe K edges for $(La_{0.6}Sr_{0.4})(Co_{0.2}Fe_{0.8})O_{3-\delta}$. *Solid State Commun.* **2012**, *152*, 278–283. [CrossRef]
69. Labhasetwar, N.; Saravanan, G.; Kumar Megarajan, S.K.; Manwar, N.; Khobragade, R.; Doggali, P.; Grasset, F. Perovskite-type catalytic materials for environmental applications. *Sci. Technol. Adv. Mater.* **2015**, *16*, 036002. [CrossRef]
70. Papazisi, K.M.; Balomenou, S.; Tsiplakides, D. Synthesis and characterization of $La_{0.75}Sr_{0.25}Cr_{0.9}M_{0.1}O_3$ perovskites as anodes for CO-fuelled solid oxide fuel cells. *J. Appl. Electrochem.* **2010**, *40*, 1875–1881. [CrossRef]
71. Civera, A.; Pavese, M.; Saracco, G.; Specchia, V. Combustion synthesis of perovskite-type catalysts for natural gas combustion. *Catal. Today* **2003**, *83*, 199–211. [CrossRef]

72. Niu, J.; Deng, J.; Liu, W.; Zhang, L.; Wang, G.; Dai, H.; He, H.; Zi, X. Nanosized perovskite-type oxides $La_{1-x}Sr_xMO_{3-\delta}$ (M = Co, Mn; x = 0, 0.4) for the catalytic removal of ethylacetate. *Catal. Today* **2007**, *126*, 420–429. [CrossRef]
73. Panagiotopoulou, P.; Kondarides, D.I.; Verykios, X.E. Mechanistic Study of the Selective Methanation of CO over Ru/TiO_2 Catalyst: Identification of Active Surface Species and Reaction Pathways. *J. Phys. Chem. C* **2011**, *115*, 1220–1230. [CrossRef]

catalysts

MDPI

Article

A New Mixed-Metal Phosphate as an Efficient Heterogeneous Catalyst for Knoevenagel Condensation Reaction

Avik Chowdhury †, Sudip Bhattacharjee †, Sayantan Chongdar, Bhabani Malakar, Anindita Maity and Asim Bhaumik *

School of Materials Sciences, Indian Association for the Cultivation of Science, 2A & 2B Raja S. C. Mullick Road, Jadavpur, Kolkata 700032, India; msac@iacs.res.in (A.C.); sudipb.hkc@gmail.com (S.B.); the.sayantan.97@gmail.com (S.C.); bhabanimalakar1997@gmail.com (B.M.); am.chem@chakdahacollege.ac.in (A.M.)

* Correspondence: msab@iacs.res.in

† These authors contributed equally to this work.

Citation: Chowdhury, A.; Bhattacharjee, S.; Chongdar, S.; Malakar, B.; Maity, A.; Bhaumik, A. A New Mixed-Metal Phosphate as an Efficient Heterogeneous Catalyst for Knoevenagel Condensation Reaction. *Catalysts* **2023**, *13*, 1053. https://doi.org/10.3390/catal13071053

Academic Editors: Georgios Bampos, Athanasia Petala and Zacharias Frontistis

Received: 11 May 2023
Revised: 25 June 2023
Accepted: 26 June 2023
Published: 29 June 2023

Abstract: The escalating demand for the cost-effective synthesis of valuable fine chemicals has fueled the search for sustainable heterogeneous catalysts. Among these catalytic reactions, Knoevenagel condensation has emerged as a very demanding reaction due to its involvement in the synthesis of new C–C bond formation. Porous metal phosphates have attracted significant attention in catalysis due to their unique surface properties. In this study, we report the synthesis of a novel porous magnesium aluminum phosphate (MALPO) material through a hydrothermal template-free approach. MALPO exhibited very promising specific surface area and hierarchical porosity. Moreover, the plate-like morphology of the material can enhance the exposure of the catalytic sites located at the surfaces, leading to enhanced catalytic activity. MALPO demonstrated excellent catalytic performance, yielding a series of Knoevenagel products with up to 99% yield. Notably, the catalyst displayed remarkable recyclability, retaining its structural integrity throughout multiple reaction cycles. The findings highlight the potential of porous mixed-metal phosphates, exemplified by MALPO, as sustainable and efficient base catalyst for the synthesis of value-added chemicals, contributing to the growing demand of the chemical industry. Further investigations are warranted to explore their catalytic potential in diverse chemical transformations and optimize their performance for large-scale operations.

Keywords: porous metal phosphate; heterogeneous catalysis; Knoevenagel condensation; base catalysis

1. Introduction

In the midst of a relentless surge for the green chemical synthesis, the chemical industry experiences an extraordinary upswing in the demand for suitable heterogeneous catalysts. Thus, the search for sustainable and cost-effective catalytic routes for synthesizing value-added chemicals has become a major concern for scientists worldwide. Knoevenagel condensation has been one of the most famous reactions in organic synthesis since its discovery (1890) [1] due to its massive importance in synthesizing valuable reactive organic building blocks [2,3]. In a typical Knoevenagel reaction a carbonyl compound reacts with active methylene groups to generate a new C–C bonds [4,5]. The reaction is found to be highly applicable for synthesizing various fine chemicals [6], hetero-Diels–Alder reactions [7,8], and carbocyclic as well as heterocyclic compounds with significant bio-active behavior [9]. Furthermore, the various kinds of intermediates, such as α, β-unsaturated esters; α-cyanocinnamates [10]; α, β-unsaturated nitriles; and cinnamic acid, involved in the Knoevenagel reaction are considered to be the major platform chemicals [11] for the pharmaceutical industry, the cosmetic industry, the production of perfumes, and the antihypertensive and polymer industries [12,13]. The Knoevenagel condensation reaction can proceed through an acid- or base-catalyzed pathway. So far, several attempts have been made to understand the mechanistic pathway for this reaction. Initially, the methodologies were developed via homogeneous routes [14,15], but in the context of the long-term usage of catalysts, the homogeneous approach has serious short comings. In this context, the heterogeneous path is considered to be the most convenient due to the scope of the recyclability of catalyst, the ease of catalyst separation, and the cost-effectiveness of the process. Several heterogeneous catalysts displayed good performances in Knoevenagel condensation reactions such as the surfactant–mesoporous silica composite [16], functionalized MCM-41 [17], indium-doped AlMCM-41 [18], mesoporous carbon nitride [19], zeolites [20,21], porous organic polymers [22], metal–organic framework [2], phosphate complexes [23], coordination polymers [24], etc. However, use of these materials as catalysts may often lead to metal contamination in the final product.

Today, porous nanomaterials play very crucial role in the field of heterogeneous catalysis due to their enhanced surface activity, pore size tenability, and ease of surface modifications. Previously, Gascon et al. reported AlMIL-53-NH_2 and IRMOF-3 in the Knoevenagel condensation of benzaldehyde with ethyl cyanoacetate [25]. On the other hand, Fischer et al. identified catalytic activities on several functionalized porous materials like Fe-MIL-101-NH_2, CAU-1-NH_2, and Al-MIL-101-NH_2 catalysts in the Knoevenagel condensation reaction of benzaldehyde with malononitrile and ethyl cyanoacetate [26]. Porous metal phosphates have emerged as a captivating class of materials [27] that has garnered significant interest within the realm of material science and engineering. These unique materials embody the fusion of advantageous properties exhibited by metals and phosphates, offering a broad spectrum of applications across diverse industries. Active catalytic sites in these porous metal phosphates are located at the surface of the interconnected pores [28] and can thus offer impressive catalytic activities.

Since the discovery of the aluminophosphate molecular sieve in 1982 by Wilson and coworkers [29], significant attention has been paid to the development of microporous aluminum phosphate molecular sieves [30–32]. Aluminum phosphate, a crystalline inorganic compound, is widely recognized for its diverse applications. Its similarities with zeolite make it a promising material in fields like gas separation [31], sensing, and heterogeneous catalysis. The framework of ALPO composed of AlO_4 and PO_4 moieties, due to the electronically neutral skeleton and the lack of sufficient acidic site results in a very weak catalytic activity [33]. Over the past few years, significant strides have been made in the synthesis and characterization of porous metal phosphates. Advanced techniques, including solvothermal and hydrothermal methods [34,35], have been harnessed to produce materials with well-defined porosity and desirable properties. Moreover, researchers have focused their efforts on designing and modifying these materials to enhance their performance and expand their potential applications. Significant focus is directed towards the development of highly active ALPO materials, while considerable attention has been dedicated to the synthesis of ALPO materials incorporating different metal ions [36].

Over past few decades, porous metal phosphates containing transition metals have garnered significant attention due to the high catalytic activity associated with the metal sites. The low-cost synthetic approach, as well as easiness in bulk synthesis, followed by significant reproducibility, make this class of materials a promising contender among the other member of the porous material family. Thus, immense effort is paying off in the fabrication of these materials, both in academia and industry. So far, a large number of different metal-incorporating aluminum phosphate catalysts have been reported. Acid properties can be significantly affected by introducing Ga, Si, and Co in the ALPO framework [37–39]. Generally, the synthesis of porous nanomaterials involves the use of structure directing agents (SDA) like amines or ionic/nonionic surfactants [38,40,41]. However, the removal of SDA can sometimes be very challenging, as the calcination process requires high temperatures, which often result in the collapse of the porous framework. Moreover, the emission of toxic gas during the calcination process is hazardous to nature. Thus, the template-free synthesis of surface-active porous metal phosphates has garnered significant attention in recent times. The scientific community is highly focused on the controlled fabrication of crystalline porous frameworks with specific morphologies. This approach holds great promise due to the wide availability of active sites, resulting in remarkable outcomes. Researchers are actively exploring the synthetic process, recognizing its potential and the immense interest associated with it. Thus, the pursuit of morphology-controlled fabrication and the accessibility of active sites in transition metal phosphates have become a subject of immense scientific interest and investigation.

Herein, we report the synthesis of a new magnesium aluminum phosphate (MALPO) through a template-free hydrothermal synthesis approach. The resulting catalyst, MALPO, exhibited a good BET surface area and hierarchical porosity. The electron microscopic analysis revealed the plate-like morphology of the material, which facilitated enhanced surface exposure and improved catalytic activity. MALPO has been employed as a heterogeneous catalyst for the Knoevenagel condensation reaction under mild reaction conditions. Remarkably, MALPO demonstrated excellent catalytic performance, resulting in high yields of the desired products and significantly reduced reaction times. Moreover, the catalyst exhibited exceptional recyclability, retaining its structural integrity throughout multiple reaction cycles. These findings highlight significant potential of MALPO as a valuable catalyst for efficient and sustainable synthesis of organic value added chemicals. The further exploration and optimization of MALPO's catalytic properties are warranted to unlock its full potential for various chemical transformations and large-scale applications.

2. Results

Herein, we haave synthesized porous magnesium aluminum phosphate (MALPO) using a template-free hydrothermal approach. The material was thoroughly characterized via different experimental tools. The bonding connectivity inside the material network was evaluated by carrying out a Fourier transform infrared (FTIR) spectroscopic analysis in a solid state by preparing the sample in the KBr pallet. The FTIR spectrum shown in Figure 1 indicates the presence of different bonding inside the material architecture. The peak at 3425 cm^{-1} could be attributed to the O–H stretching vibration [42]. On the other hand, peaks at the region of 1700–1550 cm^{-1} (1705, 1655, 1580 cm^{-1}) could be assigned to different H–O–H bending vibrations. The signals at 982, 1017, and 1053 cm^{-1} indicates the presence of phosphate groups. The 769 and 717 cm^{-1} peaks indicate the presence of metal oxygen bonds in the material [43,44].

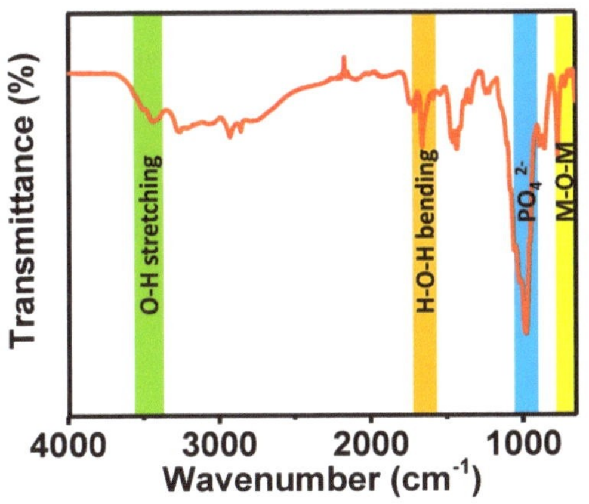

Figure 1. FTIR spectrum of MALPO.

The unit cell parameters and the crystalline phase of the MALPO was evaluated using the powder X-ray diffraction (PXRD) analysis. Prominent sharp crystalline peaks are observed at 10.08, 11.04, 12.77, 13.21, 18.96, 21.12, 23.13, 25.29, 26.72, 28.79, 30.56, 31.88, 35.48, 37.53, and 42.58 degrees of 2θ (Figure 2a). These aforementioned peaks were indexed using the Expo2014 software [45], and these are assigned as 001, 101, 201, 210, 311, 30-2, 020, 021, 51-2, 711, 003, 91-1, 82-1, 812, and 31-4 planes, respectively (Table S1, ESI). The corresponding unit cell parameters of this phase of MALPO were a = 27.652 Å, b = 7.683 Å, c = 8.866 Å, α = 90.00°, β = 98.56°, and γ = 90.00°. The unit cell volume of MALPO was calculated as 1862.59 Å^3. The space group of this triclinic phase was assigned to *P*1 (1). A structural model of MALPO (Figure 2b) was developed using VESTA 4.5.0 software [46] in order to understand the connectivity and porosity in the framework. The refined simulated PXRD of the model was generated, which matched well with the experimental PXRD pattern with a low ESD value of 0.072.

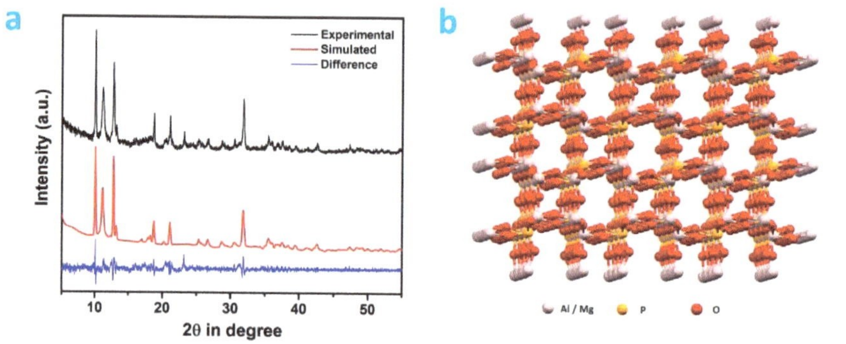

Figure 2. (**a**) Wide-angle powder XRD patterns: experimental (black), simulated (red), and difference (blue). (**b**) Ball and stick model of MALPO.

In order to investigate the porous nature of the as-prepared sample, nitrogen adsorption–desorption analysis was carried out at 77 K. The result is displayed in Figure 3. The isotherm indicated a mixture of type I and type IV isotherms, according to IUPAC convention [47]. A small rise in low pressure indicates the presence of microporosity and a steady rise in the N_2 uptake, along with a mild desorption hysteresis in the relative pressure region of 0.4 to 0.8 reflects the presence of mesoporosity. The pore size distribution, as observed in the inset Figure 3, obtained through non-local density functional theory analysis (NLDFT), reveals the presence of maximum pores with diameters of 1.6 to 2.6 nm. This finding suggests the existence of a dominant pore size within this range. The calculated BET surface area is measured to be 71 m^{2}g^{-1}. This N_2 sorption result indicates the significance of these specific pore sizes in the material, potentially influencing its properties and applications. Further investigation and characterization is warranted to explore the implications of this pore size distribution on the overall behavior and performance of the material.

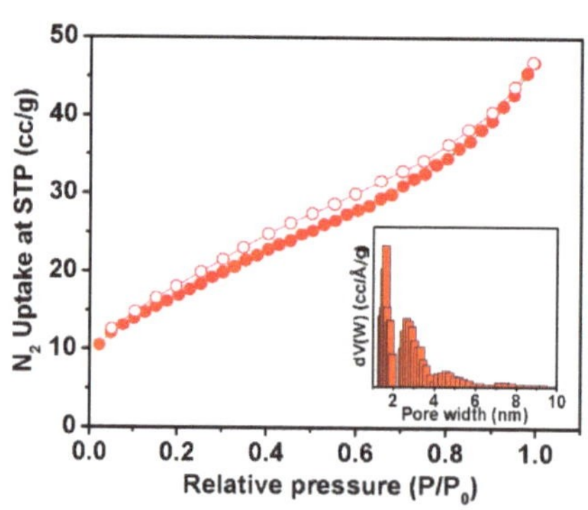

Figure 3. Nitrogen adsorption–desorption isotherm; pore size distribution curve (inset).

The MALPO's morphological characteristics were meticulously assessed utilizing ultra-high resolution transmission electron microscopy (UHR-TEM). The acquired TEM images, obtained at different resolutions, unveiled a compelling plate-like morphology exhibited by MALPO. Notably, Figure 4 vividly depicts the presence of rectangular sheets [48] in varying sizes, effectively highlighting the material's intricate structure. Remarkably, upon closer examination at a high resolution (Figure 4d), the TEM image uncovers the presence of crystalline fringes, aligning impeccably with the expected outcomes from its corresponding X-ray diffraction pattern. This observation strongly suggests the presence of a well-defined crystal lattice within the material, underscoring its inherent structural integrity. The identified two-dimensional (2D) morphological feature holds tremendous potential for enhanced surface activity [49], primarily attributable to the significantly augmented surface exposure that it offers. This unique characteristic opens up exciting prospects for applications where improved [50] surface reactivity and accessibility are critical factors.

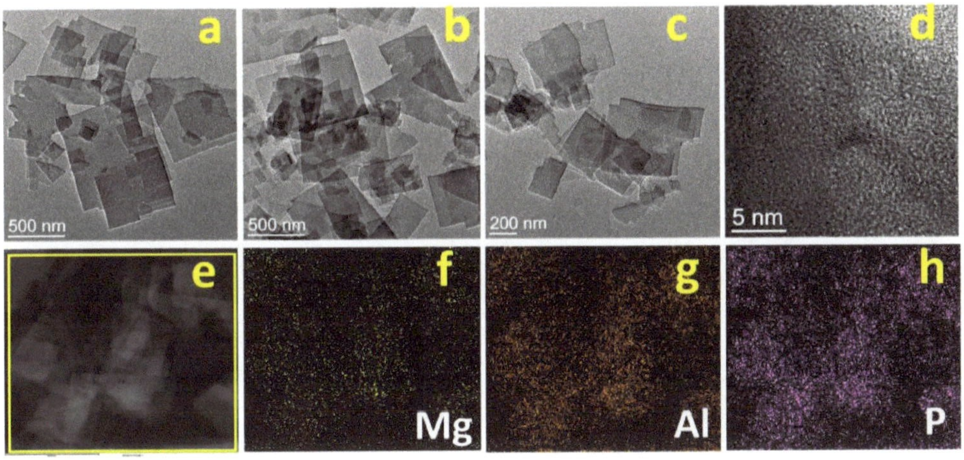

Figure 4. Transmission electron microscopy images of MALPO at different magnifications (**a**–**d**), selected area image under elemental mapping (**e**), elemental mapping of MALPO and the distribution of magnesium (Mg, (**f**)), aluminum (Al, (**g**)) and phosphorus (P, (**h**)).

In addition to the morphological evaluation, an elemental distribution (Figure 4f–h) analysis was performed using energy-dispersive X-ray spectroscopy (EDAX) on the TEM images. Notably, the results obtained demonstrate that a strikingly uniform presence off elements throughout MALPO. CO_2–TPD analysis was carried out under an inert He gas flow, and the corresponding CO_2 desorption profile of MALPO is shown in Figure 5. As can be seen from this CO_2–TPD profile, a broad CO_2 desorption peak with maxima at a very high temperature of 670 °C can be observed. The presence of no significant CO_2 desorption peaks at lower temperatures suggests that CO_2 molecules are strongly bound at the MALPO surface. The observed total basicity was 3.15 μmol g^{-1}. On the other hand, to measure the surface acidity of MALPO, we performed acid–base titration (see the Supporting Information), which suggested a total acidity of 0.71 mmol g^{-1}. The presence of defect phosphate groups are responsible for this surface acidity in MALPO.

Catalytic Activity of MALPO

Knoevenagel condensation is a classical organic transformation in which an active methylene molecule reacts with a carbonyl compound to produce an α,β-conjugated enone. As these intermediates are extensively used in perfumes, polymers, fine chemicals, cosmetics, medicines, and pharmaceuticals, many research groups are concentrating on developing heterogeneous solid catalysts for this organic reaction to generate relevant products. Basic or acidic nature-based catalysts are often used to carry out the condensation reaction, but this requires high reaction temperatures or microwave irradiation [51,52]. Although many catalytic systems have been reported for this reaction, it is very challenging to develop a catalyst which can promote this reaction without the formation of by-products due to the consecutive self-condensation and oligomerization reactions of the primary reaction product [53,54]. The type of catalyst used has a significant impact on product selectivity. In the presence of a Brønsted-base, it results in benzylidene malononitrile, while in presence of Lewis's acid, benzaldehyde reacts with ethanol solvent to produce the desired product (Figure 6). To investigate the catalytic activity of the MALPO, the Knoevenagel condensation reaction was performed by taking benzaldehyde and malononitrile as model substrates in ethanol as the solvent. The reaction yielded the desired benzylidene malononitrile through the activation of a methylene group followed by aldol condensation under the present experimental conditions. As previously discussed, ethanol is considered as the best solvent when compared to MeCN, benzene, toluene, and DCM; thus, we performed our reaction with ethanol [55]. Before confirming the catalytic activity, a blank test was performed without using catalyst, which did not provide the suitable results after 2 h of reaction (17% conversion) in ethanol at room temperature.

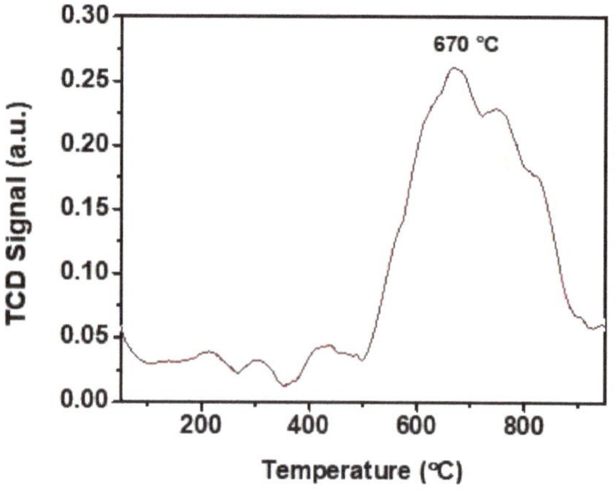

Figure 5. Temperature programmed desorption of CO_2 profile of MALPO in the temperature range 50–875 °C.

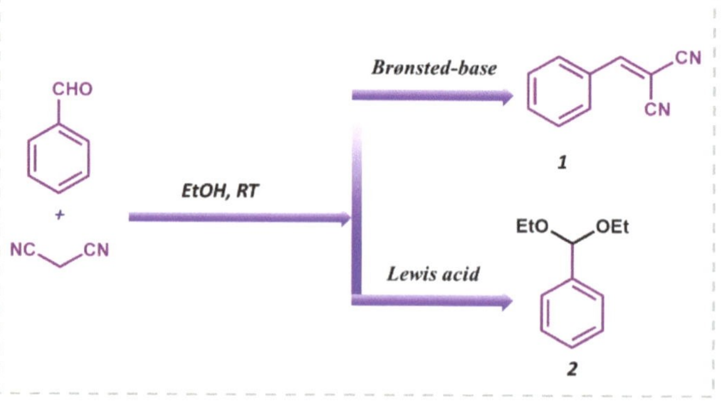

Figure 6. Knoevenagel condensation reaction between benzaldehyde and malononitrile in ethanol as the solvent to obtain benzylidene malononitrile (1) and the undesired by-product benzaldehyde diethyl acetal (2).

In the same experimental conditions, a 99% conversion of benzaldehyde was accomplished using MALPO in ethanol at room temperature after only 100 min. The results clearly reveal the contribution of catalysts towards the activation of this reaction. As we observed that the reaction proceeds smoothly in ethanol within 100 min, we further identified the actual required time for this conversion. Then, we performed the reaction at different time intervals, which is shown in Figure 7a, and while it was observed for 100 min, the reaction was completed within 60 min.

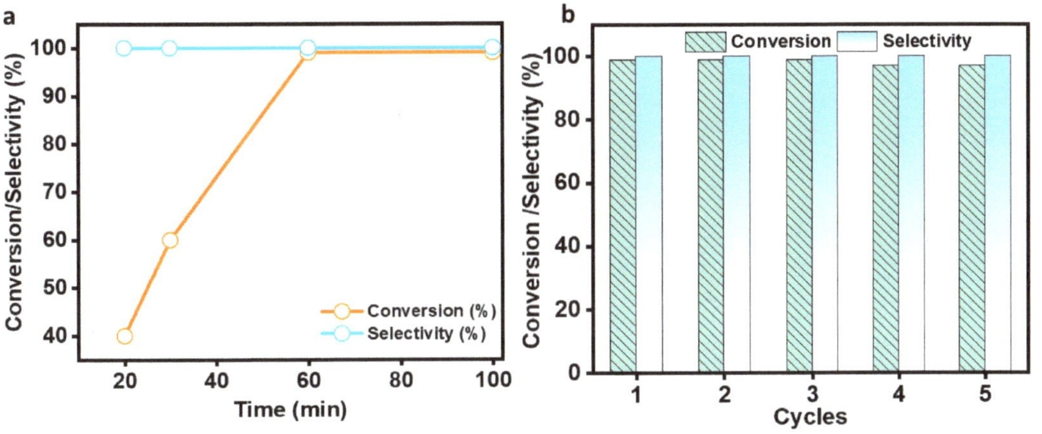

Figure 7. Conversion/selectivity plot with respect to the time (**a**) and the recyclability experiment (**b**).

To verify that the observed reaction was solely stimulated by a solid catalyst and not due to active sites leaching into the solution, a leaching experiment was conducted under identical conditions. Specifically, the benzaldehyde and malononitrile reaction was initiated in the presence of MAPLO under the same circumstances. After 30 min, an aliquot was extracted from the reaction mixture and filtered in order to eliminate the solid catalyst, allowing the resulting solution to continue reacting for an additional 30 min. The outcomes of this leaching experiment are presented in Figure S1 of the Supplementary Materials, which clearly demonstrate that the reaction rate experienced a substantial reduction in the absence of the catalyst following the filtration step. These results indicate that the presence of MALPO exclusively catalyzes the reaction, without any active sites leaching from the solid catalyst into the solution. However, the slight increase in the conversion of benzaldehyde after the catalyst's removal may be attributed to the contribution from the blank reaction, as indicated.

Having observed that the MALPO catalyst achieves an impressive 99% conversion of benzaldehyde to benzylidene malononitrile at room temperature, further experimentation at higher temperatures was deemed unnecessary. Subsequently, upon successfully optimizing the catalytic performance of MALPO for the Knoevenagel reaction, the scope of substrates was expanded to include aromatic-substituted aldehydes (Figure 8, 1a–1e) and biomass-derived heterocyclic aldehydes (Figure 8, 1f). Notably, the catalyst demonstrated the efficient conversion of various substrates, including those with electron-withdrawing groups in the para position and furfural, yielding their corresponding derivatives with high efficiency. However, para-fluoro benzaldehyde (Figure 8, 1c) proved to be an exception, as it did not undergo conversion under the same reaction conditions. The catalyst was also probed for a different active methylene group, such as ethyl cyanoacetate with benzaldehyde (Scheme 1), which showed a 94% conversion (Figure S6. ESI) regarding the condensation reaction under refluxing conditions. These findings highlight the unique catalytic prowess of MALPO and its broad substrate compatibility, thereby showcasing its potential for diverse synthetic applications, ranging from aromatic substituted aldehydes to biomass-derived heterocyclic aldehydes.

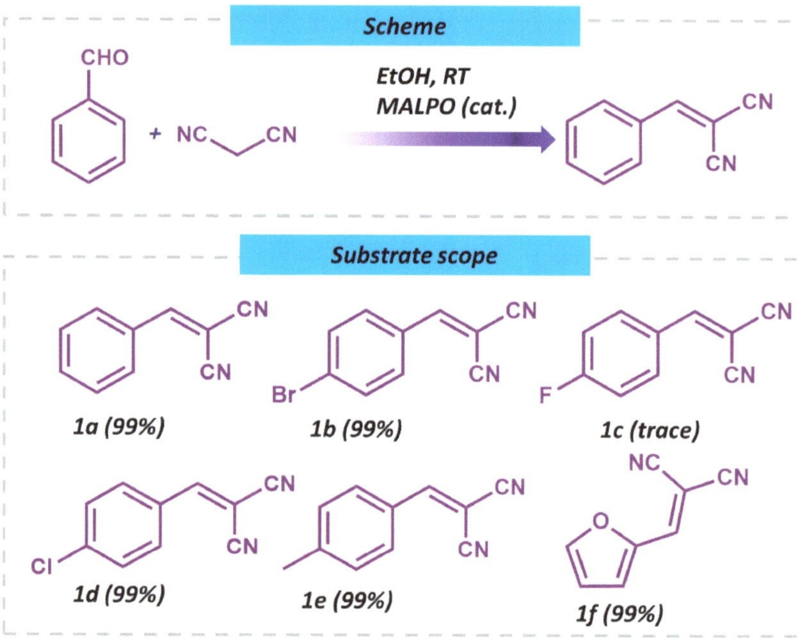

Figure 8. Scheme and substrate scopes and reaction conditions for the Knoevenagel condensation reactions for MALPO.

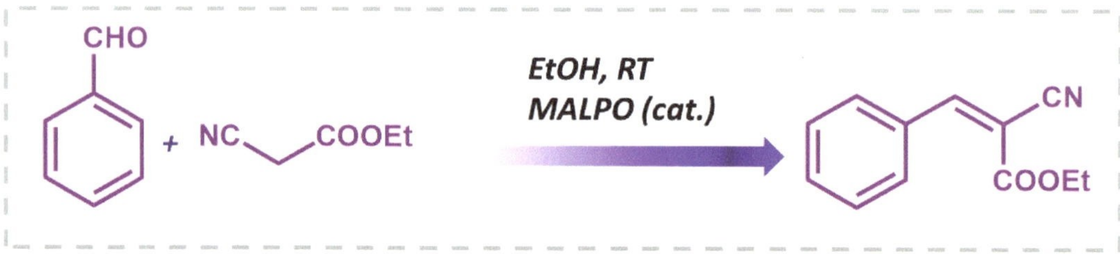

Scheme 1. Schematic representation of the Knoevenagel condensation of ethylcyanoacetate and benzaldehyde.

After the initial experiment, we ran a recyclability test for consecutive fourth cycle to verify the heterogeneity of the catalyst. The catalyst was collected after each cycle, properly washed with methanol, and activated at 120 °C for two hours to conduct the recyclability test. The catalyst was utilized for four successive cycles after it had been activated, and the conversion/selectivity plot is shown in Figure 7b. Up to the fourth cycle, no discernible difference in conversion or O_2 at acid–base paired sites on the catalyst surface that effectively catalyze the deprotonating aldol-dehydration processes was observed. The powdered X-ray diffraction pattern was taken after the fourth cycle (Figure S8, ESI) to assess the stability of the MALPO, and it reveals that there was no structural change. According to Figure 8a, the high catalytic activity of MALPO could be attributed to the basic sites, which promote the abstraction of the protons and activate the malononitrile. This active methylene attacks the -C=O bond of benzaldehyde, which further eliminate the water via aldol condensation, resulting in the final benzylidene malononitrile product.

After the Knoevenagel reaction, we further extended the procedure to produce tetrahydrobenzo[*b*]pyrans. This is a good example of tandem Knoevenagel–Michael cyclocondensation [56] reaction (Scheme 2). To perform the reaction, we chose malononitrile, aromatic aldehyde, and dimedone as model substrates in ethanol. The overall reaction was carried at room temperature, which resulted in a 99% conversion after 1 h of reaction. The reaction follows the same pathway, which follows the Knoevenagel in the first step and further reacts with dimedone to form the pyran derivative via cyclo condensation. The plausible mechanism is shown in Figure 9 In the first step, the MALPO catalyst activates the malononitrile to form the active methylene group, which reacts with the carbonyl group of the aldehyde to form the intermediate **I** via the removal of one molecule of H_2O. On the other hand, dimedone favors the enol form after tautomerization, which attacks the cyano-olefin compound **I** and which behave as a Michael acceptor to produce intermediate **II**. After the formation of intermediates **II**, it follows the cyclo-condensation via the bond shifting of -C=O to the corresponding nitrile group to form intermediate **III**. This intermediate **III** gives the final product by abstracting protons from the reaction mixture. After the successful conversion of the pyran derivatives of benzaldehyde, we examined the catalytic activity on the substrate scope for *p*-bromo derivatives, which provide 99% conversion. *p*-nitro derives also give higher yields. From the above mentioned results, it can be concluded that the catalyst is very efficient for both types of Knoevenagel condensation reactions as well as Knoevenagel–Michael cyclo-condensation. The higher activity of MALPO is due to the presence of basic sites as well as acidic sites in the framework.

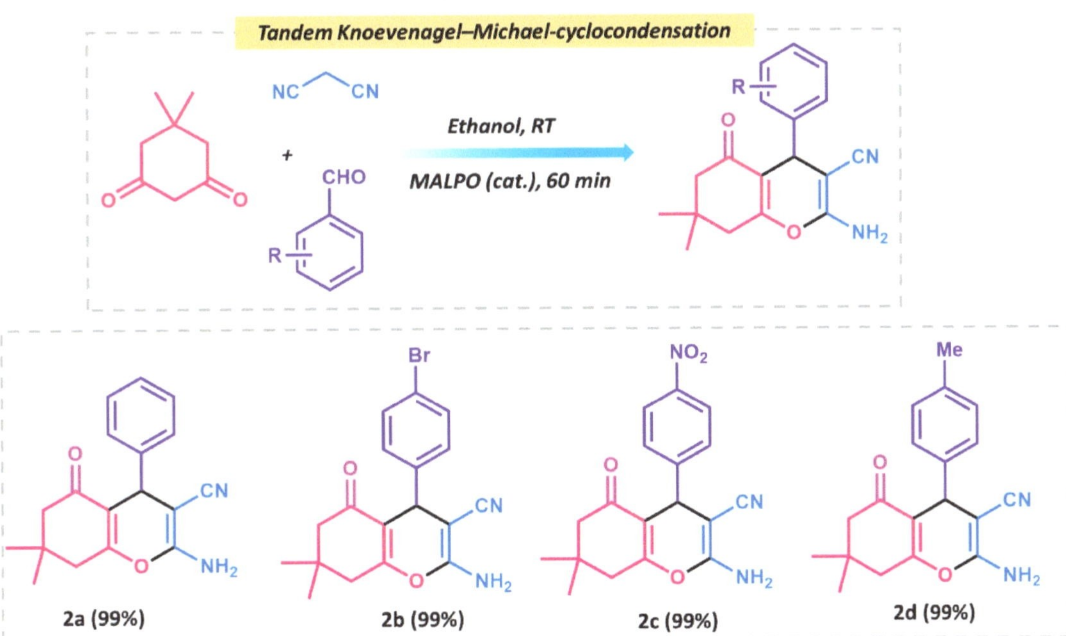

Scheme 2. Schematic representation of tandem Knoevenagel–Michael cyclocondensation reactions between dimedone, benzaldehyde, and malononitrile.

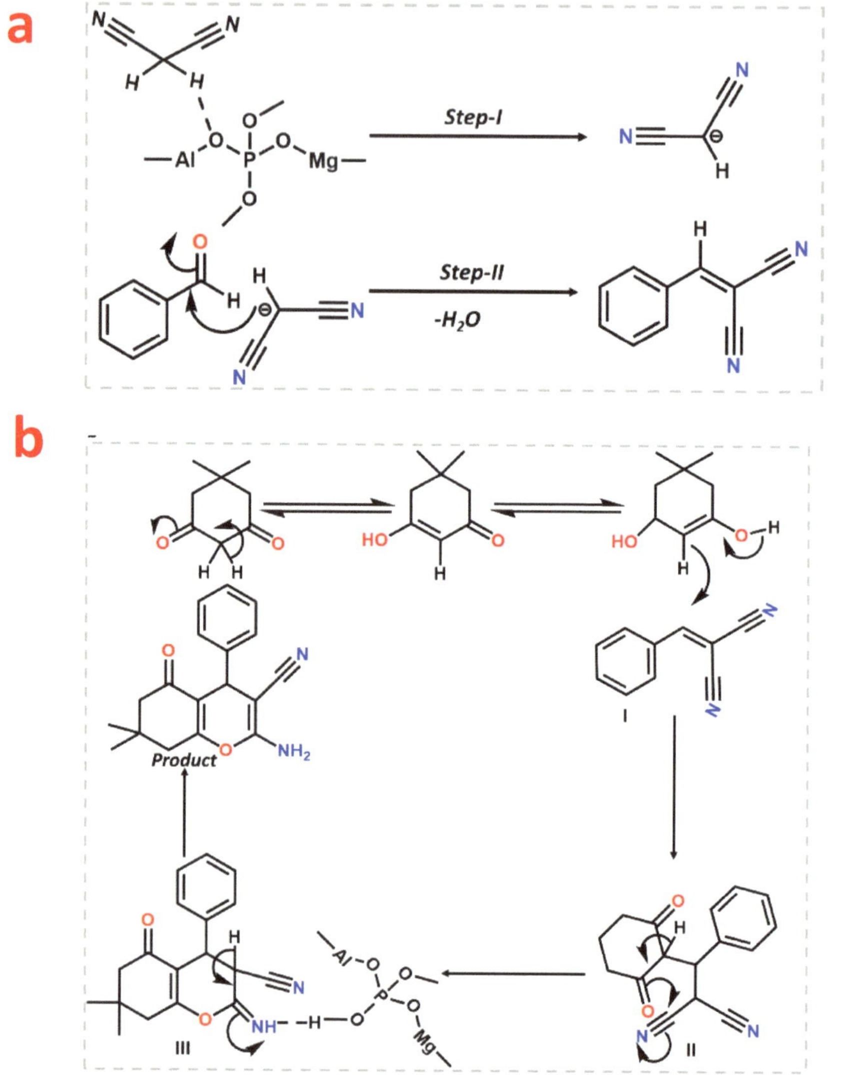

Figure 9. Probable mechanistic pathway for the Knoevenagel condensation reaction (**a**) and Knoevenagel–Michael cyclo-condensation (**b**).

3. Experimental

3.1. Characterizations

The bonding connectivity in the as-synthesized MALPO was analyzed using Fourier transform infrared (FTIR) spectroscopy and a Spectrum 100 spectrophotometer was used for the FTIR analysis (PerkinElmer, Cambridge, MA, USA). The crystalline nature of the phosphate materials was investigated using a Bruker D8 advance X-ray diffractometer, with Cu K_{α} (λ = 0.154 nm) is used as an X-ray source (Bruker AXS, Karsruhe, Germany). The surface area of the material was investigated by analyzing the Brunauer–Emmett–Teller (BET) surface through N_2 sorption analysis at 77 K using a Quantachrome Autosorb iQ surface area analyzer (Quantachrome Inc., Boynton Beach, FL, USA). In order to analyze the data, the samples were activated in a hot air oven at 80 °C, followed by the gassing of the materials at 120 °C under continuous vacuum for 3 h. The pore size distribution was investigated using the non-local density functional theory (NLDFT) method. The morphological analysis of the materials was investigated by analyzing the images obtained from ultrahigh resolution transmission electron microscopy (UHR-TEM, JEOL, Tokyo, Japan). The conformation of the catalytic products was evaluated through nuclear magnetic resonance (NMR) spectroscopy using a Bruker Avance NMR spectrometer (Bruker AXS, Germany). The elemental distribution in the material architecture was analyzed using the elemental distribution images obtained through energy dispersive spectroscopy (EDS). The JEOL 2010 TEM (JEOL, Japan), operated at 200 kV, was utilized to capture ultrahigh-resolution transmission emission microscopy (UHR-TEM) images. For the TEM analysis, the sample was dispersed in methanol and then drop-cast onto a copper grid coated with a carbon polymer. For the temperature programmed desorption analysis of CO_2 (CO_2-TPD), the MALPO sample was activated at 250 °C under a He flow of 2 h. Then, after cooling the sample to room temperature, CO_2 was purged in the absence of a carrier gas flow for 30 min. Then, the sample temperature was raised in a step-wise manner with a heating rate of 10 °C min^{-1}. The desorbed CO_2 in the temperature range 50 to 875 °C was analyzed using a thermal conductivity detector fitted in a AMI-300 Chemisorption Analyzer (Altamira Instruments, Pittsburgh, PA, USA).

3.2. Chemicals

Magnesium nitrate hexahydrate ($Mg(OH)_2$ $6H_2O$) (Spectrochem, Mumbai, India), and aluminum nitrate nonahydrate ($Al(NO_3)_3$ $9H_2O$) and 28% ammonium hydroxide solutions were purchased from Merck, Bengaluru, India. Phosphoric acid (H_3PO_4) was purchased from TCI Chemicals, India. Benzaldehyde was purchased from Spectrochem, India, Furfural was purchased from Sigma-Aldrich (St. Louis, MO, USA), and 4-chlorobenzaldehyde and 4-bromobenzaldehyde were purchased from Merck, India. Dimedone and malononitrile were purchased from Sigma-Aldrich (USA). Of the solvents, ethanol was purchased from Bengal Chemicals (Kolkata, India); acetone and methanol were purchased from Finar Chemicals (Mumbai, India); and the NMR solvent $CDCl_3$ was purchased from Sigma-Aldrich (USA). All the reagents and solvents were used without any further purification.

3.3. Synthesis of MALP

Magnesium aluminum phosphate was synthesized via a typical hydrothermal [39] method. In the synthetic procedure, 1.54 g of phosphoric acid was dissolved in 10 mL of distilled water, 1.28 g of magnesium nitrate hexahydrate ($Mg(NO_3)_2$ $6H_2O$; 0.005 mmol) and 1.87 g of aluminum nitrate nonahydrate ($Al(NO_3)_3$ $9H_2O$, 0.005 mol) were separately placed in 5 mL distilled water. The metal precursor solutions were added simultaneously drop-wise, followed by the addition of 28% ammonium hydroxide solution (NH_4OH) in order to maintain a fixed pH of 7. The final solution was kept under vigorous stirring at room temperature for another 3 h. Finally, the solution was transferred to a stainless steel hydrothermal autoclave and kept static for 72 h at 180 °C in a hot air oven. After that, the white precipitate obtained was filtered, followed by washing with water, methanol, and tetrahydrofuran. Finally, the product was dried at 80 °C under vacuum. The material was then characterized using X-ray diffraction, Fourier-transform infrared spectroscopy, transmission electron microscopy, and nitrogen sorption isotherm analysis.

3.4. Catalytic Activity of MALPO

In order to conduct the catalytic activity test for the Knoevenagel condensation reaction, we carried out the following experiment. A total of 2 mmol of aromatic aldehydes was placed in a 50 mL round-bottomed flask. Then, 2 mmol of malononitrile was added, along with the pre-activated catalyst MALPO (10 mg), followed by the addition of 10 mL ethanol. The reaction mixture was then stirred at 400 rpm for the desired time and monitored via thin-layer chromatography (TLC). After the completion of the reaction, the reaction mixture was filtered in order to separate out the solid catalyst. Thereafter, the liquid was dried under reduced pressure to the obtain solid product. The conformation of the product was evaluated via ^{1}H NMR spectroscopy, where $CDCl_3$ as used as an NMR solvent.

3.5. *Catalyst Recyclability Experiment*

To assess the reusability of MALPO, we performed catalytic recyclability tests. After the catalytic performance, the MALPO material was recovered via filtration. The catalyst was then subjected to a series of washes using distilled water, methanol, and ethanol, followed by vacuum drying. This process regenerated the catalyst, allowing us to employ it for up to four consecutive catalytic cycles. Through this evaluation, we aimed to determine the extent to which MALPO could be reused as a catalyst, providing valuable insights into its potential for sustainable and efficient applications.

4. Conclusions

In summary, herein, we successfully synthesized porous magnesium aluminum mixed-metal phosphate MALPO using ortho phosphoric acid as a phosphate source. MALPO synthesized through a template free approach was found to possess a high specific surface area and a novel triclinic crystal structure. The basic sites present in the material make it a very efficient catalyst for the heterogeneous Knoevenagel condensation reaction. The series of substrates investigated result in excellent product yields of up to 99%. Furthermore, the recyclability test carried out demonstrated the enhanced recyclability of MALPO without any loss of structural integrity, which further concludes the long term applicability of the material towards heterogeneous catalysis. In conclusion, our investigation reveals that the novel mixed-metal phosphate material MALPO holds promise as a potential catalyst for facilitating the Knoevenagel condensation reaction. This observation suggests that MALPO exhibits favorable catalytic properties and could play a crucial role in meeting the growing demand for the synthesis of value-added chemicals, involving a new C–C bond. By offering enhanced catalytic efficiency, MALPO has the potential to contribute significantly to the development and validation of advanced chemical synthesis strategies in the near future. Further research and exploration are warranted in order to fully exploit the capabilities of MALPO and optimize its performance when catalyzing the Knoevenagel condensation reaction.

Supplementary Materials: The following supporting information can be downloaded at: https://www.mdpi.com/article/10.3390/catal13071053/s1, Figure S1: Leaching experiment indicating no leaching of active sites during catalysis, Figure S2: ^{1}H NMR of 2-benzylidenemalononitrile, Figure S3: ^{1}H NMR of 2-(4-chlorobenzylidene)malononitrile, Figure S4: ^{1}H NMR of 2-(furan-2-ylmethylene)malononitrile, Figure S5: ^{1}H NMR of 2-(4-bromobenzylidene)malononitrile, Figure S6: ^{1}H NMR of ethyl (*E*)-2-cyano-3-phenylacrylate, Figure S7: ^{1}H NMR of 2-amino-7,7-dimethyl-5-oxo-4-((*p*-tolyl)-5,6,7,8-tetrahydro-4(*H*-chromene-3-carbonitrile, Figure S8: PXRD pattern of recycled catalyst, Figure S9: (a) TEM image of MALPO; (b) distribution of oxygen in MALPO. Table S1: Indexing of triclinic phase of MALPO with space group P1.

Author Contributions: Experiments, investigation, and formal analysis were carried out by A.C. and S.B. S.C. was involved in the crystal structure analysis. B.M. and A.M. were involved in the formal analysis of the catalysts and products. A.B. provided the resources, investigation, and overall supervision of this project. A.C. wrote the draft manuscript with the help of A.B. All authors have read and agreed to the published version of the manuscript.

Funding: A.C. would like to thank CSIR, New Delhi, for the Senior Research Fellowship. S.B. would like to thank IGSTC, New Delhi, for the Senior Research Fellowship. B.M. would like to thank UGC, New Delhi, for the Junior Research Fellowship. A.B. would like to acknowledge DST-SERB, New Delhi, for the core research grant (project no. CRG/2022/002812).

Data Availability Statement: Not applicable.

Acknowledgments: Authors would like to thank Keshab Maikup for helpful discussion.

Conflicts of Interest: The authors have no conflict to declare.

References

1. Knoevenagel, E. Condensation von Malonsäure mit aromatischen Aldehyden durch Ammoniak und Amine. *Ber. Dtsch. Chem. Ges.* **1898**, *31*, 2596–2619. [CrossRef]
2. Zhu, L.; Liu, X.Q.; Jiang, H.L.; Sun, L.B. Metal-Organic Frameworks for Heterogeneous Basic Catalysis. *Chem. Rev.* **2017**, *117*, 8129–8176. [CrossRef] [PubMed]
3. Jing, Y.; Meng, J.; Liu, Y.; Wan, J.P. Direct Three-Component Synthesis of α-Cyano Acrylates Involving Cascade Knoevenagel Reaction and Esterification. *Chin. J. Chem.* **2015**, *33*, 1194–1198. [CrossRef]
4. Jones, G. *Organic Reactions*; Wiley: New York, NY, USA, 1967; Volume 15, pp. 204–599.
5. Tietze, L.F.; Beifuss, U.; Trost, B.M.; Fleming, I. *An Efficient, Base-Catalyzed, Aqueous Knoevenagel Condensation for the Undergraduate Laboratory*; Pergamon Press: Oxford, UK, 1991; Volume 2, pp. 341–394.
6. Freeman, F. Properties and Reactions of Ylidene malononitriles. *Chem. Rev.* **1981**, *80*, 329–350. [CrossRef]
7. Mondal, J.; Modak, A.; Bhaumik, A. Highly efficient mesoporous base catalyzed Knoevenagel condensation of different aromatic aldehydes with malononitrile and subsequent noncatalytic Diels-Alder reactions. *J. Mol. Catal. A Chem.* **2011**, *335*, 236–241. [CrossRef]
8. Borah, H.N.; Deb, M.L.; Boruah, R.C.; Bhuyan, P.J. Stereoselective intramolecular hetero Diels–Alder reactions of 1-oxa-1, 3-butadienes: Synthesis of novel annelated pyrrolo [1, 2-a] indoles. *Tetrahedron Lett.* **2005**, *46*, 3391–3393. [CrossRef]
9. Tietze, L.F. Domino reactions in organic synthesis. *Chem. Rev.* **1996**, *96*, 115–136. [CrossRef]
10. Wan, J.P.; Jing, Y.; Liu, Y.; Sheng, S. Metal-free synthesis of cyano acrylates via cyanuric chloride-mediated three-component reactions involving a cascade consists of Knoevenagel condensation/cyano hydration/esterification. *RSC Adv.* **2014**, *4*, 63997–64000. [CrossRef]
11. Wang, H.J.; Liu, X.F.; Saliy, O.; Hu, W.; Wang, J.G. Robust Amino-Functionalized Mesoporous Silica Hollow Spheres Templated by CO_2 Bubbles. *Molecules* **2022**, *27*, 53. [CrossRef]
12. Ono, Y. Solid base catalysts for the synthesis of fine chemicals. *J. Catal.* **2003**, *216*, 406–415. [CrossRef]
13. Li, T.F.; Miras, H.N.; Song, Y.F. Polyoxometalate (POM)-Layered Double Hydroxides (LDH) Composite Materials: Design and Catalytic Applications. *Catalysts* **2017**, *7*, 260. [CrossRef]
14. Shanthan, R.P.; Venkataratnam, R.V. Zinc chloride as a new catalyst for Knoevenagel condensation. *Tetrahedron Lett.* **1991**, *32*, 5821–5822. [CrossRef]
15. Bartoli, G.; Beleggia, R.; Giuli, S.; Giuliani, A.; Marcantoni, E.; Massaccesi, M.; Paletti, M. The $CeCl_3$ $7H_2O$–NaI system as promoter in the synthesis of functionalized trisubstituted alkenes via Knoevenagel condensation. *Tetrahedron Lett.* **2006**, *47*, 6501–6504. [CrossRef]
16. Kubota, Y.; Nishizaki, Y.; Sugi, Y. High catalytic activity of as-synthesized, ordered porous silicate–quaternary ammonium composite for Knoevenagel condensation. *Chem. Lett.* **2000**, *29*, 998–999. [CrossRef]
17. Yokoi, T.; Yoshitake, H.; Tatsumi, T. Synthesis of amino-functionalized MCM-41 via direct co-condensation and post-synthesis grafting methods using mono-, di- and tri-amino-organoalkoxysilanes. *J. Mater. Chem.* **2004**, *14*, 951–957. [CrossRef]
18. Katkar, S.S.; Lande, M.K.; Arbad, B.R.; Rathod, S.B. Indium Modified Mesoporous Zeolite AlMCM-41 as a Heterogeneous Catalyst for the Knoevenagel Condensation Reaction. *Bull. Kor. Chem. Soc.* **2010**, *31*, 1301–1304. [CrossRef]
19. Ansari, M.B.; Jin, H.; Parvin, M.N.; Park, S.-E. Mesoporous carbon nitride as a metal-free base catalyst in the microwave assisted Knoevenagel condensation of ethylcyanoacetate with aromatic aldehydes. *Catal. Today* **2012**, *185*, 211–216. [CrossRef]
20. Reddy, T.I.; Verma, R.S. Rare-earth (RE) exchanged NaY zeolite promoted Knoevenagel condensation. *Tetrahedron Lett.* **1997**, *38*, 1721–1724. [CrossRef]
21. Grass, J.P.; Kluehspies, K.; Reiprich, B.; Schwieger, W.; Inayat, A. Layer-Like Zeolite X as Catalyst in a Knoevenagel Condensation: The Effect of Different Preparation Pathways and Cation Exchange. *Catalysts* **2021**, *11*, 474. [CrossRef]
22. Modak, A.; Mondal, J.; Bhaumik, A. Porphyrin based porous organic polymer as bi-functional catalyst for selective oxidation and Knoevenagel condensation reactions. *Appl. Catal. A Gen.* **2013**, *459*, 41–51. [CrossRef]
23. Bennazha, J.; Zahouilly, M.; Boukhari, A.; Hol, E.A. Investigation of the basis of catalytic activity of solid state phosphate complexes in the Knoevenagel condensation. *J. Mol. Catal. A Chem.* **2003**, *202*, 247–252. [CrossRef]
24. Karmakar, A.; Soliman, M.M.A.; Alegria, E.C.B.A.; Guedes da Silva, M.F.C.; Pombeiro, A.J.L. Polyaromatic Carboxylate Ligands Based Zn(II) Coordination Polymers for Ultrasound-Assisted One-Pot Tandem Deacetalization–Knoevenagel Reactions. *Catalysts* **2022**, *12*, 294. [CrossRef]
25. Gascon, J.; Aktay, U.; Hernandez-Alonso, M.D.; Klink, G.P.M.; Kapteijn, F. Amino-based metal-organic frameworks as stable, highly active basic catalysts. *J. Catal.* **2009**, *261*, 75–87. [CrossRef]
26. Hartmann, M.; Fischer, M. Amino-functionalized basic catalysts with MIL-101 structure. *Microporous Mesoporous Mater.* **2012**, *164*, 38–43. [CrossRef]
27. Lin, R.; Ding, Y. A Review on the Synthesis and Applications of Mesostructured Transition Metal Phosphates. *Materials* **2013**, *6*, 217–243. [CrossRef]

28. Loiseau, T.; Ferey, G. Crystalline oxyfluorinated open-framework compounds: Silicates, metal phosphates, metal fluorides and metal-organic frameworks (MOF). *J. Fluor. Chem.* **2007**, *128*, 413–422. [CrossRef]
29. Wilson, S.T.; Lok, B.M.; Messina, C.A.; Cannan, T.R.; Flanigen, E.M. Aluminophosphate Molecular Sieves: A New Class of Microporous Crystalline Inorganic Solids. *J. Am. Chem. Soc.* **1982**, *104*, 1146–1147. [CrossRef]
30. Pyke, D.R.; Whitney, P.; Houghton, H. Chemical modification of crystalline microporous aluminium phosphates. *Appl. Catal.* **1985**, *18*, 173–190. [CrossRef]
31. Ramesha, B.M.; Meynen, V. Advances and Challenges in the Creation of Porous Metal Phosphonates. *Materials* **2020**, *13*, 5366. [CrossRef]
32. Clearfield, A. Recent advances in metal phosphonate chemistry. *Curr. Opin. Solid State Mater. Sci.* **1996**, *1*, 268–278. [CrossRef]
33. Fischer, M. Porous aluminophosphates as adsorbents for the separation of CO_2/CH_4 and CH_4/N_2 mixtures—A Monte Carlo simulation study. *Sustain. Energy Fuels* **2018**, *2*, 1749–1763. [CrossRef]
34. Nan, C.; Lu, J.; Chen, C.; Peng, Q.; Li, Y. Solvothermal synthesis of lithium iron phosphate nanoplates. *J. Mater. Chem.* **2011**, *21*, 9994–9996. [CrossRef]
35. Wang, H.Y.; Cheng, H.J.; Lai, F.; Xiong, D.Y. CuAPO-5 as a Multiphase Catalyst for Synthesis of Verbenone from alpha-Pinene. *Materials* **2022**, *15*, 8097.
36. Xu, D.; Ren, J.; Yue, S.; Zou, X.; Shang, X.; Wang, X. One-Pot Synthesis of Al-P-O Catalysts and Their Catalytic Properties for O-Methylation of Catechol and Methanol. *Materials* **2021**, *14*, 5942. [CrossRef]
37. Cheng, S.; Tzeng, J.N.; Hsu, B.Y. Synthesis and Characterization of a Novel Layered Aluminophosphate of Kanemite-like Structure. *Chem. Mater.* **1997**, *9*, 1788–1796.
38. Lok, B.M.; Messina, C.R.; Patton, R.L.; Gajek, R.T.; Cannan, T.R.; Flanigan, E.M. Silicoaluminophosphate molecular sieves: Another new class of microporous crystalline inorganic solids. *J. Am. Chem. Soc.* **1984**, *106*, 6092–6093. [CrossRef]
39. Zubowa, H.L.; Richter, M.; Roost, U.; Parlitz, B.; Fricke, R. Synthesis and catalytic properties of substituted A1PO4-31 molecular sieves. *Catal. Lett.* **1993**, *19*, 67–79.
40. Sayari, A.; Moudrakovski, I.; Reddy, J.S.; Ratcliffe, C.I.; Ripmeester, J.A.; Preston, K.F. Synthesis of mesostructured lamellar aluminophosphates using supramolecular templates. *Chem. Mater.* **1996**, *8*, 2080–2088. [CrossRef]
41. Lin, X.X.; Su, M.X.; Fang, F.X.; Hong, J.F.; Zhang, Y.M.; Zhou, S.F. Hierarchically Annular Mesoporous Carbon Derived from Phenolic Resin for Efficient Removal of Antibiotics in Wastewater. *Molecules* **2022**, *27*, 6735. [CrossRef]
42. Dutta, A.; Patra, A.K.; Bhaumik, A. Porous organic–inorganic hybrid nickel phosphonate: Adsorption and catalytic applications. *Microporous Mesoporous Mater.* **2012**, *155*, 208–214. [CrossRef]
43. Thakkar, R.; Chudasama, U. Preparation and application of zirconium phosphate and its derivatives. *J. Hazard. Mater.* **2009**, *172*, 129–133. [CrossRef]
44. Ge, T.D.; Yu, L.Q.; Ni, N.R.; Dong, T.A.; Xing, T.L.; Long, H.K.; Yang, J.X. Synthesis of $LiCo1/3Ni1/3Mn1/3O_2$ as a cathode material for lithium ion battery by water-in-oil emulsion method. *Mater. Chem. Phys.* **2005**, *94*, 423–428.
45. Altomare, A.; Cuocci, C.; Giacovazzo, C.; Moliterni, A.; Rizzi, R.; Corriero, N.; Falcicchio, A. EXPO2013: A kit of tools for phasing crystal structures from powder data. *J. Appl. Crystallogr.* **2013**, *46*, 1231–1235. [CrossRef]
46. Chakroborty, D.; Chowdhury, A.; Chandra, M.; Jana, R.; Shyamal, S.; Bhunia, M.K.; Chandra, D.; Hara, M.; Pradhan, D.; Datta, A.; et al. Novel Tetradentate Phosphonate Ligand Based Bioinspired Co-Metal–Organic Frameworks: Robust Electrocatalyst for the Hydrogen Evolution Reaction in Different Mediums. *Cryst. Growth Des.* **2021**, *21*, 2614–2623. [CrossRef]
47. Kundu, S.K.; Bhaumik, A. Pyrene-Based Porous Organic Polymers as Efficient Catalytic Support for the Synthesis of Biodiesels at Room Temperature. *ACS Sustain. Chem. Eng.* **2015**, *3*, 1715–1723. [CrossRef]
48. Zhang, W.; Oulego, P.; Sharma, S.K.; Yang, X.-L.; Li, L.-J.; Rothenberg, G.; Shiju, N.R. Self-Exfoliated Synthesis of Transition Metal Phosphate Nanolayers for Selective Aerobic Oxidation of Ethyl Lactate to Ethyl Pyruvate. *ACS Catal.* **2020**, *10*, 3958–3967. [CrossRef]
49. Sharma, N.; Parhizkar, M.; Cong, W.; Mateti, S.; Kirkaland, M.A.; Puri, M.; Sutti, A. Metal ion type significantly affects the morphology but not the activity of lipase–metal–phosphate nanoflowers. *RSC Adv.* **2017**, *7*, 25437–25443. [CrossRef]
50. Li, Z.; Tang, M.; Dai, J.; Wang, T.; Wang, Z.; Bai, W.; Bai, R. Preparation of Covalent Pseudo-Two-Dimensional Polymers in Water by Free Radical Polymerization. *Macromolecules* **2017**, *50*, 4292–4299. [CrossRef]
51. Mazzotta, M.G.; Gupta, D.; Saha, B.; Patra, A.K.; Bhaumik, A.; Abu-Omar, M.M. Efficient Solid Acid Catalyst Containing Lewis and Bronsted Acid Sites for the Production of Furfurals. *ChemSusChem* **2014**, *7*, 2342–2350. [CrossRef]
52. Farhi, J.; Lykakis, I.N.; Kostakis, G.E. Metal-Catalysed A(3) Coupling Methodologies: Classification and Visualisation. *Catalysts* **2022**, *12*, 660. [CrossRef]
53. Tan, Y.; Fu, Z.; Zhang, J. A layered amino-functionalized zinc-terephthalate metal organic framework: Structure, characterization and catalytic performance for Knoevenagel condensation. *Inorg. Chem. Commun.* **2011**, *14*, 1966–1970. [CrossRef]
54. Hwang, Y.K.; Hong, D.-Y.; Chang, J.-S.; Jhung, S.H.; Seo, Y.-K.; Kim, J.; Vimont, A.; Daturi, M.; Serre, C.; Ferey, G. Amine grafting on coordinatively unsaturated metal centers of MOFs: Consequences for catalysis and metal encapsulation. *Angew. Chem. Int. Ed.* **2008**, *47*, 4144–4148. [CrossRef] [PubMed]

55. Das, A.; Anbu, N.; Dhakshinamoorthy, A.; Biswas, S. A highly catalytically active Hf(IV) metal-organic framework for Knoevenagel condensation. *Microporous Mesoporous Mater.* **2019**, *284*, 459–467. [CrossRef]
56. Basavaraja, D.; Athira, C.S.; Siddalingeshwar, V.D.; Ashitha, K.T.; Somappa, S.B. Multicomponent Synthesis of Spiro-dihydropyridine Oxindoles via Cascade Spiro-cyclization of Knoevenagel/Aza-Michael Adducts. *J. Org. Chem.* **2022**, *87*, 13556–13563.

catalysts

MDPI

Article

Electrochemical Promotion of CO_2 Hydrogenation Using Rh Catalysts Supported on O^{2-} Conducting Solid Electrolyte

Nikoleta Kokkinou, Fotios Xydas, Susanne Brosda, Georgios Kyriakou and Alexandros Katsaounis *

Department of Chemical Engineering, University of Patras, Caratheodory 1, 26504 Patras, Greece
* Correspondence: alex.katsaounis@chemeng.upatras.gr

Abstract: Electrochemical promotion was used to modify the activity and selectivity of a Rh catalyst electrode in the CO_2 hydrogenation reaction. The experiments were carried out in a temperature range of 350–430 °C at ambient pressure and at different CO_2 to H_2 gas feeding ratios (1:2 to 4:1). The only reaction products observed were CO and CH_4, both under open- and closed-circuit conditions. The CH_4 formation rate was found to increase with both positive and negative potential or current application. The CO formation rate followed the opposite trend. The selectivity to CH_4 increased under high values of hydrogen partial pressure and decreased at high pressures of CO_2. The results demonstrate how electrochemical promotion can be used to finely tune activity and selectivity for a reaction of high technical and environmental importance.

Keywords: electrochemical promotion of catalysis (EPOC); non-faradaic electrochemical modification of catalytic activity (NEMCA); CO_2 hydrogenation; rhodium (Rh); YSZ; XPS

Citation: Kokkinou, N.; Xydas, F.; Brosda, S.; Kyriakou, G.; Katsaounis, A. Electrochemical Promotion of CO_2 Hydrogenation Using Rh Catalysts Supported on O^{2-} Conducting Solid Electrolyte. *Catalysts* **2023**, *13*, 1014. https://doi.org/10.3390/catal13061014

Academic Editor: Giuseppe Bonura

Received: 18 May 2023
Revised: 6 June 2023
Accepted: 13 June 2023
Published: 16 June 2023

1. Introduction

Carbon dioxide (CO_2) is the most abundant greenhouse gas in our atmosphere with its concentration drastically increasing over time due to anthropogenic activities, most crucially the extensive use of fossil fuels for transport and power generation purposes. The ever-increasing CO_2 emissions in the atmosphere have gradually led to climate change, which is strongly associated with global warming, extreme weather conditions, and other major environmental threats [1–4]. In this regard, it is of foremost importance to globally minimize the use of fossil fuels, and accordingly, develop the necessary technologies for capturing and utilizing CO_2 to produce fuels, bulk chemicals, and value-added products. It is widely accepted that CO_2 hydrogenation may offer a relatively economic and effective way of controlling and making use of the CO_2 excess atmospheric levels [5–7].

Most studies of the catalytic hydrogenation of CO_2 have been performed in fixed-bed reactors using mainly metal catalysts (e.g., Rh, Pd, Ru, Cu, Fe, Co, Ni, Au, Ag) supported over a wide range of metal oxides (e.g., Nb_2O_3, ZrO_2, Al_2O_3, SiO_2) [8–37] and utilizing high pressures (5–70 atm) [12–15,17,24–28,30] which shift the thermodynamic equilibrium towards methanol and light hydrocarbon production. Under atmospheric pressure, CO_2 can be reduced by hydrogen to form methane (Sabatier reaction) and/or carbon monoxide (reverse water gas shift, RWGS, reaction) according to the following equations:

$$CO_2 + H_2 \rightarrow CO + H_2O, \Delta H = 41.5 \text{ kJmol}^{-1} \text{ (RWGS reaction)} \quad (1)$$

$$CO_2 + 4H_2 \rightarrow CH_4 + 2H_2O, \Delta H = -165 \text{kJmol}^{-1} \text{ (Sabatier reaction)} \quad (2)$$

The RWGS reaction is endothermic and is thermodynamically favored at high operating temperatures. Contrary to this, the Sabatier reaction leads to the formation of methane, which is an exothermic reaction and is favored at low operation temperatures. However, the low operation temperatures limit the activity and the kinetics of the reactions as well.

Therefore, it is essential to develop active and selective low temperature catalytic processes, operating at high CO_2 to H_2 feed ratios and atmospheric pressure.

The electrochemical promotion of CO_2 hydrogenation has been studied in recent years over different catalytic films including noble and non-noble metals (Ru [7,38–43] and Rh [44,45], and Ni, Co, Fe [46,47], respectively) supported on a variety of solid electrolytes including yttria-stabilized-zirconia (YSZ, an O^{2-} conductor [7,38,44,45]), H^+ conductors [40,43,46], and alkali ion conductors such as Na-β″-Al_2O_3 [40]. Under open circuit conditions, the non-promoted catalytic reaction takes place on the catalyst-working electrode. Application of a potential or current, between the catalyst film (working electrode) and a counter electrode, leads to changes of the conversion rate and product selectivity [7,38,48–56]. These non-Faradaic changes are evoked by the formation of an effective double layer due to the migration of promoting species (e.g., O^{2-} in the case of YSZ) migrating from the solid electrolyte to the metal–gas interface [7,38,49,57–59].

The CO_2 hydrogenation reaction on Rh catalyst-electrodes deposited on YSZ has been rarely investigated and mainly under conditions of excess of hydrogen (regarding Equation (1)) and cathodic polarizations. The study of the reaction in a single chamber reactor by Bebelis et al. [44] led to CO and CH_4 formation at temperatures of 346–477 °C. It was found that the rate of CH_4 formation is enhanced with positive potentials (electrophobic behavior) while the rate of CO formation is enhanced with negative potentials (electrophilic behavior). The maximum selectivity to CH_4 was up to 35%. Using a monolithic electro-promoted reactor (MEPR) [45], the reaction of CO_2 hydrogenation was studied on 22 thin Rh/YSZ/Pt plate cells at temperatures between 220 and 380 °C. The only products observed were, again, CO and CH_4. The rates of both reactions were significantly affected during polarization, while the selectivity to CH_4 remained always below 12%.

The pronounced catalytic activity of Rh-based catalysts, supported on O^{2-} conductors, under electro-promoted conditions have been associated in earlier studies with the decomposition of the rhodium oxide to metallic rhodium, which is more active especially under oxidation reaction conditions [57–59].

In the present study, the electrochemical promotion of CO_2 hydrogenation reaction was studied over a Rh catalyst-electrode deposited on an O^{2-} solid electrolyte conductor, YSZ, at atmospheric pressure, and for the first time, in a continuous single pellet flow reactor. The main scopes were in contrast to earlier studies: to perform CO_2 hydrogenation at a lower temperature range of 350–430 °C and at different CO_2 to H_2 cofeeding ratios.

2. Results and Discussion

2.1. Hydrogenation Activity Measurements

The steady-state-CO_2 conversion as a function of temperature for different CO_2 to H_2 ratios was first investigated. The results displayed in Figure 1 show that the conversion of CO_2 increases with increasing temperature reaching a maximum of 17.6% under reaction conditions of CO_2 to H_2 feeding ratio of 1:2. Under conditions of CO_2 to H_2 of 1:1 the conversion drops to 12% and is smallest with 6% and 2% at CO_2 to H_2 ratios of 2:1 and 4:1, respectively.

Figure 2 shows the CH_4 and CO production rates at steady state as a function of temperature. The rate of the methanation reaction was found to be highest for a gas feed ratio of CO_2 to H_2 of 1:2. With increasing CO_2 partial pressure in the reaction mixture, i.e., a CO_2 to H_2 ratio of 1:1, 2:1 and 4:1, the observed CH_4 formation rate is very low and almost independent of temperature. Figure 2b suggests that the RWGS reaction prevails over the Sabatier reaction. Moreover, as shown in Figure 3, CO appears to be the main product under all feeding CO_2 to H_2 ratios of this study and its selectivity is higher than 96%, while CH_4 selectivity is smaller than 4%. The reaction rate values of the RWGS reaction are more than one order of magnitude higher than those of the methanation one.

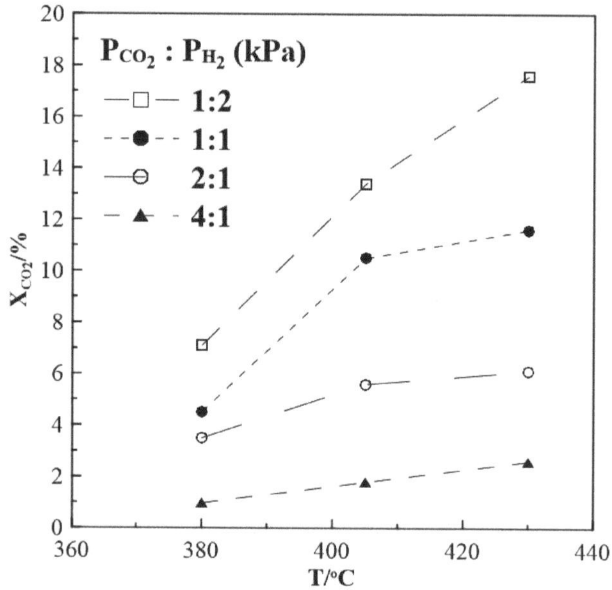

Figure 1. Steady-state effect of temperature on CO_2 conversion. Different gas feeding ratios of CO_2 to H_2. F_T = 50 cm^3 min^{-1}.

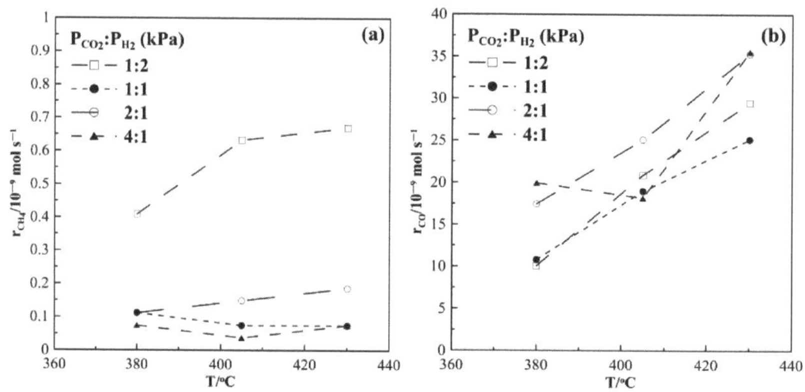

Figure 2. Steady-state effect of temperature on (**a**) CH_4 formation rate and (**b**) CO formation rate. Different gas feeding ratios of CO_2 to H_2. F_T = 50 cm^3 min^{-1}.

Figures 4 and 5 depict the transient effect of constant applied positive (Figure 4) and negative (Figure 5) current on the catalytic rate and turnover frequency (TOF) for the formation of CH_4 and CO at a CO_2 to H_2 ratio of 1:2 at 380 °C. Initially, as presented in Figure 4, at t < 0, the circuit is open and the steady-state formation rates of CH_4 and of CO are equal to 0.38 × 10^{-9} mols^{-1} and 9.2 × 10^{-9} mols^{-1} respectively. At t = 0, a constant anodic current (I = +0.8 mA) is applied between the catalyst and counter electrode, which causes an applied potential of U_{WR} = +1.2 V. Oxygen ions, O^{2-}, are transferred from the YSZ support to the Rh catalyst-electrode at a rate of I/2F equal to approximately 10^{-3} s^{-1}. The rate of CH_4 increases and approaches a new steady-state value (r_{CH4} = 0.62 × 10^{-9} mols^{-1}). This increase of the CH_4 catalytic rate (Δr = 0.24 × 10^{-9} mols^{-1}) is 1.7 times greater than the initial rate achieved under open circuit conditions. The CO formation rate decreases under anodic current application to 8.0 × 10^{-9} mols^{-1}, resulting in a rate enhancement value of ρ = 0.84. After current interruption, the CH_4 rate of formation returns to its initial open circuit value (reversible behavior), while the CO formation rate returns only to a value above the initial open circuit rate (non-reversible behavior). A very similar trend is observed under cathodic current application, as shown in Figure 5. The decrease in CO formation rate is less pronounced with a rate enhancement ratio, ρ, (see Equation (5) in Section 3.3) equal to 0.94. After current interruption, the rate does not return to its initial open circuit value. Under cathodic current application, CH_4 formation rate is increased and a rate enhancement ρ-value of 2.7 is estimated.

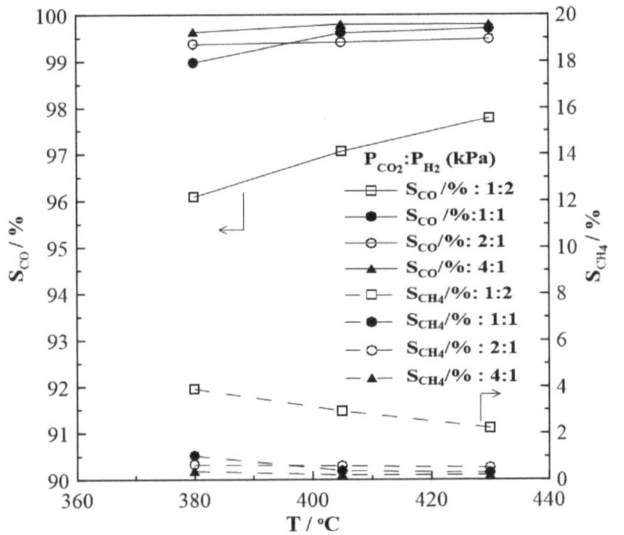

Figure 3. Steady-state effect of temperature on the selectivity to CO and CH_4 formation under open circuit conditions and at different gas feeding ratios of CO_2 to H_2. F_T = 50 cm^3 min^{-1}.

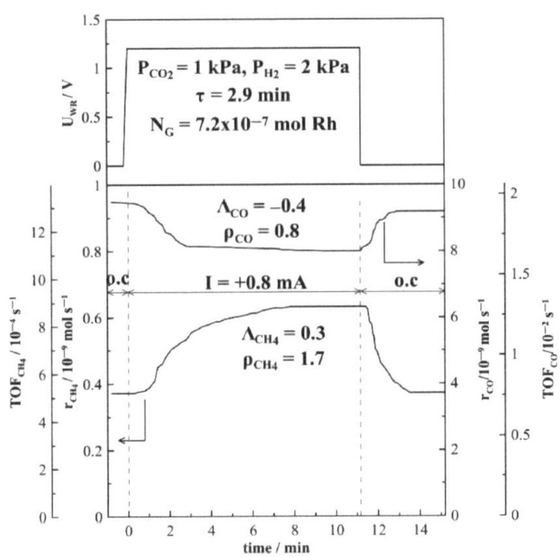

Figure 4. Transient effect of constant applied current (I = +0.8 mA) on the formation rate of CH_4 and CO at T = 380 °C and CO_2 to H_2 ratio of 1:2. F_T = 50 cm^3 min^{-1}.

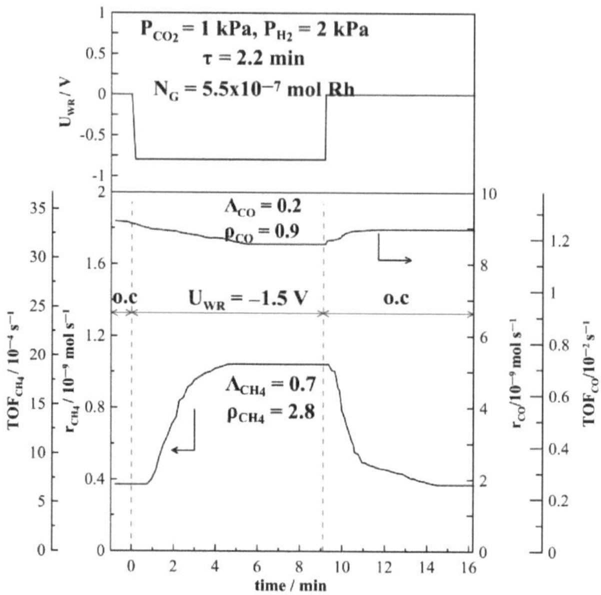

Figure 5. Transient effect of constant applied current (I = −0.8 mA) on the formation rate of CH_4 and CO at T = 380 °C and CO_2 to H_2 ratio of 1:2.

The time constant, τ, given in both Figures 4 and 5, is defined as the time required for the rate increase Δr to reach 63% of its new steady-state value during a galvanostatic transient [38,59]. When an O^{2-} ion conductor is used, the magnitude of τ can be generally predicted by

$$\tau \approx \frac{2FN_G}{I} \tag{3}$$

where N_G (mol) is the reactive oxygen uptake on the metal catalyst. The time constant τ expresses the time required to form a monolayer of $O^{\delta-}$ on the metal surface, while N_G expresses, approximately, the surface mols of metal—here, Rh. The average active catalyst surface area, N_G, which is equal to 7.62×10^{-7} mol Rh, has been calculated based on 6 galvanostatic transients. The estimated value of N_G agrees well with the values found in previous studies of Rh catalyst electrodes supported on YSZ [44,45,56,59]. The average N_G value obtained was used to calculate turnover frequencies (TOFs) for CH_4 and CO formation. TOFs for CH_4 formation are found to be small and in the order of 10^{-4} s^{-1}, which is in good agreement with the literature data [21].

The non-reversible behavior of CO formation has been further investigated by repeated current and/or potential application, which have generally shown that upon current and/or potential interruption the rate of CO formation does not return to its initial open circuit value. Figure 6 summarizes the results for potentiostatic transient operation at different CO_2 to H_2 feed ratios. The bottom part of the figure shows that the increase in CH_4 formation rate under anodic and cathodic polarization (+ and −1.5 V), which is more pronounced at high H_2 to CO_2 feeding ratios, is reversible. However, the CO formation rate, as shown in the top part of Figure 6, exhibits a "permanent", non-reversible NEMCA behavior [60,61]. The guide (dotted) lines clearly show that this non-reversible effect of current and/or potential application is more pronounced at higher CO_2 to H_2 feed ratios.

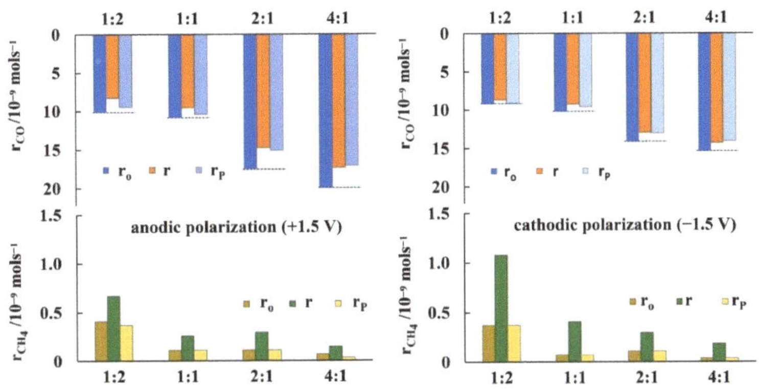

Figure 6. Transient effect of constant applied potential (**left** +1.5 V, **right** −1.5 V) on the formation rates of CH_4 and CO at T = 380 °C and at different ratios of CO_2 to H_2. Dotted lines in the sketch for the CO formation rates point guidance for the non-reversible behavior of the open circuit rate after potential interruption, for details see text. F_T = 50 cm^3 min^{-1}. R_o is the formation rate under open circuit conditions, r is the formation rate under potentiostatic polarization, and r_p corresponds to the formation rate after potential interruption.

The "permanent" rate enhancement ratio γ, was defined for the first time by Comninellis et al. [61] as

$$\gamma = r_p / r_o \tag{4}$$

where r_p is the "permanent" promoted catalytic rate after current or potential interruption, and r_o is the unpromoted rate (i.e., the open-circuit catalytic rate). Figure 7 displays the non-reversibility magnitude of CO formation for different gas feed ratios. Under reaction conditions of a CO_2 to H_2 ratio of 1:2, the γ/ρ ratio is higher than 1, expressing, that the CO formation rate observed under potential application partially returns to its initial open circuit value. However, γ/ρ ratio equals to 1 for a reaction mixture with a CO_2 to H_2 ratio of 4:1 and a significant "permanent" NEMCA behavior is observed.

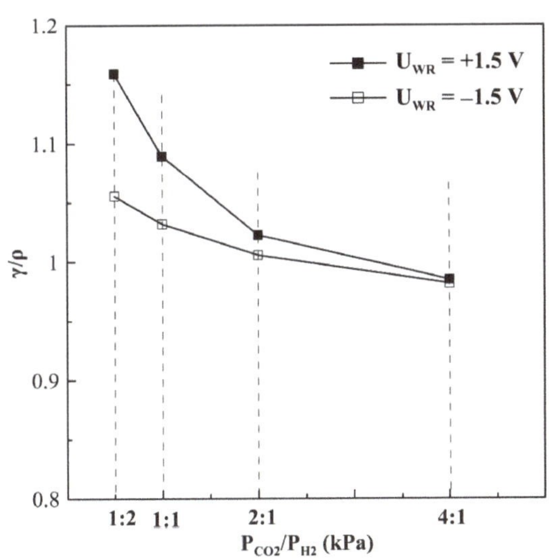

Figure 7. Ratio of "permanent" rate enhancement γ to rate enhancement ρ for CO formation at different gas feeding ratios, T = 380 °C.

In the general case of an oxidation reaction on catalysts deposited on O^{2-} conductors, like YSZ, the ionic species migrating in the solid electrolyte is participating in the electrochemical reaction leading to distinct values of apparent Faradaic efficiency, $|\Lambda| > 1$ (see Equations (6)–(8), Section 3.3). In the present case, however, in which oxygen is not a reactant, any positive current or positive potential-induced catalytic rate change denotes electrochemical promotion, even when $|\Lambda| < 1$ [39].

Figure 8 summarizes the results of steady-state potential application on the rate of CH_4 formation at T = 430 °C, for which the highest rate enhancement ratio, ρ has been found. The observed currents do not depend on gas feed conditions, which means they are almost equal at a given potential and do not depend on the CO_2 to H_2 ratios. This observation is one of the very first hints that the overall oxidation state of the film remains relatively unchanged during the course of the reaction. The rate of CH_4 production displays an inverted volcano type behavior, i.e., it increases with increasing and decreasing catalyst potential as shown in Figure 8a. Maximum changes in the CH_4 rate are observed under reaction conditions of a CO_2 to H_2 ratio of 1:2, but the highest rate enhancement ratios are achieved at conditions of CO_2:H_2 = 1:1, with ρ reaching values of up to 11. If the partial pressure of CO_2 is further increased, i.e., CO_2 to H_2 ratios of 2:1 and 4:1, methane formation is found to be small with values below 0.6×10^{-9} $mols^{-1}$. Despite this, however, methane formation can be electrochemically promoted with ρ-values of up to 5. Product selectivity towards CH_4 is electrochemically promoted under anodic and cathodic polarization; furthermore, it is highest under reaction conditions of a CO_2 to H_2 ratio of 1:2, but does not exceed 10%, as seen in Figure 9.

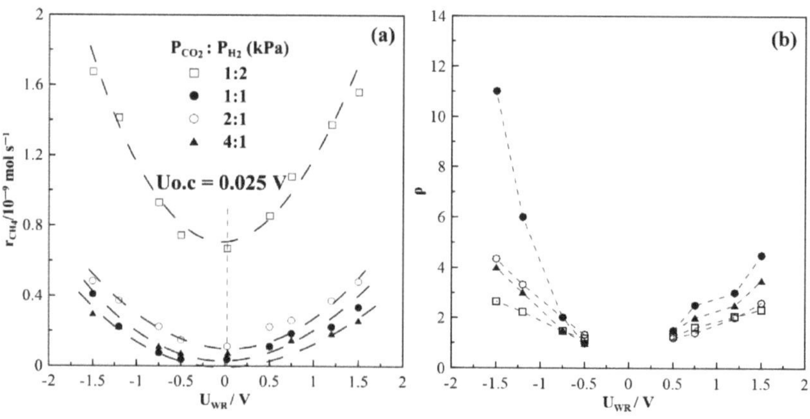

Figure 8. Steady-state effect of potential on (**a**) CH_4 formation rate and (**b**) the corresponding rate enhancement ratio, T = 430 °C, different gas feeding ratios, F_T = 50 cm^3 min^{-1}.

2.2. Catalyst Characterization

The morphology and chemical state of the Rh/YSZ film was assessed with XRD (X-ray diffraction), SEM (scanning electron microscopy) and XPS (X-ray photoelectron spectroscopy). The characterization studies were carried out on 2 different samples, designated as fresh and used, which correspond to the freshly reduced (i.e., after treatment in 15% H_2 in He) and a post-experiment (i.e., after exposure to different CO_2: H_2 gas feed compositions) Rh/YSZ catalyst electrode, respectively.

The diffractograms displayed in Figure 10 show no major differences in the phase distribution of the fresh and used sample. More specifically both diffractograms show clear reflections at 2θ = 30.3°, 35.0°, 50.4°, 59.9°, 62.9°, 74.1°, and 81.9°, which correspond to the (111), (200), (220), (311), (222), (400), (331) planes of YSZ [62–67]. In addition, both diffractograms show clear reflections at 41.5°, 48.4°, and 70.7° which correspond to the (111), (200), and (220) phases Rh particles in their metallic (Rh^0) state [66]. No reflections corresponding to RhO_2 or Rh_2O_3 were detected in the XRD measurements, suggesting that the chemical state of the Rh film is mostly metallic before and after use.

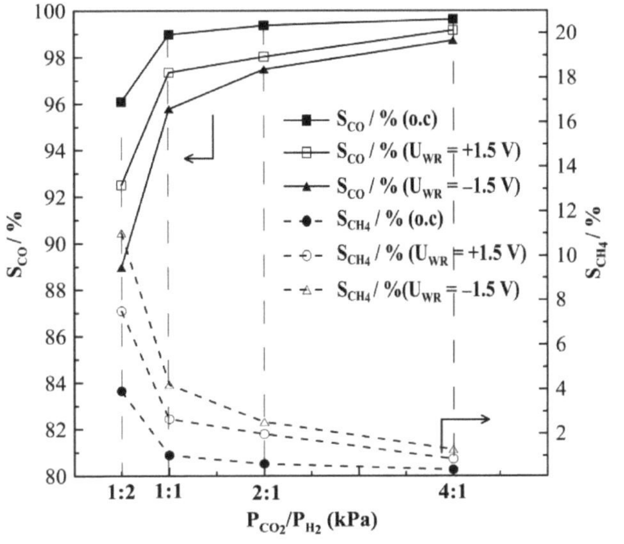

Figure 9. Steady-state effect of gas feeding ratios of CO_2 to H_2 on the selectivity to CO and CH_4 formation on Rh catalyst under open and closed-circuit conditions, T = 380 °C.

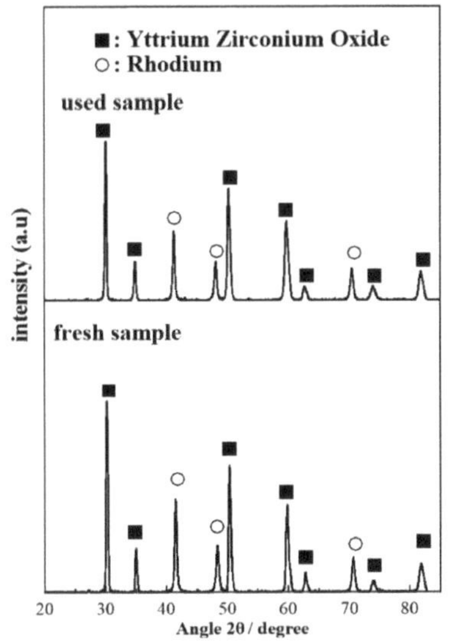

Figure 10. XRD patterns of the freshly reduced and used (after exposing to experimental operating conditions) sample.

The mean Rh crystallite size of the fresh and used samples were calculated by means of the Scherrer equation [65] by assuming a spherical crystallite shape and averaging the crystalline domain diameter obtained for the (111), (200), and (220) reflections. The size of the Rh crystallites in the fresh sample (19.9 ± 0.8 nm) was found to be slightly smaller as compared to the used sample (24 ± 1 nm), which suggests that some degree of crystallite size increase during the course of the electrocatalytic reaction due to agglomeration.

Previous XRD studies on Rh/YSZ systems have shown that after reduction and prior to experimental measurements the metallic phase of Rh is mainly present in the sample with only tracers of metal oxides being detectable [66]. Using metalorganic paste (Engelhard 8826) as a precursor, Jimenez et al. reported rhodium crystallite sizes equal to 58.5 ± 0.3 nm, calculated by the XRD patterns of the reduced sample [66]. The observed size variation and the Rh^0: Rh oxide ratio were attributed to the different conditions of calcination and reduction conditions of the working Rh-catalyst electrode.

Top view and cross section SEM images of the fresh and used catalyst film are shown in Figures 11 and 12, respectively. Overall, the sample preparation method followed in this work has led to a continuous rhodium film, sufficiently porous which facilitates the efficient access of both reactants and products on the Rh active sites, and well attached to the YSZ support to ensure an electrochemically active interface. The apparent average particle size is estimated to be 40 nm, which is roughly twice the crystallite size calculated from the XRD spectra. It is worth mentioning here that particle size estimation from the SEM images is limited by the resolution of the instrument. Both the morphology and structure of the Rh film appears to be unchanged comparing the images from the sample prior experiment and after exposing it to reaction conditions. The thickness of the catalyst film was estimated from the cross-section micrograph and was found to be in the range of 6.5–10 μm, which is a common thickness for films prepared by applying metalorganic pastes [38].

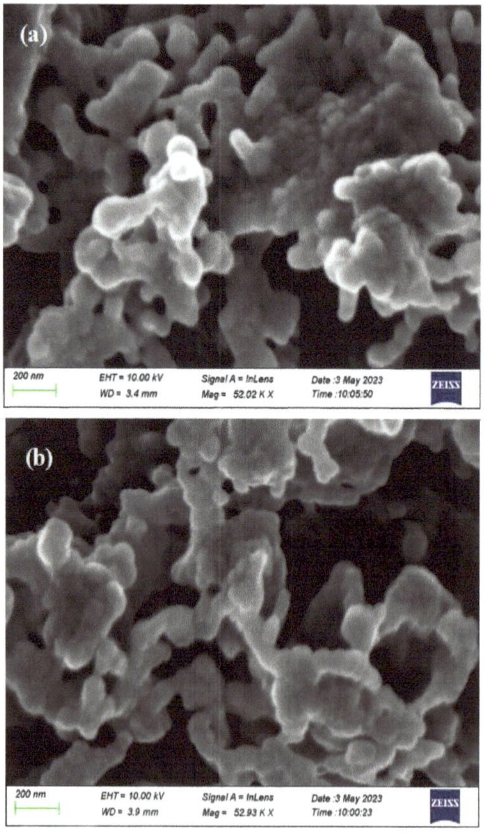

Figure 11. SEM micrographs of Rh/YSZ from top view (**a**) fresh sample and (**b**) used sample.

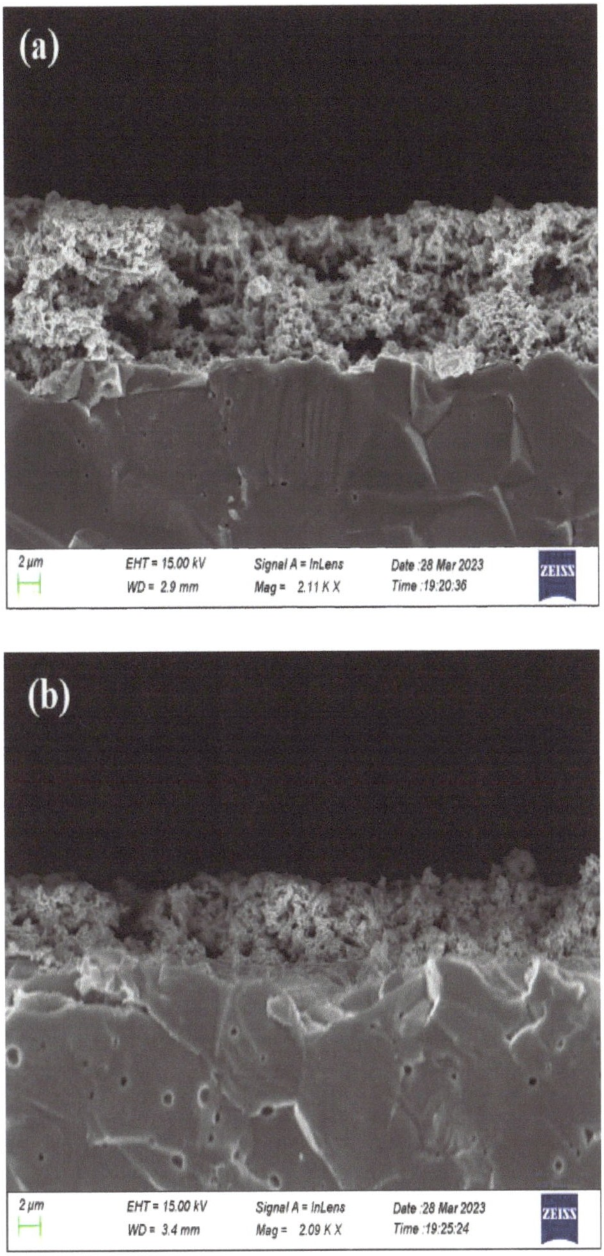

Figure 12. SEM micrographs of Rh/YSZ from cross section (**a**) fresh sample and (**b**) used sample.

Figure 13a shows the Rh 3d spectra of the freshly reduced and post-reaction (used) sample. The Rh $3d_{5/2}$ peak of the fresh sample is centred at approximately 307.2 eV, which indicates that the Rh catalyst film is mostly in the metallic state, without excluding the possibility of some minor Rh oxide quantities being also present [66]. The post-reaction data suggest that the surface chemical state of the Rh electrocatalyst film has not changed significantly. A rather small shift of ~0.2 eV towards lower binding energy is observed suggesting that a minor oxide component may be present in the fresh sample which has been reduced after reaction leading to the apparent shift. To clarify this point, the raw Rh3d spectra were deconvoluted using an asymmetric peak shape for the metallic state and Gaussian–Lorentzian peak shape for the oxidized state. Figure 13a shows that the metallic $Rh3d_{5/2}$ component (solid line) is centred at ~307 eV while the oxidized $Rh3d_{5/2}$ component (dashed line) is centred at ~308 eV. The ratio of metallic Rh to oxidized Rh is 1:9, confirming that Rh film is nearly metallic both before and after reaction. It is worth noting here that the post-reaction spectra have been acquired after testing the catalyst under reaction conditions of CO_2 to H_2 of 1:4. The XPS results are in close agreement with the XRD data presented above (Figure 10), with the two techniques suggesting the absence of any substantial quantity of Rh oxide on the surface or bulk of the Rh film. Minor quantities of carbon were present on both the fresh and used samples (Figure 13b). More crucially, the amount of carbon was not found to increase in the post-reaction sample, which indicates that there is no carbon deposition on the catalyst surface during the reaction, especially under feed conditions of P_{CO2}:P_{H2} ratios of 4:1.

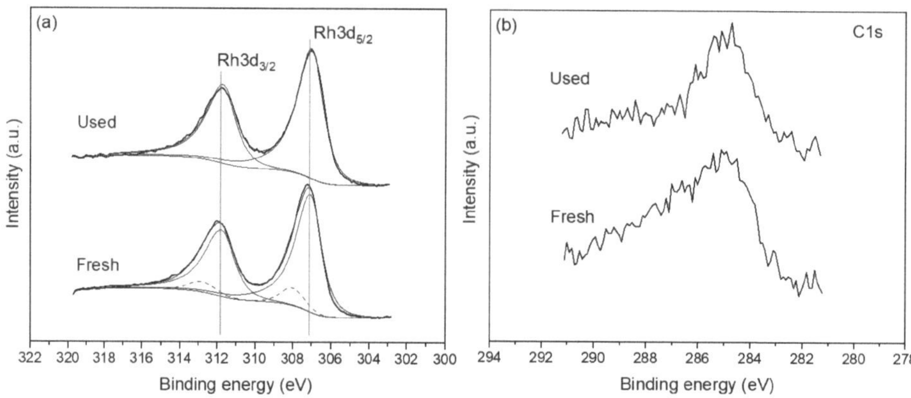

Figure 13. XPS measurements of the freshly reduced and post-reaction used catalyst sample. (**a**) Rh 3d and (**b**) C 1s regions.

3. Experimental

3.1. Catalyst Preparation

A disc of 8% mol Y_2O_3-stabilized ZrO_2 (YSZ), with 17 mm diameter and 1.5 mm thickness, respectively, was utilized as solid electrolyte. Gold (Au) reference and counter electrodes were deposited on one side of the electrolyte, using Au paste (Metalor, AU 201 Gold Resinate, Birmingham, UK) and calcination at 450 °C for 30 min following a sintering at 700 °C for 1 h, with a heating rate of 10 °C min^{-1}.

The working electrode (Rh catalyst electrode) was deposited on the opposite side of the YSZ solid electrolyte by application of thin coatings of Rh paste (Engelhard Rhodium Resinate 8826, Cinderford, UK). Calcination was carried out in air at 550 °C for 3 h with a heating rate of 10 °C min^{-1}. Prior electrochemical experiments, the Rh catalyst electrode was reduced at 440 °C for 2 h under a flow of 15% H_2 in He. The total mass of the Rh catalyst electrode was 1.8 mg.

3.2. Reactor Operation

The electrocatalytic experiments were carried out in a continuous single pellet flow reactor, which has been discussed previously [38–40]. The feed gas composition and total flow rates were controlled by a set of electronic flow meters (Brooks Instrument, Hatfield, PA, USA). Reactants were certified standards of 5% CO_2 in He and 15% H_2 in He. Pure He (99.999%) was fed to further adjust the total flow rate and the inlet gas composition at the desired values. Electrochemical experiments were performed at ambient pressure. Feed reactant partial pressure was varied between 1 to 4 kPa CO_2 and 1 to 7 kPa H_2.

The concentrations of reactants and products were analyzed with an IR CO_2-CO-CH_4 gas analyzer (Futzi Electric ZRE, Shinagama City, Tokyo, Japan). Constant currents and potentials were applied using an AMEL 2053 galvanostat-potensiostat. CO_2, CO, and CH_4 were further analyzed by gas chromatography (Gas Chromatographer, Shimadzu 2014, Kyoto, Japan) with TCD and FID detectors, using a Porapak QS packed column.

Blank experiments have been carried out to confirm the non-catalytic behavior of the Au counter and reference electrodes under open- and closed-circuit conditions. For this purpose, an electrochemical cell of the type Au/YSZ/Au (replacing the Rh working electrode with Au) was investigated under reaction conditions of this study (T = 400 °C, F_T = 100 cm^3 min^{-1} and different H_2 to CO_2 gas feeding ratios). The obtained CO formation rate on Au was small with values of $r_{CO} < 10^{-9}$ mols^{-1} and it was not affected by potential application. CH_4 formation on Au was not observed.

3.3. Electrochemical Promotion Parameters Computation

Figure 14 shows a schematic representation of the (Rh | YSZ | Au) electrochemical cell under (a) open circuit and (b) closed circuit conditions. WE correspond to the working electrode (Rh deposited catalytic active film), CE to the counter, and RE to the reference electrodes both made out of Au in order to be catalytically inactive. All 3 electrodes are deposited on yttria-stabilized zirconia (YSZ) which serves as a solid O^{2-} ion conductor at operating temperatures of 380 to 430 °C. Under open circuit conditions (i.e., no current passing through the cell), H_2 and CO_2 are co-fed over the conductive catalyst film leading to the formation of CO and CH_4, expressed by the unpromoted catalytic rate, r_o. Under, for example, anodic polarization, the application of a positive potential between the counter and working electrode, O^{2-} ions from the solid electrolyte are migrating to the catalyst-film forming at the metal gas interface an effective double layer, that alters the catalytic activity, and leading to the electropromoted rate, r.

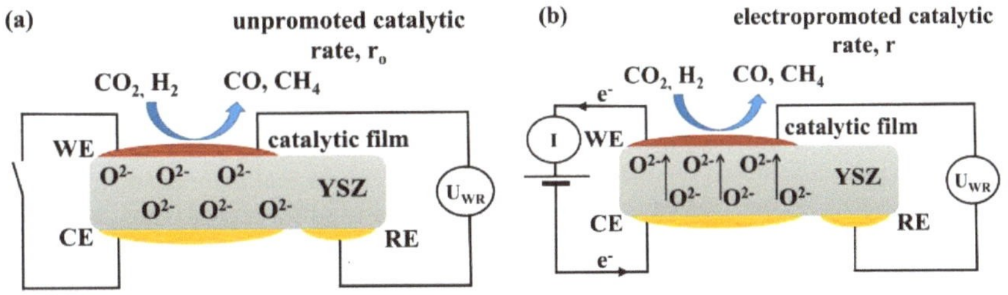

Figure 14. Schematic presentation of the electrochemical cell under (**a**) open and (**b**) closed circuit conditions, WE corresponds to the working electrode, CE to counter, and RE to reference electrode.

The non-Faradaic behavior is defined by two basic parameters, the rate enhancement ratio, ρ,

$$\rho = \frac{r}{r_o} \quad (5)$$

where r is the electropromoted catalytic rate under polarization and r_o is the unpromoted rate (i.e., the open-circuit catalytic rate). The apparent Faradaic efficiency, Λ, defined from:

$$\Lambda = \frac{\Delta r}{I/nF} \quad (6)$$

where Δr is the current- or potential-induced observed change in catalytic rate ($\Delta r = r - r_o$), I is the applied current, n is the number of exchanged electrons, and F is Faraday's constant.

In this study, where CH_4 and CO were the only products of the CO_2 hydrogenation reaction, the Faradaic efficiency for each formation reaction can be defined from the follow equations:

$$\Lambda_{CO} = \frac{2\Delta r_{CO}\left(\text{in}\,\frac{\text{mol}}{\text{s}}\right)}{I/F} \quad (7)$$

$$\Lambda_{CH_4} = \frac{8\Delta r_{CH_4}\left(\text{in}\,\frac{\text{mol}}{\text{s}}\right)}{I/F} \quad (8)$$

3.4. Catalyst Characterization

Wide-angle X-ray diffraction (XRD) patterns were recorded using a Bruker D8 Advance system equipped with Cu–Ka radiation (λ = 1.5418 Å). The voltage and lamp current were adjusted to 40 kV and 40 mA, respectively. The angle range 2θ was 20–90° and scan speed was 0.3 s/step (Siemens D-500, Frankfurt, Germany).

High resolution field-emission scanning electron microscopy (FE-SEM) was employed using a Zeiss SUPRA 35VP system operating at 10–15 kV voltage range (Carl Zeiss AG, Jena, Germany).

Ex situ X-ray photoelectron spectroscopy (XPS) measurements of the fresh Rh catalyst sample (after H_2 reduction) and post-experiment sample were carried out in an ultrahigh-vacuum (UHV) system described in detail elsewhere [68]. Measurements were carried out using non-monochromatic AlKα radiation (1486.6 eV) and a Leybold LH EA11 energy analyzer (Leybold, Germany), which was operated at a constant pass energy (100 eV). The analyzed sample area was a 2 × 5 mm^2 rectangle. All spectra were corrected for charge transfer using the C1s peak at 284.8 eV for adventitious carbon

4. Conclusions

In this work, the electrochemical promotion of CO_2 hydrogenation was studied over a Rh/YSZ electrode in a continuous single pellet flow reactor for P_{CO2}/P_{H2} gas feed mixtures ranging from 1:2 to 4:1. Despite the large body of literature on the use of Rh/YSZ electrodes for other reactions such as NO reduction and ethylene oxidation under various experimental set-ups, CO_2 hydrogenation has been rarely investigated on Rh/YSZ using laboratory reactors. There are only a couple of references discussing data under reaction conditions of a narrow CO_2 to H_2 ratio of 1–1.5 or 1:1.8 and cathodic polarizations. The results of the present study show that potential or current application can strongly affect the CO_2 hydrogenation reaction, both in regard to the catalytic rates obtained but also selectivity to reactions products. In the temperature range of 350 to 430 °C, under atmospheric total pressure and with a total flowrate of F_T = 50 cm^3 min^{-1}, the only products found were CH_4 and CO. Increasing or decreasing the catalyst potential by up to 1.5 V leads to an inverted volcano type behavior of the methanation rate (the formation rate increases with anodic and cathodic polarization), while the rate of CO production follows a volcano type behavior (the formation rate decreases with anodic and cathodic polarization). The observed trends are confirmed by kinetic measurements and agree with the rules governing the EPOC phenomenon as described in bibliography.

The observed rate changes were non-faradaic and after current interruption, the catalytic rate of CH_4 formation returned to its initial open circuit value, demonstrating the reversibility of the phenomenon. However, the CO formation rate, exhibited a partial non-reversible behavior, which is more pronounced under a gas feed composition of CO_2 to H_2 ratio of 4:1., i.e., in excess of CO_2. Permanent EPOC behavior was discussed in view of the stabilization of the Rh metallic phase under anodic polarization. Our preliminary XPS studies show that for pre- and post-reaction samples the Rh metal phase is mainly present, even after exposing the Rh film to higher CO_2 to H_2 ratios. This confirms that, at first glance, the partially permanent EPOC behavior in our case is due to the electrochemically induced stabilization of the Rh metallic phase.

Author Contributions: Conceptualization, A.K. and S.B.; methodology, S.B., G.K. and A.K.; investigation, N.K. and F.X.; resources, N.K., F.X. and S.B.; data curation, N.K. and F.X.; writing—original draft preparation, N.K. and S.B.; writing—review and editing, S.B., G.K. and A.K.; supervision, S.B., G.K. and A.K.; funding acquisition, G.K. and A.K. All authors have read and agreed to the published version of the manuscript.

Funding: NK thanks the National Scholarship Foundation (IKY) for financial support through the Program code: 5113934.

Data Availability Statement: All the experimental data presented in the current manuscript.

Acknowledgments: Authors are thankful to Panagiota Natsi for XRD patterns, and FORTH ICE-HT for SEM images.

Conflicts of Interest: The authors declare no conflict of interest.

References

1. Kinney, P.L. Interactions of Climate Change, Air Pollution, and Human Health. *Curr. Environ. Health Rep.* **2018**, *5*, 179–186. [CrossRef] [PubMed]
2. Berry, H.L.; Waite, T.D.; Dear, K.B.G.; Capon, A.G.; Murray, V. The Case for Systems Thinking about Climate Change and Mental Health. *Nat. Clim. Chang.* **2018**, *8*, 282–290. [CrossRef]
3. Kattel, S.; Liu, P.; Chen, J.G. Tuning Selectivity of CO_2 Hydrogenation Reactions at the Metal/Oxide Interface. *J. Am. Chem. Soc.* **2017**, *139*, 9739–9754. [CrossRef] [PubMed]
4. Knutson, T.R.; Tuleya, R.E. Impact of CO_2-Induced Warming on Simulated Hurricane Intensity and Precipitation: Sensitivity to the Choice of Climate Model and Convective Parameterization. *J. Clim.* **2004**, *17*, 3477–3495. [CrossRef]
5. Schwiderowski, P.; Ruland, H.; Muhler, M. Current Developments in CO_2 Hydrogenation towards Methanol: A Review Related to Industrial Application. *Curr. Opin. Green Sustain. Chem.* **2022**, *38*, 100688. [CrossRef]
6. Xu, D.; Wang, Y.; Ding, M.; Hong, X.; Liu, G.; Tsang, S.C.E. Advances in Higher Alcohol Synthesis from CO_2 Hydrogenation. *Chem* **2021**, *7*, 849–881. [CrossRef]
7. Chatzilias, C.; Martino, E.; Zagoraios, D.; Kyriakou, G.; Katsaounis, A. Electrochemical Promotion of Catalysis for CO_2 Valorization. In *Recent Advances in Electrochemical Promotion of Catalysis*; Modern Aspects of Electrochemistry; Vernoux, P., Vayenas, C.G., Eds.; Springer: Cham, Switzerland, 2022; Volume 61.
8. Nozaki, F.; Sodesawa, T.; Satoh, S.; Kimura, K. Hydrogenation of Carbon Dioxide into Light Hydrocarbons at Atmospheric Pressure over Rh/Nb_20_5 or $Cu/Si0_2$-Rh/Nb_20_5. *J. Catal.* **1987**, *104*, 339–346. [CrossRef]
9. Solymosi, F.; Pasztor, M. Analysis of the R-Spectral Behavior of Adsorbed CO Formed in $H_2 + CO_2$ Surface Interaction over Supported Rhodium. *J. Catal.* **1987**, *104*, 312–322. [CrossRef]
10. Lapidus, A.L.; Gaidai, N.A.; Nekrasov, N.V.; Tishkova, L.A.; Agafonov, Y.A.; Myshenkova, T.N. The Mechanism of Carbon Dioxide Hydrogenation on Copper and Nickel Catalysts. *Pet. Chem.* **2007**, *47*, 91–98. [CrossRef]
11. Borodko, Y.; Somorjai, G.A. Catalytic Hydrogenation of Carbon Oxides-a 10-Year Perspective. *Appl. Catal. A Gen.* **1999**, *186*, 355–362. [CrossRef]
12. Kusama, H.; Kitamura Bando, K.; Okabe, K.; Arakawa, H. Effect of Metal Loading on CO_2 Hydrogenation Reactivity over Rh/SiO_2 Catalysts. *Appl. Catal.* **2000**, *197*, 255–268. [CrossRef]
13. Gasser, D.; Baiker, A. Hydrogenation of Carbon Dioxide over Copper-Zirconia Catalysts Prepared by In-Situ Activation of Amorphous Copper-Zirconium Alloy. *Appl. Catal.* **1989**, *48*, 279–294. [CrossRef]
14. Amenomiya, Y. Methanol synthesis from CO + H_2. II Copper-based binary and ternary catalysts. *Appl. Catal.* **1987**, *30*, 57–68. [CrossRef]
15. Sahibzada, M.; Chadwick, D.; Metcalfe, I.S. Hydrogenation of Carbon Dioxide to Methanol over Palladium-Promoted $Cu/ZnO/A1_2O_3$ Catalysts. *Catal. Today* **1996**, *29*, 367–372. [CrossRef]
16. Solymosi, F.; Erdhelyi, A.; Bansagi, T. Methanation of CO, on Supported Rhodium Catalyst. *J. Catal.* **1981**, *68*, 371–382. [CrossRef]
17. Yang, R.; Zhang, Y.; Tsubaki, N. Dual Catalysis Mechanism of Alcohol Solvent and Cu Catalyst for a New Methanol Synthesis Method. *Catal. Commun.* **2005**, *6*, 275–279. [CrossRef]
18. Schild, C.; Wokaun, A.; Baiker, A. On the Mechanism of CO and CO_2 Hydrogenation Reactions on Zirconia-Supported Catalysts: A Diffuse Reflectance FTIR Study Part II. Surface Species on Copper/Zirconia Catalysts: Implications for Methanol Synthesis Selectivity. *J. Mol. Catal.* **1990**, *63*, 243–254. [CrossRef]
19. Vannice, M.A. The Catalytic Synthesis of Hydrocarbons from HJCO Mixtures over the Group VIII Metals I. The Specific Activities and Product Distributions of Supported Metals. *J. Catal.* **1975**, *31*, 449–461. [CrossRef]
20. Araki, R.; Ponec, V. Methanation of Carbon Monoxide on Nickel and Nickel-Copper Alloys. *J. Catal.* **1976**, *44*, 439–448. [CrossRef]
21. Panagiotopoulou, P.; Kondarides, D.I.; Verykios, X.E. Selective Methanation of CO over Supported Noble Metal Catalysts: Effects of the Nature of the Metallic Phase on Catalytic Performance. *Appl. Catal. A Gen.* **2008**, *344*, 45–54. [CrossRef]
22. Henderson, A.; Worley, S.D.; Peebles, D.E. An Infrared Study of the Hydrogenation of Carbon Dioxide on Supported Rhodium Catalys An inverse spillover effect. *J. Phys. Chem.* **1985**, *89*, 392–394. [CrossRef]

23. Solymosi, F.; Tombacz, I.; Koszta, J. Effects of Variation of Electric Properties of TiOp Support on Hydrogenation of CO and CO_2 over Rh Catalysts. *J. Catal.* **1985**, *95*, 578–586. [CrossRef]
24. Brown Bourzutschky, J.A.; Homs, N.; Bell, A.A.T. Hydrogenation of 002 and 002/00 Mixtures over Copper-Containing Catalysts. *J. Catal.* **1990**, *124*, 73–85. [CrossRef]
25. Chanchlani, K.G.; Hudgins, R.R.; Silveston, P.L. Methanol Synthesis from H_2, CO, and 002 over Cu/ZnO Catalysts. *J. Catal.* **1992**, *136*, 59–75. [CrossRef]
26. Nitta, Y.; Suwata, O.; Ikeda, Y.; Okamoto, Y.; Imanaka, T. Copper-Zirconia Catalysts for Methanol Synthesis from Carbon Dioxide: Effect of ZnO Addition to Cu-ZrO_2 Catalysts. *Catal. Lett.* **1994**, *26*, 345–354. [CrossRef]
27. Arena, F.; Italiano, G.; Barbera, K.; Bordiga, S.; Bonura, G.; Spadaro, L.; Frusteri, F. Solid-State Interactions, Adsorption Sites and Functionality of Cu-ZnO/ZrO_2 Catalysts in the CO_2 Hydrogenation to CH_3OH. *Appl. Catal. A Gen.* **2008**, *350*, 16–23. [CrossRef]
28. Reubroycharoen, P.; Vitidsant, T.; Yoneyama, Y.; Tsubaki, N. Development of a New Low-Temperature Methanol Synthesis Process. *Catal. Today* **2004**, *89*, 447–454. [CrossRef]
29. Hori, Y.; Wakebe, H.; Tsukamoto, T.; Koga, O. Adsorption of CO Accompanied with Simultaneous Charge Transfer on Copper Single Crystal Electrodes Related with Electrochemical Reduction of CO_2 to Hydrocarbons. *Surf. Sci.* **1995**, *335*, 258–260. [CrossRef]
30. Ando, H.; Xu, Q.; Fujiwara, M.; Matsumura, Y.; Tanaka, M.; Souma, Y. Hydrocarbon Synthesis from CO_2 over Fe $\pm$ Cu. *Catal. Today* **1998**, *45*, 229–234. [CrossRef]
31. Trovarelli, A.; Mustazza, C.; Dolcetti, G.; Kagpar, J.; Graziani, M. Carbon dioxide hydrogenation on rhodium supported on transition metal oxides. Effect of reduction temperature on product distribution. *Appl. Catal.* **1990**, *65*, 129–142. [CrossRef]
32. Marwood, M.; Doepper, R.; Renken, A. In-Situ Surface and Gas Phase Analysis for Kinetic Studies under Transient Conditions: The Catalytic Hydrogenation of CO_2. *Appl. Catal.* **1997**, *151*, 223–246. [CrossRef]
33. Falconer, J.L.; Zagli, A.E. Adsorption and Methanation of Carbon Dioxide on a Nickel/Silica Catalyst. *J. Catal.* **1980**, *62*, 280–285. [CrossRef]
34. Biloen, P.; Helle, J.N.; Van Den Berg, F.G.A.; Sachtler, W.M.H. On the Activity of Fischer-Tropsch and Methanation Catalysts: A Study Utilizing Isotopic Transients. *J. Catal.* **1983**, *81*, 450–463. [CrossRef]
35. Coenen, J.W.E.; Van Nisselrooy, P.F.M.T.; De Croon, M.H.J.M.; Van Dooren, P.F.H.A.; Van Meerten, R.Z.C. The dynamics of methanation of carbon modoxide on Nickel catalysts. *Appl. Catal.* **1986**, *25*, 1–8. [CrossRef]
36. Zheng, Y.; Zhang, J.; Ma, Z.; Zhang, G.; Zhang, H.; Fu, X.; Ma, Y.; Liu, F.; Liu, M.; Huang, H. Seeded Growth of Gold–Copper Janus Nanostructures as a Tandem Catalyst for Efficient Electroreduction of CO_2 to C_{2+} Products. *Small* **2022**, *18*, 2201695. [CrossRef]
37. Zhong, Y.; Kong, X.; Song, Z.; Liu, Y.; Peng, L.; Zhang, L.; Luo, X.; Zeng, J.; Geng, Z. Adjusting Local CO Confinement in Porous-Shell Ag@Cu Catalysts for Enhancing C-C Coupling toward CO_2 Eletroreduction. *Nano Lett.* **2022**, *22*, 2554–2560. [CrossRef]
38. Vayenas, C.G.; Bebelis, S.; Pliangos, C.; Brosda, S.; Tsiplakides, D. *Electrochemical Activation of Catalysis Promotion, Electrochemical Promotion, and Metal-Support Interactions*; Springer: New York, NY, USA, 2002.
39. Theleritis, D.; Souentie, S.; Siokou, A.; Katsaounis, A.; Vayenas, C.G. Hydrogenation of CO_2 over Ru/YSZ Electropromoted Catalysts. *ACS Catal.* **2012**, *2*, 770–780. [CrossRef]
40. Kalaitzidou, I.; Makri, M.; Theleritis, D.; Katsaounis, A.; Vayenas, C.G. Comparative Study of the Electrochemical Promotion of CO_2 Hydrogenation on Ru Using Na^+, K^+, H^+ and O^{2-} Conducting Solid Electrolytes. *Surf. Sci.* **2016**, *646*, 194–203. [CrossRef]
41. Panaritis, C.; Michel, C.; Couillard, M.; Baranova, E.A.; Steinmann, S.N. Elucidating the Role of Electrochemical Polarization on the Selectivity of the CO_2 Hydrogenation Reaction over Ru. *Electrochim. Acta* **2020**, *350*, 136405. [CrossRef]
42. Chatzilias, C.; Martino, E.; Tsatsos, S.; Kyriakou, G.; Katsaounis, A.; Vayenas, C.G. Kinetic study of CO_2 hydrogenation on Ru/YSZ catalyst using a monolithic electropromoted reactor (MEPR). *Chem. Eng. J.* **2022**, *430*, 132967. [CrossRef]
43. Zagoraios, D.; Panaritis, C.; Krassakopoulou, A.; Baranova, E.A.; Katsaounis, A.; Vayenas, C.G. Electrochemical Promotion of Ru Nanoparticles Deposited on a Proton Conductor Electrolyte during CO_2 Hydrogenation. *Appl. Catal. B* **2020**, *276*, 119148. [CrossRef]
44. Bebelis, S.; Karasali, H.; Vayenas, C.G. Electrochemical Promotion of CO_2 Hydrogenation on Rh/YSZ Electrodes. *J. Appl. Electrochem.* **2008**, *38*, 1127–1133. [CrossRef]
45. Papaioannou, E.I.; Souentie, S.; Hammad, A.; Vayenas, C.G. Electrochemical Promotion of the CO_2 Hydrogenation Reaction Using Thin Rh, Pt and Cu Films in a Monolithic Reactor at Atmospheric Pressure. *Catal. Today* **2009**, *146*, 336–344. [CrossRef]
46. Kotsiras, A.; Kalaitzidou, I.; Grigoriou, D.; Symillidis, A.; Makri, M.; Katsaounis, A.; Vayenas, C.G. Electrochemical Promotion of Nanodispersed Ru-Co Catalysts for the Hydrogenation of CO_2. *Appl. Catal. B* **2018**, *232*, 60–68. [CrossRef]
47. Panaritis, C.; Zgheib, J.; Ebrahim, S.A.H.; Couillard, M.; Baranova, E.A. Electrochemical In-Situ Activation of Fe-Oxide Nanowires for the Reverse Water Gas Shift Reaction. *Appl. Catal. B* **2020**, *269*, 118826. [CrossRef]
48. Nicole, J.; Comninellis, C.; Tsiplakides, D.; Pliangos, C.; Verykios, X.E.; Vayenas, C.G. Electrochemical Promotion and Metal-Support Interactions. *J. Catal.* **2001**, *204*, 23–34. [CrossRef]
49. Stoukides, M.; Vayenas, C.G. The Effect of Electrochemical Oxygen Pumping on the Rate and Selectivity of Ethylene Oxidation on Polycrystalline Silver. *J. Catal.* **1981**, *70*, 137–146. [CrossRef]
50. Vayenas, C.G.; Koutsodontis, C.G. Non-Faradaic Electrochemical Activation of Catalysis. *J. Chem. Phys.* **2008**, *128*, 182506. [CrossRef]

51. Vayenas, C.G.; Bebelis, S.; Ladas, S. Dependence of Catalytic Rates on Catalyst Work Function. *Nature* **1990**, *343*, 625–627. [CrossRef]
52. Bebelis, S.; Vayenas, C.G. Non-Faradaic Electrochemical Modification of Catalytic Activity 1. The Case of Ethylene Oxidation on Pt. *J. Catal.* **1989**, *118*, 125–146. [CrossRef]
53. Vayenas, C.G.; Bebelis, S.; Despotopoulou, M. Non-Faradaic Electrochemical Modification of Catalytic Activity 4. The Use of B"-Al_2O_3 as the Solid Electrolyte. *J. Catal.* **1991**, *128*, 415–435. [CrossRef]
54. Ladas, S.; Bebelis, S.; Vayenas, C.G. Work Function Measurements on Catalyst Films Subject to in Situ Electrochemical Promotion. *Surf. Sci.* **1991**, *251–252*, 1062–1068. [CrossRef]
55. Neophytides, S.G.; Tsiplakides, D.; Stonehart, P.; Jaksic, M.M.; Vayenas, C.G. Electrochemical Enhancement of a Catalytic Reaction in Aqueous Solution. *Nature* **1994**, *370*, 45–47. [CrossRef]
56. Pliangos, C.; Raptis, C.; Badas, T.; Tsiplakides, D.; Vayenas, C.G. Electrochemical Promotion of a Classically Promoted Rh Catalyst for the Reduction of NO. *Electrochim. Acta* **2000**, *46*, 331–339. [CrossRef]
57. Vernoux, P.; Gaillard, F.; Bultel, L.; Siebert, E.; Primet, M. Electrochemical Promotion of Propane and Propene Oxidation on Pt/YSZ. *J. Catal.* **2002**, *208*, 412–421. [CrossRef]
58. Katsaounis, A.; Teschner, D.; Zafeiratos, S. The Effect of Polarization and Reaction Mixture on the Rh/YSZ Oxidation State During Ethylene Oxidation Studied by Near Ambient Pressure XPS. *Top. Catal.* **2018**, *61*, 2142–2151. [CrossRef]
59. Constantinou, I.; Archonta, D.; Brosda, S.; Lepage, M.; Sakamoto, Y.; Vayenas, C.G. Electrochemical Promotion of NO Reduction by C_3H_6 on Rh Catalyst-Electrode Films Supported on YSZ and on Dispersed Rh/YSZ Catalysts. *J. Catal.* **2007**, *251*, 400–409. [CrossRef]
60. Souentie, S.; Xia, C.; Falgairette, C.; Li, Y.D.; Comninellis, C. Investigation of the "Permanent" Electrochemical Promotion of Catalysis (P-EPOC) by Electrochemical Mass Spectrometry (EMS) Measurements. *Electrochem. Commun.* **2010**, *12*, 323–326. [CrossRef]
61. Falgairette, C.; Jaccoud, A.; Fóti, G.; Comninellis, C. The Phenomenon of "Permanent" Electrochemical Promotion of Catalysis (P-EPOC). *J. Appl. Electrochem.* **2008**, *38*, 1075–1082. [CrossRef]
62. Kokka, A.; Petala, A.; Panagiotopoulou, P. Support Effects on the Activity of Ni Catalysts for the Propane Steam Reforming Reaction. *Nanomaterials* **2021**, *11*, 1948. [CrossRef]
63. Yentekakis, I.V.; Goula, G.; Panagiotopoulou, P.; Kampouri, S.; Taylor, M.J.; Kyriakou, G.; Lambert, R.M. Stabilization of Catalyst Particles against Sintering on Oxide Supports with High Oxygen Ion Lability Exemplified by Ir-Catalyzed Decomposition of N_2O. *Appl. Catal. B* **2016**, *192*, 357–364. [CrossRef]
64. Blomberg, S.; Lundgren, E.; Westerström, R.; Erdogan, E.; Martin, N.M.; Mikkelsen, A.; Andersen, J.N.; Mittendorfer, F.; Gustafson, J. Structure of the Rh_2O_3(0001) Surface. *Surf. Sci.* **2012**, *606*, 1416–1421. [CrossRef]
65. Holzwarth, U.; Gibson, N. The Scherrer Equation versus the "Debye-Scherrer Equation". *Nat. Nanotechnol.* **2011**, *6*, 534. [CrossRef] [PubMed]
66. Jiménez-Borja, C.; De Lucas-Consuegra, A.; Sapountzi, F.; Dorado, F.; Katsaounis, A.; Valverde, J.L. Oscillatory Behavior of Rh/YSZ under Electropromoted Conditions. *Chem. Phys. Lett.* **2012**, *519–520*, 89–92. [CrossRef]
67. Hamdy, M.S.; Alhanash, A.M.; Benaissa, M.; Alsalme, A.; Alharthi, F.A.; Al-Zaqri, N. Rhodium Nanoparticles Incorporated Mesoporous Silica as an Active Catalyst for Cyclohexene Hydrogenation under Ambient Conditions. *Catalysts* **2020**, *10*, 925. [CrossRef]
68. Tsatsos, S.; Kyriakou, G. Copper Growth on a Stepped Nickel Surface: Electronic and Geometric Effects on CO Reactivity. *J. Phys. Chem.* **2023**, *127*, 6337–6346. [CrossRef]

catalysts

MDPI

Article

$Nd_{2-x}Sr_xNiO_4$ Solid Solutions: Synthesis, Structure and Enhanced Catalytic Properties of Their Reduction Products in the Dry Reforming of Methane

Oleg A. Shlyakhtin [1,*], Grigoriy M. Timofeev [1], Sergey A. Malyshev [1,2], Alexey S. Loktev [3,4,5], Galina N. Mazo [1], Tatiana Shatalova [1,2,6], Veronika Arkhipova [4], Ilya V. Roslyakov [5,6] and Alexey G. Dedov [3,4,5]

1 Department of Chemistry, M.V. Lomonosov Moscow State University, 119991 Moscow, Russia
2 Department of Materials Sciences, Shenzhen MSU-BIT University, Shenzhen 518172, China
3 A.V. Topchiev Institute of Petrochemical Synthesis, Russian Academy of Sciences, 119991 Moscow, Russia
4 Department of General and Applied Chemistry, Gubkin Russian State University of Oil and Gas, 119991 Moscow, Russia
5 N.S. Kurnakov Institute of General and Inorganic Chemistry, Russian Academy of Sciences, 119991 Moscow, Russia
6 Department of Materials Sciences, M.V. Lomonosov Moscow State University, 119991 Moscow, Russia
* Correspondence: oleg@inorg.chem.msu.ru

Abstract: Solid solutions $Nd_{2-x}Sr_xNiO_{4\pm\delta}$ (x = 0, 0.5, 1, 1.2, 1.4) with a K_2NiF_4 structure can be obtained from freeze-dried precursors. The end members of this series can be obtained at T ≥ 1000 °C only, while complex oxides with x = 1; 1.5 are formed at T ≥ 700 °C. Thermal analysis revealed the two stages of $Nd_{2-x}Sr_xNiO_{4\pm\delta}$ thermal reduction in a 10%H_2/Ar gas mixture that was completed at 900 °C. For x < 0.2, the reduction products demonstrated an exsolution-like morphology with Ni nanoparticles allocated at the surface of oxide grains. As-obtained nanocomposites with x = 0 and x > 1 revealed the outstanding catalytic activity and selectivity in the dry reforming of the methane (DRM) reaction at 800 °C with CH_4 conversion close to the thermodynamic values. The appearance of two different maxima of the catalytic properties of Ni/(Nd_2O_3,$SrCO_3$) nanocomposites could be affiliated with the domination of the positive contributions of Nd_2O_3 and $SrCO_3$, respectively.

Keywords: metal–oxide nanocomposites; multicomponent catalysts; complex oxide precursors; K_2NiF_4 structure; high temperature reduction; exsolution; dry reforming of methane

Citation: Shlyakhtin, O.A.; Timofeev, G.M.; Malyshev, S.A.; Loktev, A.S.; Mazo, G.N.; Shatalova, T.; Arkhipova, V.; Roslyakov, I.V.; Dedov, A.G. $Nd_{2-x}Sr_xNiO_4$ Solid Solutions: Synthesis, Structure and Enhanced Catalytic Properties of Their Reduction Products in the Dry Reforming of Methane. *Catalysts* **2023**, *13*, 966. https://doi.org/10.3390/catal13060966

Academic Editors: Georgios Bampos, Athanasia Petala and Zacharias Frontistis

Received: 6 May 2023
Revised: 29 May 2023
Accepted: 30 May 2023
Published: 2 June 2023

1. Introduction

Through the dry reforming of methane (DRM), the reaction of methane with CO_2 is one of the most prospective methods of methane conversion into synthesis gas and is a useful raw material for the large-scale chemical industry. The advantage of DRM in the steam reforming of methane currently available on large-scale deals with a more balanced composition of Syngas is that it allows its direct processing into petrochemical products. Another advantage of DRM is that it deals with the utilization of significant amounts of CO_2 byproducts which are often formed in other petrochemical and other industrial processes. In addition, more and more attention is being paid to the dry reforming of methane to a synthesis gas as an effective way to utilize these two main greenhouse gases [1–5]. The DRM process is also rather beneficial for utilization and conversion into value-added products with the growing amounts of biogas consisting mostly of CH_4 and CO_2 [6,7].

The main processes that occurred during DRM is shown as follows [1,2,8–10]:

$$CH_4 + CO_2 \leftrightarrow 2CO + 2H_2;\ \Delta H_{298K} = +247\ \text{kJ mol}^{-1} \quad (1)$$

$$CO_2 + H_2 \leftrightarrow CO + H_2O;\ \Delta H_{298K} = +41\ \text{kJ mol}^{-1} \quad (2)$$

$$2CH_4 + CO_2 \leftrightarrow C_2H_6 + CO + H_2O; \Delta H_{298K} = +106 \text{ kJ mol}^{-1} \quad (3)$$

$$2CO \leftrightarrow C(s) + CO_2; \Delta H_{298K} = -171 \text{ kJ mol}^{-1} \quad (4)$$

$$CH4 \leftrightarrow C(s) + 2H_2; \Delta H_{298K} = +75 \text{ kJ mol}^{-1} \quad (5)$$

$$CO_2 + 2H_2 \leftrightarrow C + 2H_2O; \Delta H298K = -90 \text{ kJ mol}^{-1} \quad (6)$$

$$CO + H_2 \leftrightarrow H_2O + C; \Delta H_{298K} = -131.3 \text{ kJ mol}^{-1} \quad (7)$$

The main reaction (1) of the DRM process was strongly endothermic due to the large amount of energy necessary to break down methane and CO_2. According to thermodynamic calculations, the equilibrium was shifted to CO and H_2 at T > 640 °C. The amount of these reaction products increased with the temperature and in the lack of catalyst approaches 100% at T = 1000–1100 °C. The actual composition of the reaction mixture depended on the reaction rate and the contribution of the side reactions (2)–(5). The amount of hydrogen in the reaction products could be reduced by a reverse water-gas shift reaction (2). The reaction of methane and CO_2 could also cause the formation of hydrocarbons through a reaction (3) through one of the main obstacles to the large-scale application of DRM were reactions (4) and (5) and, to a lesser extent, (6) and (7), which caused the formation of coke at the surface of the catalysts and resulted in their deterioration [11].

Due to these features, DRM is usually performed as a catalytic process; the possibility of realizing it with a high yield of products at reasonable temperatures depends greatly on the selection of efficient catalysts. This selection deals with features of the DRM reaction mechanism at the atomic level. According to the existing models, the DRM reaction in the catalytic mode could be divided into four elementary steps: (1) the dissociative adsorption of methane at the metal surface and fast desorption of H_2; (2) the dissociative adsorption of CO_2 at the metal and metal–oxide interface and fast desorption of CO; (3) formation of the surface hydroxyl groups by hydrogen and oxygen spillover at the surface of the catalyst; (4) the oxidation of methyl-like CH_x groups by the surface oxygen and hydroxyls, resulting in the formation and desorption of CO and H_2 [10,12,13].

The highest activity and selectivity in DRM among the base metals was demonstrated by the nickel catalysts, especially at T < 900 °C. The main problem with their industrial application is related to intense coking which results in the progressive deactivation of the catalyst. However, recent studies of the coking processes demonstrated that their intensity could be significantly reduced by the particle size control and selection of oxide substrates. Taking into account the atomic scale model of the DRM process, the oxide substrate is now considered one of the key elements of metal–oxide composite catalysts, as the sorption and activation of CO_2 usually occur at the oxide components of catalysts. It was found that the surface acidity/basicity of the oxide components was rather important for this process, which made it preferable to the application of alkaline earth and rare earth oxides, such as the elements of DRM catalysts [8]. Several other, more complex metal–oxide interactions could also be important in DRM catalysis. It was found, for example, that the methane activation barriers were drastically different for Co and Ni surfaces of pure metals and for Co/CeO_{2-x} (111) and Ni/CeO_{2-x} (111) surfaces [10,14,15]. These features make the processes of metal–oxide interaction in the catalytic systems one of the hottest topics in modern studies of heterogeneous catalysts.

Until now, most of the metal-oxide catalysts have been prepared by the incipient wetness technique when the porous oxide substrate is soaked with a solution containing active metal, followed by drying, thermal decomposition, and the chemical reduction in the metal-containing component. This method ensures the relatively uniform distribution of catalytically active components across the surface of the supporting material. Along with the preferences for their chemical composition described before, these materials should possess a high surface area, extended porosity, and excellent thermal stability. A significant interaction of the active metal with the oxide surface is considered a positive factor here, as the anchoring of active particles to the surface of the supporting material promotes the prevention of sintering the nanoparticles of active metals in the reaction environment which is the second most important reason for catalyst deactivation. However, this method is really efficient when the composite metal-oxide catalyst consists of a single active component and a single supporting material only [11].

Due to the multistage mechanism of the DRM process, the modern Ni-based DRM catalysts usually consist of several components promoting the most efficient realization of the particular stages of the process [10]. To implement the efficient transport of the adsorbed reaction species between these components, it is necessary to provide a number of tight multiphase contacts of the various catalyst components. One of the best ways to ensure that multiphase contacts are based on a reduction in the complex oxide precursors containing all the necessary cations. The application of single oxide precursors ensures the reproducibility not only of the phase composition but also the morphology of the catalyst [1].

Currently, the most often used precursors are complex oxides with q perovskite structure, though, in the case of nickel-based catalysts, the number of suitable perovskites is rather limited. The largest number of studies in this field deal with $LaNiO_3$ and its solid solutions. The preliminary chemical reduction in this complex oxide, often called the activation of the catalyst, results in the formation of a Ni/La_2O_3 nanocomposite when the nanoparticles of the Ni metal are allocated at the surface of lanthanum oxide. The processing of native $LaNiO_3$ at DRM conditions also results in the fast decomposition of $LaNiO_3$, which usually occurs through the formation of the $La_2O_2CO_3$ intermediate. Despite the same phase composition of reduction products in both cases, different reaction routes cause a different Ni particle size and a different character of its spatial distribution. Taking into account the reversibility of $La_2O_2CO_3$ decomposition and the closeness of its decomposition temperature and usual processing conditions of the DRM process (T = 700–800 °C; CO_2-containing atmosphere), the actual composition of the nanocomposite catalyst, in this case, is rather $Ni/La_2O_2CO_3$ than Ni/La_2O_3. The temperature of the reductive decomposition of $LaNiO_3$ and, hence, the morphology of its decomposition products could be substantially modified by the cationic substitution both to La- and to Ni-sublattices of the perovskite structure by Ca, Sr, Ce, and by Fe, Mn, respectively. The easy resynthesis of $LaNiO_3$ from the spent catalyst at moderate temperatures in the air promotes its oxidative regeneration, followed by its further reduction to $Ni/(La_2O_3, La_2O_2CO_3)$ nanocomposites [16,17].

The number of the available Ni-containing complex oxides with the K_2NiF_4 structure is much larger, which opens a broader selection of the possible metal-oxide compositions that can be obtained by their thermal reduction [18]. Our previous studies demonstrated the efficiency of the metal–oxide nanocomposites obtained by the thermal reduction of $(Ln,Ca)_2(Ni,Co)O_4$ as catalysts for the partial oxidation and dry reforming of methane [19–21]. The essential morphological feature of as-obtained composites is the formation of the Ni metal nanoparticles at the surface of larger agglomerates of the oxide-based reduction products. This type of composite is usually observed before in the course of several processes of a partial reduction in complex oxides (redox exsolution) and is characterized by the enhanced strength of the metal particle connection with a surface of oxide particle [22–24]. The DRM catalysts based on these composites and obtained by both a partial or complete reduction in oxides were found prone not only to the sintering of active metal particles in the course of the DRM process but also to intense coking at their surface: one of the main reasons for the deactivation of DRM catalysts [25,26].

The best catalysts of the DRM at T < 800 °C among these products were the $Ni/(Nd_2O_3, CaO)$ nanocomposites obtained by the thermal reduction of $(Nd,Ca)_2NiO_4$ [20]. To find the most efficient one among them, it was necessary to vary their composition and compare their properties. However, the homogeneity range of these solid solutions was found to be rather limited ($0 < x < 0.5$), which, in turn, limited the variation range of these metal–oxide composites by the Nd/Ca ratio. It is known that the variation range was wider for $Nd_{2-x}Sr_xNiO_4$ solid solutions [27], and their reduction products were also expected to be good DRM catalysts. Similarly to $LaNiO_3$-derived catalysts, SrO in the Ni/(Nd_2O_3, SrO) composite could transform into $SrCO_3$ in the DRM environment. Apart from $La_2O_2CO_3$, the information on the $SrCO_3$ effect on the properties of DRM catalysts was rather limited. However, the high thermal stability of $SrCO_3$, especially in a CO_2-containing DRM environment, could promote the morphological stabilization of the composite Ni/(Nd_2O_3, $SrCO_3$) catalyst at elevated temperatures.

Due to these reasons, the present study was aimed at the synthesis of $Nd_{2-x}Sr_xNiO_4$ complex oxides, the refinement of the homogeneity ranges of these solid solutions with K_2NiF_4 structure in the air, the study of their thermal reduction processes, an analysis of the composition and morphology of their reduction products and evaluation of their catalytic properties in the DRM reaction.

2. Results and Discussion

In order to investigate the thermal decomposition features of the freeze-dried precursors, TG measurements with a simultaneous DSC signal detection were carried out for two samples of significantly different compositions. It was observed that both Nd_2NiO_4 and $Nd_{0.6}Sr_{1.4}NiO_4$ precursors exhibited rather similar behavior in the course of their TG-DSC examination (Figures S1 and S2). Their TG profiles contained three distinct temperature ranges that likely corresponded to the different chemical reactions that occurred in precursors. The first step at ~150–250 °C could be attributed to the elimination of chemically bound water presented as –OH species or H_2O molecules incorporated into the structure of inorganic hydrates. The second step at 250–400 °C corresponded to the decomposition of Nd, Sr, and Ni nitrates, and the last step was likely to be the complete combustion of PVA fragments by atmospheric O_2 or/and active oxygen species released during the decomposition of nitrates at temperatures near the combustion reaction. The combustion nature of this process was confirmed by a strong exothermic effect at 400–450 °C, which was detected by the DSC method. No effects at temperatures above 500 °C were observed in the TG-DSC profiles except a slight mass loss at 850–900 °C for the $Nd_{0.6}Sr_{1.4}NiO_4$ precursor. This could have been elucidated by the decomposition of the minor amount of $SrCO_3$ formed by the partial interaction of SrO with CO_2 gas released in the course of PVA combustion. Thus, the intensive formation of the target K_2NiF_4 complex oxides under isothermal temperature conditions was expected at T > 500–600 °C after the completion of all decomposition/combustion processes in freeze-dried precursors.

To study the complex oxide formation, the thermal decomposition of the freeze-dried precursors was performed in the air at T = 600–1200 °C for 6 h. A careful comparison of the XRD patterns of the decomposition products revealed the considerable dependence of the phase formation conditions on the $Nd_{2-x}Sr_xNiO_4$ composition (Figure 1 and Figures S3–S7).

In the case of Nd_2NiO_4, the samples obtained at 600–700 °C consisted of a mixture of cubic Nd_2O_3 and $NdNiO_3$ with an orthorhombic perovskite structure. These observations were in good accordance with the fact that neodymium-nickel perovskite can be obtained at lower temperatures under an air atmosphere [28]. Further heating provided a solid-state interaction between Nd_2O_3 and $NdNiO_3$, forming the target Nd_2NiO_4 compound at T > 900 °C.

In the case of $Nd_{1.5}Sr_{0.5}NiO_4$ and $NdSrNiO_4$ compositions, the thermal decomposition of the freeze-dried precursors led to the intense dominant formation of the target K_2NiF_4 phases even at 600–700 °C; only trace amounts of the $SrCO_3$ carbonate was observed by XRD. All the samples obtained at T ≥ 800 °C were single-phase complex nickelates. Such low temperatures demanded $Nd_{1.5}Sr_{0.5}NiO_4$ and $NdSrNiO_4$ formation to be explained by the drastic stabilization of the K_2NiF_4 structure due to the substitution of Nd^{3+} with a larger Sr^{2+} cation.

	Nd_2NiO_4	$Nd_{1.5}Sr_{0.5}NiO_4$	$NdSrNiO_4$	$Nd_{0.8}Sr_{1.2}NiO_4$	$Nd_{0.6}Sr_{1.4}NiO_4$
1100 °C	Nd_2NiO_4	$Nd_{1.5}Sr_{0.5}NiO_4$	$NdSrNiO_4$	$Nd_{0.8}Sr_{1.2}NiO_4$	$Nd_{0.6}Sr_{1.4}NiO_4$
1000 °C	Nd_2NiO_4	$Nd_{1.5}Sr_{0.5}NiO_4$	$NdSrNiO_4$	$Nd_{0.8}Sr_{1.2}NiO_4$	$Nd_{0.6}Sr_{1.4}NiO_4$
900 °C	Nd_2NiO_4	$Nd_{1.5}Sr_{0.5}NiO_4$	$NdSrNiO_4$	$Nd_{0.8}Sr_{1.2}NiO_4$	RP P
800 °C	RP c-Nd_2O_3 P	$Nd_{1.5}Sr_{0.5}NiO_4$	$NdSrNiO_4$	RP $SrCO_3$	RP P
700 °C	P c-Nd_2O_3	$Nd_{1.5}Sr_{0.5}NiO_4$	RP $SrCO_3$	RP $SrCO_3$	RP $SrCO_3$
600 °C	P c-Nd_2O_3	RP $SrCO_3$	RP $SrCO_3$	RP $SrCO_3$	RP $SrCO_3$

Figure 1. The phase composition of $Nd_{2-x}Sr_xNiO_4$ samples obtained at various temperatures; RP—oxide with K_2NiF_4 structure, P—oxide with perovskite structure.

According to the XRD data, the thermolysis products at 600–700 °C contained phases with a K_2NiF_4 structure mixed with considerable amounts of strontium carbonate. Apparently, these K_2NiF_4 products were stable complex nickelates with a moderate Sr content; the excess of the last one was presented as a $SrCO_3$ phase. Interestingly, further heating to 800–900 °C led to the formation of mixtures of K_2NiF_4 and orthorhombic perovskite phases. A similar feature was observed for Nd-Ca cobaltite in [29], where the formation of the target K_2NiF_4 complex oxide at moderate temperatures was accompanied by the kinetically controlled formation of an intermediate phase with a perovskite structure. For Nd-Sr nickelates, the observed features could be explained by the formation of Sr-rich (Nd,Sr)NiO_3 perovskites with a Ni oxidation state close to +3 followed by their solid state interaction with the "low temperature" Nd-rich $(Nd,Sr)_2NiO_4$ phases at 900–1000 °C, resulting in the target $Nd_{0.8}Sr_{0.2}NiO_4$ and $Nd_{0.6}Sr_{1.4}NiO_4$ nickelates.

Thus, despite such diverse solid-state reaction pathways, all the discussed nickelates could be securely synthesized by the freeze-drying method at T = 1000–1200 °C in the air. In order to ensure a high level of crystallographic ordering in such multicomponent solid solutions and to make a reliable comparison between the samples with various Sr content, a series of $Nd_{2-x}Sr_xNiO_4$ samples (x = 0; 0.5; 1; 1.2; 1.4) obtained at 1200 °C was chosen for further investigation. According to the XRD study, their crystal structure was reliably assigned to the perovskite-related K_2NiF_4 type. Similarly to [20], the orthorhombically distorted Fmmm structure was observed for Nd_2NiO_4 (x = 0); no additional reflections of the other K_2NiF_4 polymorphs were observed. On the other hand, for the Sr-containing solid solutions ($0.5 \leq x \leq 1.4$), a tetragonal I4/mmm structure was verified (Figure 2). It clearly indicated the stabilization effect. A profile analysis of these complex oxides was performed by the Le Bail method from powder XRD data; the resulting unit cell parameters are given in Table 1.

The variation in the K_2NiF_4 lattice parameters values on the Sr content was found similar to that found in [27]. It is noteworthy that the observed changes could be attributed to both Nd/Sr and O stoichiometry which provided a notable deviation from the linear Vegard's rule (Figure S8). The absence of these changes for the sample with $x > 1.4$ corresponded quite well with the previous conclusion on the allocation of the limit of the $Nd_{2-x}Sr_xNiO_{4\pm\delta}$ homogeneity range at $1.4 < x < 1.6$.

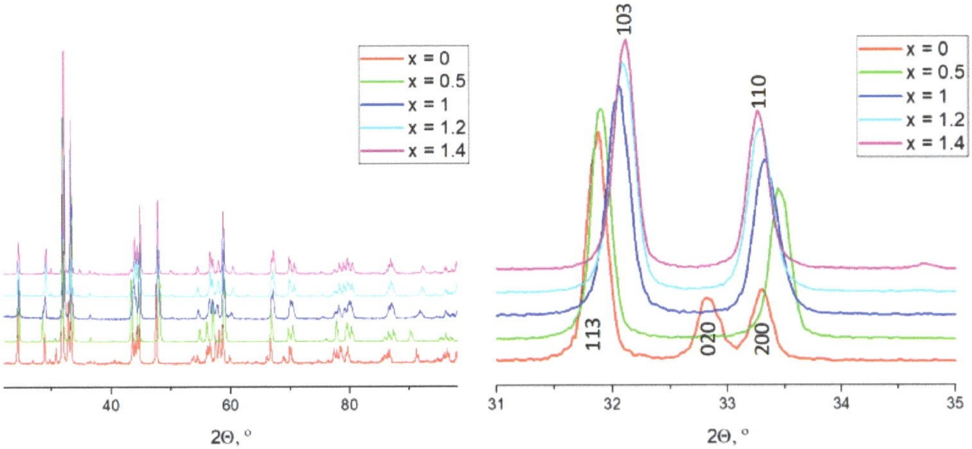

Figure 2. XRD patterns of $Nd_{2-x}Sr_xNiO_4$ obtained at T = 1200 °C.

Table 1. The unit cell parameters of $Nd_{2-x}Sr_xNiO_{4\pm\delta}$.

Composition	Space Group	a, Å	b, Å	c, Å	V, Å^3
$Nd_2NiO_{4\pm\delta}$	Fmmm	5.3789(3)	5.4514(3)	12.3724(8)	362.79(5)
$Nd_{1.5}Sr_{0.5}NiO_{4\pm\delta}$	I4/mmm	3.776(2)	3.776(2)	12.482(1)	177.9(2)
$NdSrNiO_{4\pm\delta}$	I4/mmm	3.7886(2)	3.7886(2)	12.320(1)	176.8(2)
$Nd_{0.8}Sr_{1.2}NiO_{4\pm\delta}$	I4/mmm	3.798(2)	3.798(2)	12.261(1)	176.8(2)
$Nd_{0.6}Sr_{1.4}NiO_{4\pm\delta}$	I4/mmm	3.804(2)	3.804(2)	12.252(6)	177.3(2)
$Nd_{0.4}Sr_{1.6}NiO_{4\pm\delta}$	I4/mmm	3.8027(3)	3.8027(3)	12.266(11)	177.38(3)

The thermal analysis of the reduction processes of the as-obtained complex oxides in the Ar/H_2 mixture revealed two different stages of this process for all samples in the study (Figure 3A–C), as it was observed before for other K_2NiF_4 complex oxides [19–21]. According to the XRD analysis results of the reduction products, the partial reduction in the complex oxide precursor at the first stage caused its transformation into another complex oxide with an individual crystal structure. The detection of the Ni^0 phase possibly formed during its reduction was complicated by the strong overlapping of its reflections with the reflections of complex oxide structures. Complete decomposition of this intermediate at the second stage of the reduction resulted in the formation of Ni^0, Nd_2O_3, and SrO, which could be transformed into $Sr(OH)_2$ during further cooling in the air (Figure 3D). An increase in the Sr content in the initial complex oxide was accompanied by a systematic increase in the complete reduction temperature.

The microstructure of the complete reduction products of $Nd_{2-x}Sr_xNiO_{4\pm\delta}$ was similar to the microstructure in the reduction products of their Ca-substituted counterparts, as studied in more detail before. According to [19,20], the coarser grains were likely formed by Nd and Sr oxides that were coated with spherical nanoparticles of the Ni metal. Due to a wider range of Sr substitution in Nd_2NiO_4 compared to Ca, it was possible to observe in this case that the increase in Sr content was accompanied by a significant coarsening of the exsolved Ni particles from ~20 nm for x = 0 to 40–60 nm for x $\geq$ 1, probably due to a higher reduction temperature (Figure 4C,D).

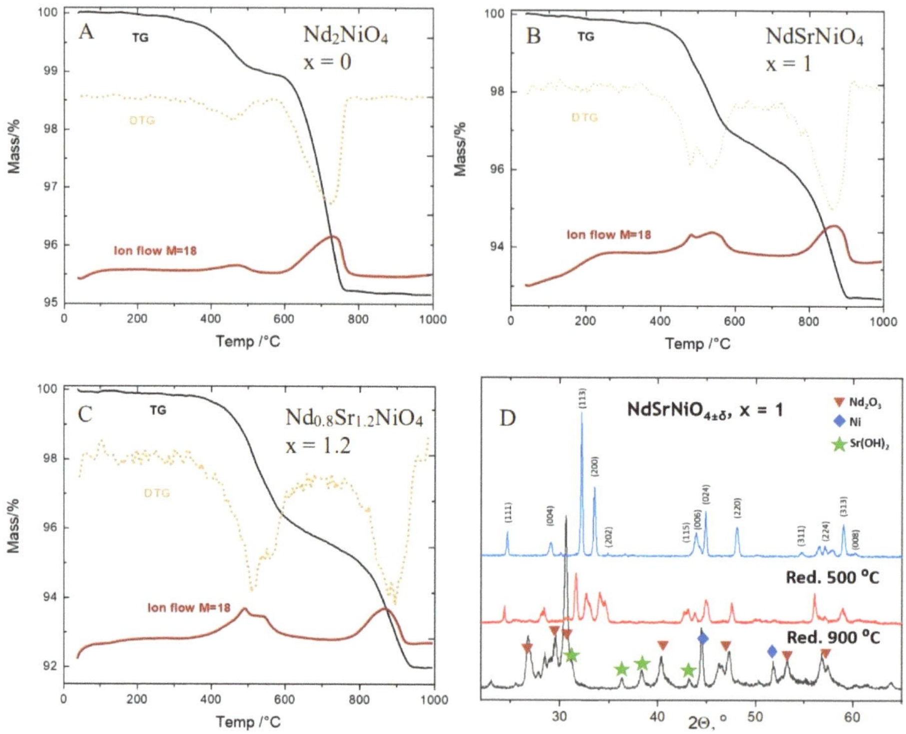

Figure 3. TG, DTG, and H_2O evolution curves for the reduction of Nd_2NiO_4 (**A**), $NdSrNiO_4$ (**B**), $Nd_{0.8}Sr_{1.2}NiO_4$ (**C**) in Ar/H_2 mixture; XRD patterns of $NdSrNiO_4$ and its reduction products at 500 °C and 900 °C (**D**).

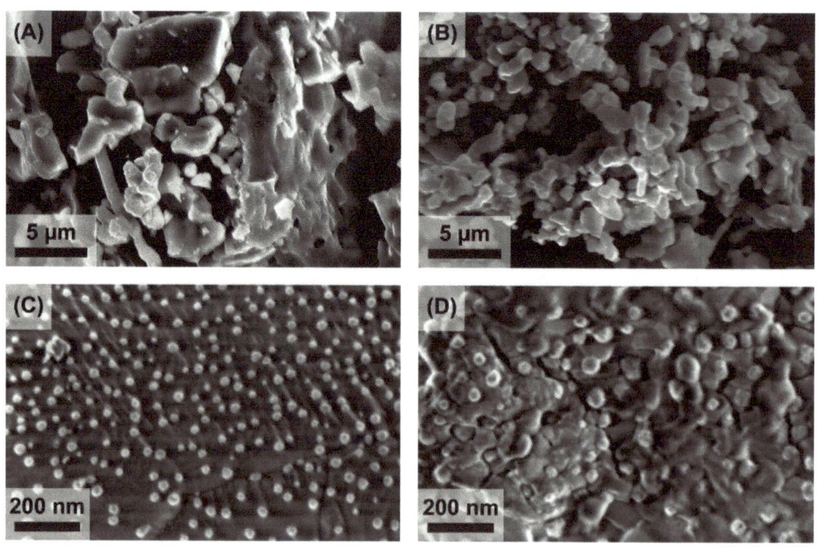

Figure 4. SEM micrographs of Nd_2NiO_4 (**A**), $NdSrNiO_4$ (**B**), and the products of their thermal reduction at 900 °C ((**C**,**D**), respectively).

Taking into account the results of our previous studies, it was expected that the reduction products of $Nd_{2-x}Sr_xNiO_{4\pm\delta}$ with the morphology described before would demonstrate significant catalytic properties in the DRM process. Due to the similarity of the catalytic processes, the DRM catalysts often demonstrated significant catalytic activity and selectivity in the partial oxidation of the methane (POM) reaction: another widely discussed alternative to the currently applied steam reforming process [2,19–21]. However, the production of syngas by DRM at T = 700–800 °C over the Ni-based catalysts could be more efficient than POM. According to [20], a lower CO yielded in POM at T < 800 °C over Ni/(Nd_2O_3,CaO) nanocomposites dealt with Ni oxidation to NiO, the catalyst of the total oxidation of methane and CO to CO_2. The oxidation of the part of methane to CO_2 led to a notable decrease in syngas selectivity. Similar processes were expected to be obtained in $Nd_{2-x}Sr_xNiO_{4\pm\delta}$-derived nanocomposites in the POM environment, but their appearance during DRM was less probable due to the lack of oxidants.

An experimental evaluation of their catalytic performance confirmed a considerable catalytic activity and selectivity in DRM for all metal–oxide composites in a study at T $\geq$ 700 °C (Figure 5, Figures S9 and S10). However, the catalytic behavior of these composites demonstrated several important anomalies. Similar metal–oxide composites obtained from K_2NiF_4-like precursors usually demonstrate a monotonous dependence of their catalytic properties in DRM both on the temperature and on the rare earth to alkaline earth ratio in the composite [19,20]. In the present case, the samples in the study could be divided into two different groups according to the behavior of their catalytic properties-Sr-free and Sr-containing composites. The catalytic activity and selectivity of the only member of the first group, the Nd_2NiO_4 reduction product (x = 0), similarly to [19,20], increased monotonously with the temperature demonstrating maximum values at the maximum temperature of the study (T = 900 °C) and approaching their thermodynamic limits (Figure 5A and Figure S9B). Meanwhile, the maximum values of activity and selectivity in all the Sr-containing composites (0.5 $\leq$ x $\leq$ 1.4) were observed at T = 800 °C while further heating to T = 900 °C, contrary to the thermodynamically predicted behavior, caused a considerable decrease in their catalytic performance. Another unusual feature in this series of composites dealt with a significant drop in catalytic activity at 0 < x $\leq$ 0.5. However, the further substitution of Nd with Sr resulted in the systematic recovery of the catalytic activity and selectivity of composites with their maximal values at T = 800 °C for the samples with x = 1.2–1.4.

The XRD analysis of spent catalysts revealed that they consisted of Ni and Nd_2O_3 for x = 0 and Ni, Nd_2O_3, and $SrCO_3$ in all other cases (Figure 6A), as SrO converted easily into $SrCO_3$ in the DRM reaction environment [30]. Hence, the most probable reasons for the different behaviors of these two types of DRM catalysts were likely affiliated with the appearance of $SrCO_3$ in the composite catalysts and with specific properties of strontium carbonate. According to the existing models of DRM catalysis, the catalytic properties of the Nd_2NiO_4 reduction products were determined by the combination of the methane sorption and catalytic activity of nickel metal and the CO_2 sorption activity of the neodymium oxide. As the amount of Ni metal was almost the same in all samples in the study, the higher catalytic activity of the composites with a larger Sr content could be affiliated with more intense CO_2 sorption due to a higher basicity catalyst surface. However, a decrease in the sorption activity at higher temperatures was likely more pronounced for $SrCO_3$ than for Nd_2O_3, thus promoting a definite decrease in the other catalytic properties of Sr-containing composites at T > 800 °C.

The reasons for the observed drop in the catalytic properties at 0 < x $\leq$ 0.5 and, in general, the non-monotonous dependence of the catalytic properties of Ni/($Nd_{2-x}Sr_x$)Oy composites on x were likely due to a competition between the two catalytic pathways, via Nd_2O_3 and via $SrCO_3$. It should also be noted that the catalytic activity of samples (x = 1.2;1.4) with a larger size of Ni particles was the same or even higher than the corresponding activity of the Sr-free (x = 0) sample with much smaller Ni exclusions (Figure 4C,D). It appeared contradictory to the common trend of the better catalytic performance of the smaller particles of a catalyst. This discrepancy could also be affiliated with different mechanisms of the DRM catalysis on Nd_2O_3 and $SrCO_3$ substrates when more efficient CO_2 activation with $SrCO_3$ became more important than the smaller activity of coarser Ni particles on the $SrCO_3$ substrate.

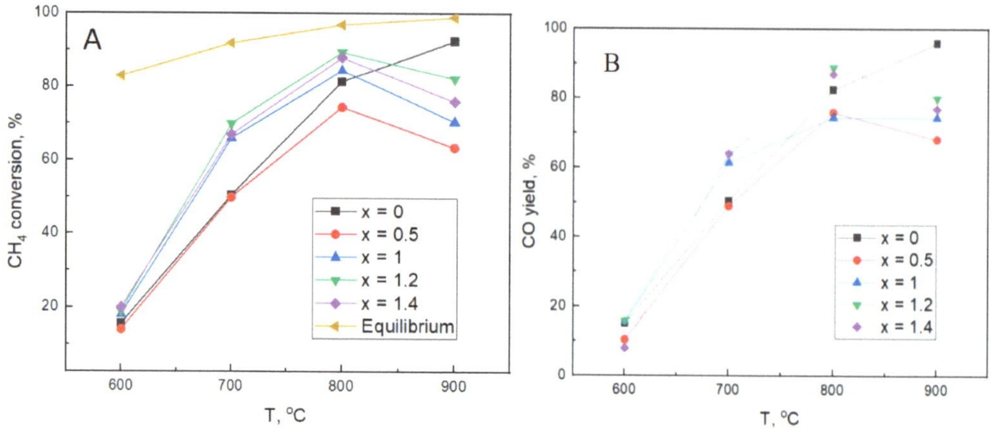

Figure 5. CH_4 conversion (**A**) and CO yield (**B**) over the reduction products of various composites. The calculated values of CH_4 conversion are given according to [13,31–33].

(A)

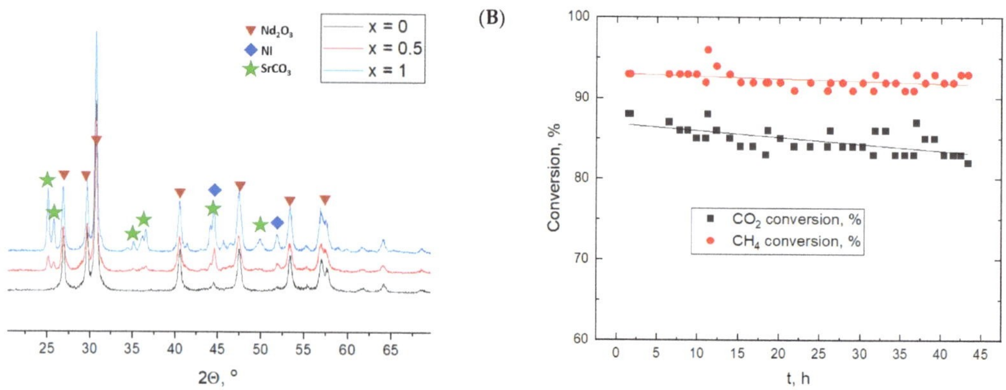

Figure 6. XRD patterns of spent catalysts (**A**) and the results of the stability test of the metal–oxide nanocomposite with x = 1.4 at T = 800 °C (**B**).

These interesting and unusual features demanded more detailed studies using modern instrumental methods. It should be noted, however, that the values of the activity and selectivity of as-obtained DRM catalysts were comparable with the results obtained by other groups [3,4,10,30,31]. An essential feature of the $Nd_{2-x}Sr_xNiO_{4\pm\delta}$ reduction products was good reproducibility (Figure S10) and the enhanced stability of their catalytic properties in the DRM environment. This was illustrated by the results of the 45h stability test, which revealed only the minimum variation in the catalytic properties (Figure 6B). This stability could be affiliated with an absence of significant coking at the surface for the catalyst as one of the main reasons of catalyst deactivation during the DRM process (Figure S11).

3. Materials and Methods

$Nd_{2-x}Sr_xNiO_4$ (x = 0 ÷ 1.4) samples were obtained by the modified freeze-drying synthesis technique. To obtain multicomponent starting solutions, preliminarily calcined Nd_2O_3 and $SrCO_3$ were dissolved in the warm diluted acetic acid, then a $Ni(NO_3)_2x6H_2O$ aqueous solution was added. To stabilize the frozen solution during freeze drying, an aqueous solution (5 wt. %) of polyvinyl alcohol was added to all solutions under intense stirring. The freeze-drying of flash-frozen solutions was performed at P = 0.7 mbar for 2 days. The thermal decomposition of the freeze-dried precursors was performed in the air at 800–1200 °C for 6 h. A reduction in as-obtained $Nd_{2-x}Sr_xNiO_4$ powders was performed in an H_2:Ar = 1:10 gas mixture at different temperatures for 1 h, followed by slow cooling to room temperature under a reducing atmosphere.

XRD analysis of the powders was performed using a Rigaku D/MAX-2500PC diffractometer (Rigaku, Tokyo, Japan) with Cu K_α radiation generated on a rotating Cu anode (40 kV, 250 mA). The profile analysis of the powder XRD data was carried out by the Le Bail method using the Jana 2006 program package. The TG-DSC thermal analysis of samples combined with a simultaneous QMS evolved gas analysis was performed in the air or in a 10% H_2/Ar gas mixture by STA 409PC/PG (NETZSCH) at T = 40 °C ÷ 1000 °C and a 10 K/min heating rate. The morphology of the powders was studied using a Carl Zeiss NVision 40 scanning electron microscope (Carl Zeiss SMT AG, Oberkochen, Germany).

The catalytic tests of the DRM reaction were carried out in a quartz glass flow fixed-bed reactor (18 mm internal diameter, 300 mm length) with a 0.2 g sample of the catalyst (100–250 mesh fraction) at atmospheric pressure in the absence of dilution with inert gas. This catalyst was first heated in a hydrogen flow at 10 K/min to 900 °C. Then, the gas stream was switched to a mixture of CH_4/CO_2 = 1/1 at GHSV values of 16 L/g h. No dilution in the feed flow by the inert gas was applied. The catalytic experiments were performed consecutively at 900, 800, 700, and 600 °C by holding at the preselected temperatures for 1–5 h at a steady state. After the analysis, the furnace was switched off, and the catalyst was cooled to room temperature over 3–4 h in pure N_2. The details of the testing technique and GC analysis can be found elsewhere [34].

The methane conversion (X), product selectivity (S), and yield (Y) of the products were defined as follows:

$$X(CH_4,\%) = \frac{\text{moles of } CH_4 \text{ converted}}{\text{moles of } CH_4 \text{ in feed}} \times 100 \quad (8)$$

$$S(CO,\%) = \frac{\text{moles of CO in products}}{\text{moles of } CH_4{+}CO_2 \text{ converted}} \times 100 \quad (9)$$

$$S(H_2,\%) = \frac{\text{moles of } H_2 \text{ produced}}{2 \times \text{ moles of } CH_4 \text{ converted}} \times 100 \quad (10)$$

$$Y(\text{products},\%) = \frac{X(CH_4,\%) \times S(\text{products},\%)}{100} \quad (11)$$

$$\text{C balance } (\%) = \frac{\text{moles of C in products}}{\text{moles of C in feed}} \times 100 \quad (12)$$

The number of moles in the feed gases and gaseous products of the reactions was calculated based on the measured volumetric velocity of the feed gases and the products formed, as well as chromatography data, which made it possible to fully take into account the stoichiometry of the reaction and the corresponding volume expansion of the reacted gas mixture.

4. Conclusions

A full scale of the solid solutions $Nd_{2-x}Sr_xNiO_{4\pm\delta}$ ($0 \leq x \leq 1.4$) was obtained during these studies; the boundaries of their homogeneity ranged by the Nd/Sr ratio were refined.

The processes of a high-temperature reduction in these complex oxides were investigated; the composition and the morphology of as-obtained metal-oxide nanocomposites were determined. It was found that several Ni/(Nd_2O_3,$SrCO_3$) nanocomposites demonstrated outstanding activity and selectivity in the DRM process at T = 800 °C and the significant stability of their catalytic properties on stream. A positive effect of $SrCO_3$ on the catalytic properties of Ni/(Nd_2O_3,$SrCO_3$) nanocomposites was detected.

Supplementary Materials: The following supporting information can be downloaded at: https://www.mdpi.com/article/10.3390/catal13060966/s1. Figure S1: TG (green) and DSC (blue) profiles of freeze-dried Nd_2NiO_4 precursor, Figure S2: TG (green) and DSC (blue) profiles of freeze-dried $Nd_{0.6}Sr_{1.4}NiO_4$ precursor, Figure S3: XRD patterns of thermal decomposition products of Nd_2NiO_4 (x = 0) precursors at different temperatures, Figure S4: XRD patterns of thermal decomposition products of $Nd_{1.5}Sr_{0.5}NiO_4$ (x = 0.5) precursors at different temperatures, Figure S5: XRD patterns of thermal decomposition products of $NdSrNiO_4$ (x = 1) precursors at different temperatures, Figure S6: XRD patterns of thermal decomposition products of $Nd_{0.8}Sr_{1.2}NiO_4$ (x = 1.2) precursors at different temperatures, Figure S7: XRD patterns of thermal decomposition products of $Nd_{0.6}Sr_{1.4}NiO_4$ (x = 1.4) precursors at different temperatures, Figure S8: The dependence of $Nd_{2-x}Sr_xNiO_{4\pm\delta}$ lattice parameters on Sr content, Figure S9: CO_2 conversion (A) and H_2 yield (B) over the reduction products of various composites. The calculated values of H_2 yield are given according to [13,31–33], Figure S10: The results of testing the composites with various Sr content in the DRM process, Figure S11: SEM micrographs of the spent catalyst (x = 1.4).

Author Contributions: Conceptualization, O.A.S. and A.G.D.; methodology, O.A.S., S.A.M., A.S.L. and G.N.M.; validation, O.A.S., A.S.L. and G.N.M.; investigation, S.A.M., G.M.T., V.A., T.S. and I.V.R.; data curation, S.A.M. and A.S.L.; formal analysis, S.A.M., A.S.L.; software, S.A.M. and G.M.T.; resources, O.A.S. and A.S.L.; writing—original draft, O.A.S. and G.M.T.; Writing—review and editing, O.A.S., S.A.M., G.N.M. and A.S.L.; visualization, S.A.M. and G.M.T.; project administration, O.A.S. and A.G.D.; funding acquisition, O.A.S.; supervision, A.G.D. All authors have read and agreed to the published version of the manuscript.

Funding: This study is supported by the Russian Science Foundation (grant 22-23-00460). DRM experiments were carried out within the state funding of TIPS RAS. The Moscow State University program of development is gratefully acknowledged for the partial support of instrumental studies. SEM images were obtained at the IGIC RAS Joint Research Center for Physical Methods of Research.

Data Availability Statement: The data presented in this study are available in the present paper and supplementary materials.

Conflicts of Interest: The authors declare no conflict of interest.

References

1. Bhattar, S.; Abedin, M.A.; Kanitkar, S.; Spivey, J.J. A review on dry reforming of methane over perovskite derived catalysts. *Catal. Today* **2021**, *365*, 2–23. [CrossRef]
2. Moiseev, I.I.; Loktev, A.S.; Shlyakhtin, O.A.; Mazo, G.N.; Dedov, A.G. New approaches to the design of nickel, cobalt, and nickel–cobalt catalysts for partial oxidation and dry reforming of methane to synthesis gas. *Petrol. Chem.* **2019**, *59* (Suppl. S1), S1–S27. [CrossRef]
3. Wang, L.; Wang, F. Design Strategy, Synthesis, and Mechanism of Ni Catalysts for Methane Dry Reforming Reaction: Recent Advances and Future Perspectives. *Energy Fuels* **2022**, *36*, 5594–5621. [CrossRef]
4. Ranjekar, A.M.; Yadav, G.D. Dry reforming of methane for syngas production: A review and assessment of catalyst development and efficacy. *J. Indian Chem. Soc.* **2021**, *98*, 100002. [CrossRef]
5. Zhenghong, B.; Fei, Y. Catalytic conversion of biogas to syngas via dry reforming process. *Adv. Bioenergy* **2018**, *3*, 43–76.
6. Gao, Y.; Jiang, J.; Meng, Y.; Yan, F.; Aihemaiti, A. A review of recent developments in hydrogen production via biogas dry reforming. *Energy Convers. Manag.* **2018**, *171*, 133–155. [CrossRef]
7. Rosha, P.; Rosha, A.K.; Ibrahim, H.; Kumar, S. Recent advances in biogas upgrading to value added products: A review. *Int. J. Hydrogen Energy* **2021**, *46*, 21318–21337. [CrossRef]
8. Abdulrasheed, A.; Abdul Jalil, A.; Gambo, Y.; Ibrahim, M.; Hambali, H.U.; Shahul Hamid, M.Y. A review on catalyst development for dry reforming of methane to syngas: Recent advances. *Renew. Sustain. Energy Rev.* **2019**, *108*, 175–193. [CrossRef]
9. Aramouni, N.A.K.; Touma, J.G.; Tarboush, B.A.; Zeaiter, J.; Ahmad, M.N. Catalyst design for dry reforming of methane: Analysis review. *Renew. Sustain. Energy Rev.* **2018**, *82*, 2570–2585. [CrossRef]
10. Li, M.; Sun, Z.; Hu, Y.H. Catalysts for CO_2 reforming of CH_4: A review. *J. Mater. Chem. A* **2021**, *9*, 12495. [CrossRef]

11. Arora, S.; Prasad, R. An overview on dry reforming of methane: Strategies to reduce carbonaceous deactivation of catalysts. *RSC Adv.* **2016**, *6*, 108668. [CrossRef]
12. Papadopoulou, C.; Matralis, H.; Verykios, X. Utilization of Biogas as a Renewable Carbon Source: Dry Reforming of Methane. In *Catalysis for Alternative Energy Generation*; Guczi, L., Erdôhelyi, A., Eds.; Springer: New York, NY, USA, 2012; pp. 57–127.
13. Le Sache, E.; Reina, T.R. Analysis of Dry Reforming as direct route for gas phase CO_2 conversion. The past, the present and future of catalytic DRM technologies. *Prog. Energy Combust. Sci.* **2022**, *89*, 100970. [CrossRef]
14. Liu, Z.; Lustemberg, P.; Gutierrez, R.A.; Carey, J.J.; Palomino, R.M.; Vorokhta, M.; Grinter, D.C.; Ramirez, P.J.; Matolin, V.; Nolan, M.; et al. In Situ Investigation of Methane Dry Reforming on Metal/Ceria(111) Surfaces: Metal–Support Interactions and C-H Bond Activation at Low Temperature. *Angew. Chem. Int. Ed.* **2017**, *56*, 13041–13046. [CrossRef] [PubMed]
15. Liu, Z.; Grinter, D.C.; Lustemberg, P.G.; Nguyen-Phan, T.-D.; Zhou, Y.; Luo, S.; Waluyo, I.; Crumlin, E.J.; Stacchiola, D.J.; Zhou, J.; et al. Dry Reforming of Methaneon a Highly-Active Ni-CeO_2 Catalyst: Effects of Metal-Support Interactions on C-H Bond Breaking. *Angew. Chem. Int. Ed.* **2016**, *55*, 7455–7459. [CrossRef] [PubMed]
16. Batiot-Dupeyrat, C.; Gallego, G.A.S.; Mondragon, F.; Barrault, J.; Tatibouët, J.-M. CO_2 reforming of methane over $LaNiO_3$ as precursor material. *Catal. Today* **2005**, *107–108*, 474–480. [CrossRef]
17. Bonmassar, N.; Bekheet, M.F.; Schlicker, L.; Gili, A.; Gurlo, A.; Doran, A.; Gao, Y.; Heggen, M.; Bernardi, J.; Klötzer, B.; et al. In Situ-Determined Catalytically Active State of $LaNiO_3$ in Methane Dry Reforming. *ACS Catal.* **2020**, *10*, 1102–1112. [CrossRef]
18. Tarutin, A.P.; Lyagaeva, J.G.; Medvedev, D.A.; Bi, L.; Yaremchenko, A. Recent advances in layered $Ln_2NiO_{4+\delta}$ nickelates: Fundamentals and prospects of their applications in protonic ceramic fuel and electrolysis cells. *J. Mater. Chem. A* **2021**, *9*, 154–195. [CrossRef]
19. Shlyakhtin, O.A.; Malyshev, S.A.; Loktev, A.S.; Mazo, G.N.; Garshev, A.V.; Chumakov, R.G.; Dedov, A.G. Synthesis and decomposition of $Nd_{2-y}Ca_yCo_{1-x}Ni_xO_4$: The effect of resynthesis on the catalytic performance of decomposition products in the partial oxidation of methane. *ACS Appl. Energy Mater.* **2021**, *4*, 7661–7673. [CrossRef]
20. Malyshev, S.A.; Shlyakhtin, O.A.; Loktev, A.S.; Mazo, G.N.; Timofeev, G.M.; Mukhin, I.E.; Kaplin, I.Y.; Svetogorov, R.D.; Valeev, R.G.; Dedov, A.G. Exsolution-like synthesis of Ni/(Nd_2O_3,CaO) nanocomposites from $Nd_{2-x}Ca_xNiO_4$ precursors for catalytic applications. *J. Solid State Chem.* **2022**, *312*, 123267. [CrossRef]
21. Malyshev, S.A.; Shlyakhtin, O.A.; Loktev, A.S.; Mazo, G.N.; Timofeev, G.M.; Mukhin, I.E.; Svetogorov, R.D.; Roslyakov, I.V.; Dedov, A.G. Ni/(R_2O_3,CaO) nanocomposites produced by the exsolution of $R_{1.5}Ca_{0.5}NiO_4$ nickelates (R = Nd, Sm, Eu): Rare earth effect on the catalytic performance in the dry reforming and partial oxidation of methane. *Materials* **2022**, *15*, 7265. [CrossRef]
22. Kwon, O.; Joo, S.; Choi, S.; Sengodan, S.; Kim, G. Review on exsolution and its driving forces in perovskites. *J. Phys. Energy* **2020**, *2*, 032001. [CrossRef]
23. Tsekouras, G.; Miller, D.N.; Ménard, H.; Irvine, J.T.S. In situ growth of nanoparticles through control of non-stoichiometry. *Nat. Chem.* **2013**, *5*, 916–923.
24. Shah, S.; Sayono, S.; Junzunza, J.; Pan, R.; Xu, M.; Pan, X.; Gilliard-AbdulAziz, K.L. The effects of stoichiometry on the properties of exsolved Ni-Fe alloy nanoparticles for dry methane reforming. *AIChE J.* **2020**, *66*, e17078. [CrossRef]
25. Neagu, D.; Oh, T.S.; Miller, D.N.; Ménard, H.; Bukhari, S.M.; Gamble, S.R.; Gorte, R.J.; Vohs, J.M.; Irvine, J.T.S. Nano-socketed nickel particles with enhanced coking resistance grown in situ by redox exsolution. *Nat. Commun.* **2015**, *6*, 8120. [CrossRef] [PubMed]
26. Steiger, P.; Nachtegaal, M.; Kröcher, O.; Ferri, D. Reversible segregation of Ni in $LaFe_{0.8}Ni_{0.2}O_{3\pm\delta}$ during coke removal. *ChemCatChem* **2018**, *10*, 4456–4464. [CrossRef]
27. Takeda, Y.; Nishijima, M.; Imanishi, N.; Kanno, R.; Yamamoto, O.; Takano, M. Crystal chemistry and transport properties of $Nd_{2-x}A_xNiO_4$ (A = Ca, Sr, or Ba, $0 \leq x \leq 1.4$). *J. Solid State Chem.* **1992**, *96*, 72–83. [CrossRef]
28. Blasco, J.; Garsia, J. A comparative study of of the crystallographic, magnetic and electrical properties of the $Nd_{1-x}La_xNiO_{3-\delta}$ system. *J. Phys. Condens. Matter* **1994**, *6*, 10759–10772. [CrossRef]
29. Malyshev, S.A.; Shlyakhtin, O.A.; Mazo, G.N.; Garshev, A.V.; Mironov, A.V.; Loktev, A.S.; Dedov, A.G. Comparative analysis of $NdCaCoO_4$ phase formation from cryogel and from solid state precursors. *J. Sol-Gel Sci. Technol.* **2017**, *81*, 372–377. [CrossRef]
30. Omata, K.; Nukui, N.; Hottai, T.; Showa, Y.; Yamada, M. Strontium carbonate supported cobalt catalyst for dry reforming of methane under pressure. *Catal. Commun.* **2004**, *5*, 755–758. [CrossRef]
31. Han, K.; Wang, S.; Hu, N.; Shi, W.; Wang, F. Alloying Ni–Cu Nanoparticles Encapsulated in SiO_2 Nanospheres for Synergistic Catalysts in CO_2 Reforming with Methane Reaction. *ACS Appl. Mater. Interfaces* **2022**, *14*, 23487–23495. [CrossRef]
32. Shi, Y.; Han, K.; Wang, F. Ni–Cu Alloy Nanoparticles Confined by Physical Encapsulation with SiO_2 and Chemical Metal–Support Interaction with CeO_2 for Methane Dry Reforming. *Inorg. Chem.* **2022**, *61*, 15619–15628. [CrossRef] [PubMed]
33. Cai, X.; Hu, Y.H. Advances in catalytic conversion of methane and carbon dioxide to highly valuable products. *Energy Sci. Eng.* **2019**, *7*, 4–29. [CrossRef]
34. Gavrikov, A.V.; Loktev, A.S.; Ilyukhin, A.B.; Mukhin, I.E.; Bykov, M.A.; Maslakov, K.I.; Vorobei, A.M.; Parenago, O.O.; Sadovnikov, A.A.; Dedov, A.G. Supercritical fluid-assisted modification combined with the resynthesis of $SmCoO_3$ as an effective tool to enhance the long-term performance of $SmCoO_3$-derived catalysts for the dry reforming of methane to syngas. *Dalton Trans.* **2022**, *51*, 18446–18461. [CrossRef] [PubMed]

catalysts

MDPI

Article

Electrocatalytic Hydrogen Evolution Reaction from Acetic Acid over Gold Immobilized Glassy Carbon Surface

Basmah H. Alshammari [1], Humayra Begum [2], Fatma A. Ibrahim [3], Mohamed S. Hamdy [3], Tahamida A. Oyshi [2], Nazia Khatun [4] and Mohammad A. Hasnat [2,*]

1 Chemistry Department, Faculty of Science, University of Hail, P.O. Box 2440, Hail 81451, Saudi Arabia
2 Electrochemistry & Catalysis Research Laboratory (ECRL), Department of Chemistry, School of Physical Sciences, Shahjalal University of Science and Technology, Sylhet 3114, Bangladesh
3 Catalysis Research Group (CRG), Department of Chemistry, College of Science, King Khalid University, P.O. Box 9004, Abha 61413, Saudi Arabia
4 Industrial Physics Division, Bangladesh Council of Scientific and Industrial Research (BCSIR), Dhaka 1205, Bangladesh; naziabcsir@gmail.com
* Correspondence: mah-che@sust.edu

Abstract: A hydrogen fuel cell is a highly promising alternative to fossil fuel sources owing to the emission of harmless byproducts. However, the operation of hydrogen fuel cells requires a constant supply of highly pure hydrogen gas. The scarcity of sustainable methods of producing such clean hydrogen hinders its global availability. In this work, a noble Au-atom-decorated glassy carbon electrode (Au/GCE) was prepared via a conventional electrodeposition technique and used to investigate the generation of hydrogen from acetic acid (AA) in a neutral electrolyte using 0.1 M KCl as the supporting electrolyte. Electrochemical impedance spectroscopy (EIS), open circuit potential measurement, cyclic voltammetry (CV), and rotating disk electrode voltammetry (RDE) were performed for the characterization and investigation of the catalytic properties. The constructed catalyst was able to produce hydrogen from acetic acid at a potential of approximately −0.2 V vs. RHE, which is much lower than a bare GCE surface. According to estimates, the Tafel slope and exchange current density are 178 mV dec^{-1} and 7.90×10^{-6} A cm^{-2}, respectively. Furthermore, it was revealed that the hydrogen evolution reaction from acetic acid has a turnover frequency (TOF) of approximately 0.11 s^{-1}.

Keywords: acetic acid; hydrogen evolution reaction; hydrogen fuel cell; rotating disk voltammetry; gold electrode; turnover frequency

Citation: Alshammari, B.H.; Begum, H.; Ibrahim, F.A.; Hamdy, M.S.; Oyshi, T.A.; Khatun, N.; Hasnat, M.A. Electrocatalytic Hydrogen Evolution Reaction from Acetic Acid over Gold Immobilized Glassy Carbon Surface. *Catalysts* **2023**, *13*, 744. https://doi.org/10.3390/catal13040744

Academic Editors: Georgios Bampos, Athanasia Petala and Zacharias Frontistis

Received: 15 February 2023
Revised: 5 April 2023
Accepted: 10 April 2023
Published: 13 April 2023

1. Introduction

Energy demand is surging due to population expansion and the introduction of technological conveniences. It is predicted that by 2040, the energy demand will increase by 30%. Most of this energy is produced from fossil fuels, which are non-renewable and detrimental to the environment due to the release of harmful gasses. The existing fossil fuel reserve is already under strain due to the rising energy demand, which is leading it to be exhausted more quickly and release more harmful polluting gases into the atmosphere. As a result, developing environmentally benign as well as self-sustainable fuel sources is crucial and therefore has grown to become one of the most alluring research areas in the present era. In this regard, many alternative energy sources have been reported, including solar energy, ocean energy, hydroelectric power, and wind energy, but none of them have the potential to completely supplant fossil fuel [1–4]. Moreover, the complete utilization of these fuel sources is limited by their uneven distribution.

The conversion of chemical energy to electric energy in an electrochemical fuel cell has been considered a promising technique for providing an alternative fuel source due to its eco-friendliness, lower toxicity, and ability to produce electricity with zero carbon dioxide emission [5]. In an electrochemical fuel cell, the chemical energy of fuel materials such as hydrogen, methanol, and oxygen is converted to electrical energy via a pair of redox reactions. Potential applications range from operating large-scale power stations to small-scale laptops, mobiles, and other electrical devices. In particular, hydrogen has demonstrated great potential as an energy carrier owing to its ability to have a high energy density, greater than almost all hydrocarbons, as well as its ability to produce harmless byproducts (only water) [6–8]. The challenge for the widespread application of hydrogen fuel cells is the availability of highly pure hydrogen. Commercially, hydrogen is obtained from the steam reforming process in which natural gas is heated with steam where hydrogen gas is produced along with toxic carbon monoxide [9,10]. This process utilizes fossil fuel sources for heating and releases toxic gases into the environment. Moreover, the steam reforming process does not produce hydrogen in a pure state but generates a mixture of gasses including CO_2, CH_4, etc. [11]. Although the hydrogenase enzyme can produce hydrogen using earth-abundant metals such as iron and nickel, its commercial application is limited due to the low stability of the enzymes outside their native environment and the complex preparation of the catalysts [12]. Two other alternative techniques for generating hydrogen in a highly pure state are photochemical and electrochemical water-splitting reactions [13–15]. However, photochemical water splitting requires a comparatively larger electrode area than electrochemical water splitting, imposing restrictions on choosing viable catalysts with desirable construction sites. Due to its greater adaptability and flexibility, electrochemical water splitting is more desirable. Once more, hydrogen evolution can be driven by electrochemical water splitting because water is renewable and abundant [16,17]. Until now, Pt has been the most efficient catalyst for HER due to its ability to reduce protons at a low overpotential [18–22]. The widespread use of hydrogen fuel cells has remained unexplored even decades after its invention due to the high cost and scarcity of Pt.

Thus, it is exceedingly necessary to develop cost-effective, efficient, and green catalysts for HER. For this purpose, metal phosphides, chalcogenides, carbides, borides, and nitrides have been investigated extensively, yet the problems of high metal content, convoluted operation, and long-term durability have not been overcome yet [23–28]. On the other hand, various nanostructures, heterostructures, and carbon composites have shown interesting activity, but the preparation procedure is still complex [29–32].

The majority of hydrogen evolution reactions have been reported in acidic and alkaline media, while the study of HER in neutral media is very limited [33–35]. However, extreme pH conditions are not only harmful to fuel cell components but also degrade the electrocatalysts. Moreover, since the pH of the natural water reserves on our planet, including the ocean and rivers, is close to neutral, the development of HER in neutral media would be helpful for powering devices sailing in seawater. HER in a neutral media occurs as follows [36,37]:

$$H_2O + e^- \rightarrow H_{ads} + OH^- \rightarrow \text{VolmerStep} \tag{1}$$

$$H_2O + e^- + H_{ads} \rightarrow H_2 + OH^- \rightarrow \text{HeyrovskyStep} \tag{2}$$

$$H_{ads} + H_{ads} \rightarrow H_2 \rightarrow \text{TafelStep} \tag{3}$$

Nevertheless, due to the low proton concentration in a neutral solution, HER follows sluggish kinetics compared to those of acidic and alkaline media, and water molecules must act as a reactant. Several catalysts such as amorphous MoS_x, CoP, and FeP have been shown to exhibit good activity for HER in neutral media, but the synthesis procedure of that catalyst is still difficult [38–40].

Pyrolysis of biomass from agricultural waste produces bio-oil in which acetic acid is a major component, constituting around 25% [41]. Furthermore, acetic acid is a great hydrogen carrier that is inflammable and safer than other types of hydrocarbons. The generation of hydrogen from acetic acid can be of great significance, as it not only provides clean fuel but also assists in recycling agricultural waste. Previously, V.S Thoi et al. reported a molecular catalyst for the generation of hydrogen from acetic acid using non-aqueous media [42]. The preparation of molecular catalysts is a complex procedure and the use of organic media during electrolysis is a convoluted process.

Here, we have investigated a green method for the generation of hydrogen from acetic acid in a KCl medium using a Au-modified glassy carbon electrode. To make the catalysts cost-effective, we have decorated the glassy carbon surface with an extremely thin layer of gold. To the best of our knowledge, no one has ever produced hydrogen from acetic acid using gold-modified glassy carbon, despite substantial research on HER in gold electrodes using H_2SO_4 being available.

2. Experimental Section

2.1. Chemicals

Analytical-grade chemicals were purchased and used as received without additional purification. KCl salt was obtained from Merck, Darmstadt, Germany. Alumina powder, sulfuric acid, acetic acid, and tetra chloroauric acid ($HAuCl_4$) were purchased from Sigma-Aldrich. All the solutions were prepared using Millipore Milli-Q water (resistivity < 18 MΩ cm and micro-organic concentration < 3 ppb). All the experiments were conducted in clean and N_2-saturated conditions.

2.2. Fabrication of Electrode and Electrochemical Measurements

All electrochemical experiments, such as cyclic voltammetry (CV), open circuit potential (OCP) measurements and electrochemical impedance (EIS) measurements, were conducted with PGSTAT 128N (Autolab, Utrecht, The Netherlands) and CHI 660 (Bee Cave, TX, USA) potentiostats using a typical three-electrode arrangement. In the case of EIS measurements, the frequency range was maintained from 0.1 MHz to 0.1 Hz, with 50 measuring points.

A glassy carbon electrode (GCE) with an exposed geometric surface area of ca. 0.0314 cm^2 was cleaned and modified with Au atoms using the electrodeposition technique. Ag/AgCl saturated with KCl and Pt wire electrodes were used as reference and counter electrodes, respectively. The GCE surface was first polished mechanically with fine alumina powder (down to 0.3 μm) until it turned a shiny black. Then, the polished electrode was rinsed with water and sonicated in ethanol and water for 10 min each to remove the adsorbed particles. Finally, the GCE surface was subjected to electrochemical cleaning in a N_2-saturated 0.1 M H_2SO_4 solution from 0.1 to +1.0 V at a scan rate of 0.1 V s^{-1} for 50 cycles. The clean electrode was then modified with Au by cycling the potential in a 0.01 M $HAuCl_4$ solution from 0 V to −1.0 V with a scan rate of 0.1 V s^{-1} for two cycles.

To unveil the catalysis of the hydrogen evolution reaction, cyclic voltammograms were recorded at potentials lower than the open circuit potential in a KCl medium by varying the acetic acid concentration (1–7 mM) and scan rate.

The hydrodynamic analysis and linear sweep voltammograms (LSVs) were recorded in a 0.1 M KCl solution with 5 mM acetic acid at a 0.05 V s^{-1} scan rate at different rotation rates (400–1200 rpm) using a Modulated Speed Rotator (PINE Incorp., Los Angeles, CA, USA).

In this article, all the potentials were converted into a reversible hydrogen electrode (RHE) using the following expression (4):

$$E_{RHE} = E_{Ag/AgCl} + E^o{}_{Ag/AgCl} + (0.059 \times pH) \tag{4}$$

where $E^o{}_{Ag/AgCl} = 0.1976\ V$.

The prepared Au/GCE surface was morphologically characterized using a scanning electron microscope (SEM) (JSM-7610F, JEOL Ltd., Tokyo, Japan). The X-ray photoelectron spectroscopy (XPS)-related analysis was executed using a Thermo Scientific (Waltham, MA, USA) K-Alpha KA1066 spectrometer to unveil the composition of the Au/GCE surface. A monochromatic AlKα X-ray radiation source was used as the excitation source, where the beam-spot size was kept at 300.0 μm. The spectra were recorded in terms of fixed analyzer transmission mode, where the pass energy was kept at 200.0 eV. The recording of the XPS spectra was executed at pressures of less than 10^{-8} Torr.

3. Results and Discussion

3.1. Surface Characterization

To reveal the surface feature of the Au/GCE electrode, a field-emission scanning electron microscopic (SEM) image was recorded. Figure 1 depicts the SEM image of the Au/GCE surface, which indicates that the Au particles were deposited successfully on the GCE surface. Additionally, the Au particles present on the surface are in a globular shape.

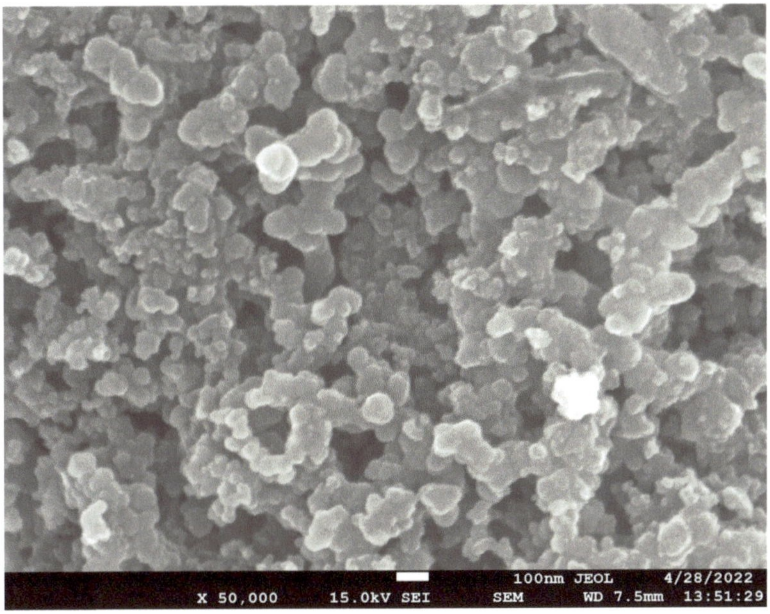

Figure 1. SEM image of the Au/GCE surface.

To understand the characteristics of the modified Au/GCE electrode, X-ray photoelectron spectroscopy (XPS) was conducted. Figure 2 depicts the XPS spectrum of Au on the GCE surface. In Figure 2a, the two intense peaks at 83.34 eV and 87.05 eV represent Au $4f_{7/2}$ and Au $4f_{5/2}$, respectively. The difference between the peaks is noted to be 3.71 eV, which suggests that the surface-located Au particles are metallic. Figure 2b demonstrates that the O 1s spectrum of the modified electrode at approximately 531.34 eV is due to the existence of oxygen in the environment.

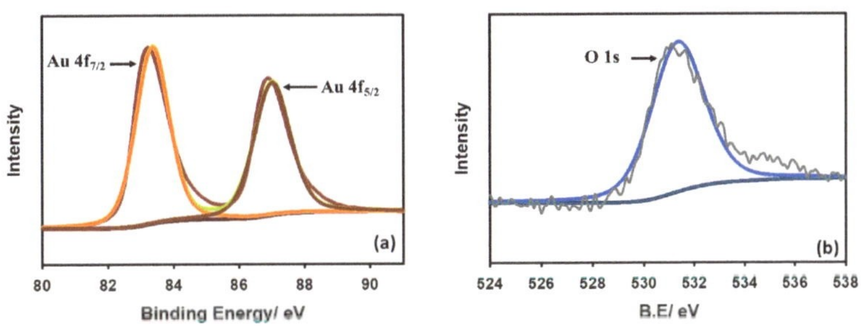

Figure 2. XPS spectra of Au on GCE surface (**a**) Au 4f and (**b**) O 1s.

3.2. Electrochemical Characterization

The electrochemical characterization of the Au/GCE electrode was performed by electrochemical impedance spectroscopy (EIS) and open circuit potential (OCP) measurements. Furthermore, the results were compared with those of unmodified GCE electrodes in a similar experimental condition. From the comparative analysis, the modification of GCE with the gold film was confirmed. The conventional display of Nyquist plots recorded using GCE and Au/GCE electrodes is shown in Figure 3 with 5 mM acetic acid in 0.1 M KCl at ca. −0.2 V and −0.6 V. It can be noted from Figure 3a that at −0.2 V in the Nyquist plot with bare GCE, there is a distorted semicircle with an extremely high charge transfer resistance, while in Au/GCE, the Nyquist plot exhibits a Warburg impedance curve. On the other hand, both figures show a Warburg impedance at −0.6 V, indicating that the reduction process is diffusion controlled (Figure 3b). The charge transfer resistance is lower in the case of Au/GCE compared to that of a bare GCE at each potential, indicating that the reduction process is more feasible on the Au/GCE surface than on the bare GCE surface. An Au/GCE electrode was perhaps naturally more negative than a GCE, hence the Au/GCE electrode could attract positively charged protons toward the electrode surface due to Coulombic attraction. The equivalent circuit is shown in the inset of Figure 3b and the charge transfer resistances in each electrode are compared in Table 1.

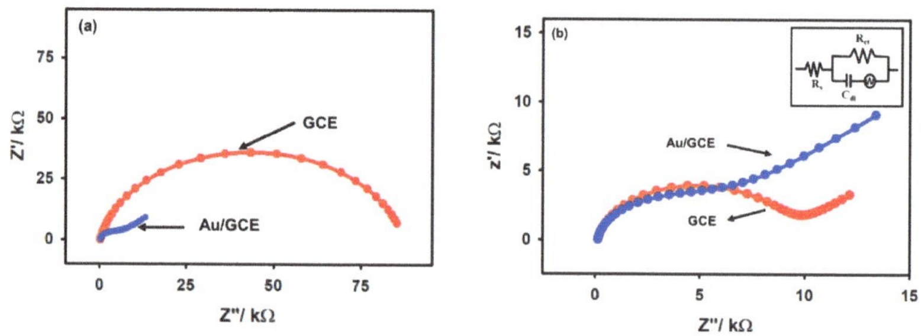

Figure 3. The comparative EIS spectra of bare GCE and Au/GCE at the potentials (**a**) −0.2 V and (**b**) −0.6 V recorded with 5 mM acetic acid in a 0.1 M KCl N_2-saturated solution; inset shows the equivalent circuit.

Table 1. Values of OCP and charge transfer resistance of a bare GCE and Au/GCE electrode recorded in a N_2-saturated 0.1 M KCl solution.

Electrode	OCP/V		R_s/Ω	$R_{ct}/k\Omega$
	Without AA	With 5 mM AA		
GCE	0.651	0.542	126	9.78
Au/GCE	0.622	0.675	130	4.70

Next, to justify the OCP observation, the linear polarization curves were recorded with both GCE and Au/GCE electrodes in the presence and absence of 5 mM acetic acid in a KCl medium. It can be noted from Figure 4a that in the absence of acetic acid, the Au/GCE electrode's OCP value is less positive compared to a bare GCE electrode. This observation shows that the Au/GCE electrode has greater capability to attract positively charged cations due to Coulombic interactions. This fact is also consistent with the results obtained from the EIS experiments. However, when the linear polarization curve was recorded in the presence of 5 mM acetic acid, it was observed that the Au/GCE surface became more positive than that of the bare GCE surface (Figure 4b). Thus, it can be assumed that, in the presence of acetic acid, the negatively charged Au/GCE surface provides electrons to reduce the positively charged protons and becomes more positive itself. The OCP values are summarized in Table 1. Figure 4c compares only Au/GCE electrodes with and without acetic acid. The shifting of the OCP value towards more positive potential in the presence of acetic acid also supports the above argument.

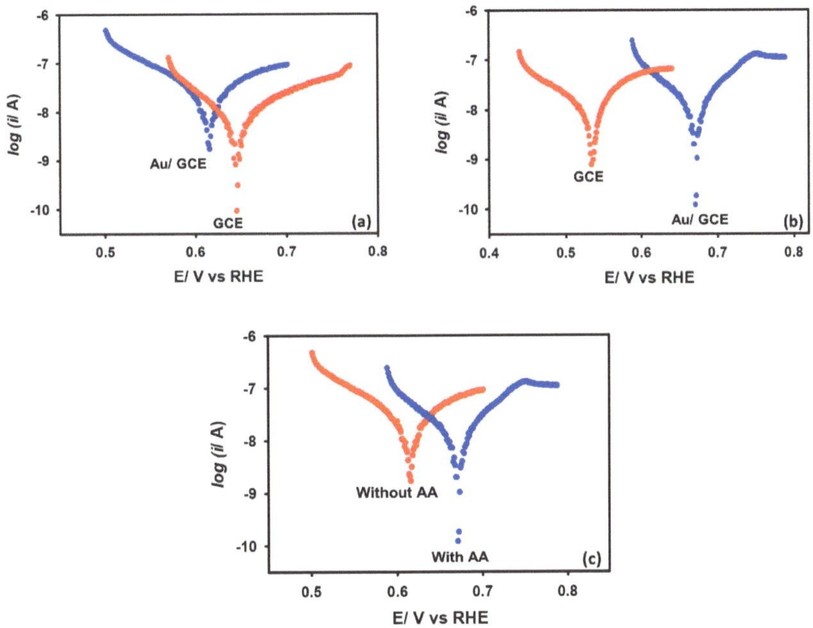

Figure 4. Linear polarization curves were recorded with (**a**) bare GCE and Au/GCE electrodes without AA, (**b**) bare GCE and Au/GCE electrodes with 5 mM AA, and (**c**) a Au/GCE electrode with and without AA in a N_2-saturated KCl solution. Scan rate: 0.1 V s^{-1}.

3.3. HER Studies

3.3.1. Catalysis

Figure 5a depicts the cyclic voltammograms recorded in a 0.1 M KCl solution with and without 5 mM acetic acid at a 0.1 V s^{-1} scan rate with a Au/GCE electrode. As shown, no reduction peak is seen between the potential range of 0 V and −0.5 V when acetic acid is not present. The reduction peak beyond −0.5 V is due to the hydrogen evolution from water. When 5 mM acetic acid is added to the electrolyte solution, an intense reduction peak is obtained at −0.2 V. Thus, this reduction peak is attributed to the reduction of acetic acid. Two reasons can be assumed for the occurrence of this wave: (1) the reduction of acetic acid to aldehyde or alcohol, or (2) the reduction of the proton to generate hydrogen. To confirm the identity of this peak, 5 mM sodium acetate is added to the KCl solution, but no such peak is found at this region. The peak is therefore proven to be related to proton reduction or hydrogen generation from acetic acid, not the reduction of the acetate ion. This is also evident from the appearance of small bubbles at the electrode surface during the CV measurement. Note that in bare GCE, proton reduction from acetic acid occurs at −0.6 V, while in Au/GCE the potential shifted to −0.2 V with an almost 48% higher current. This positive shift of potential is very important as it indicates that less energy is required to generate hydrogen from acetic acid with Au/GCE than that of bare GCE. After it is confirmed that the reduction wave is from proton reduction, some kinetics analysis is performed with the Au/GCE electrode.

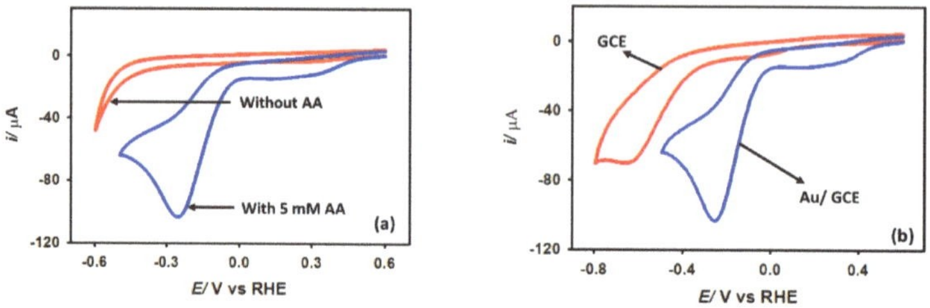

Figure 5. Cyclic voltammograms of (**a**) Au/GCE surface in KCl solution in the presence and absence of 5 mM CH_3COOH (**b**) bare GCE and Au/GCE surface in presence of 5 mM acetic acid at 0.1 V s^{-1} scan rate.

The experiment was repeated using a Au electrode. It was found that a gold electrode reduces protons under the same experimental conditions with insignificant activity (Figure S1, Supplementary information) in the KCl medium. Significant H^+ reduction performance was obtained in a 0.1 M KCl medium while a Au/GCE electrode was employed. It was prepared by depositing Au particles on the GCE surface from a 0.01 M $HAuCl_4$ solution by potential cycling between 0 and −1.0 V (vs. Ag/AgCl (sat. KCl) two times with a scan rate of 0.1 V s^{-1} (Figure S2, Supplementary Information). The Au/GCE electrode prepared with more than two cycles produces large capacitive currents without a significant improvement in Faradaic currents. Hence, all the rest of the experiments were performed by fabricating a Au/GCE electrode with controlled Au deposition on the GCE surface for two potential cycles only.

3.3.2. Effect of Acetic Acid Concentration

The variation of peak current (i_p) and peak potential (E_p) against the concentration of acetic acid at the Au/GCE electrode is shown in Figure 6a. The plot shows that as the AA concentration rises, the peak current progressively increases, and the peak potential slightly shifts in the negative direction. This is due to the concentration overpotential resulting from the increased AA concentration. The kinetic order (β) of the reaction is often used to disclose the mechanistic route of an electrochemical process. The value of β can be computed from the dependencies of peak currents on the concentration of the electro-active species with the assumption that at the peak potentials, the log i_p vs. log C curve gives straight lines as per Equation (5):

$$log\, i = log k + \beta log[C] \rightarrow \tag{5}$$

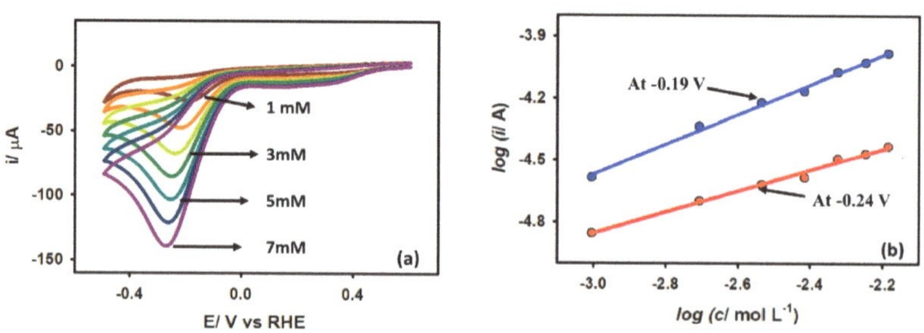

Figure 6. (**a**) CVs of Au/GCE electrode with varying concentration of acetic acid at 0.1 V s^{-1} scan rate. (**b**) Linear regression plots of log i_p vs. log C on Au/GCE electrode at −0.19 V and −0.24 V.

Here, k is the reaction rate constant. From Figure 6b, using Equation (5) the order of the reaction was found to be 0.56 and 0.62 for Au/GCE at potentials −0.19 V and −0.24 V, respectively, which means that the reaction followed half-order kinetics.

3.3.3. Effect of Scan Rate

Cyclic voltammograms were obtained for a Au/GCE electrode in a 0.1 M KCl medium with 5 mM acetic acid with variable scan rates in order to further study the kinetics. It is observed from Figure 7a that when the scan rate is raised from 0.05 V s^{-1} to 0.7 V s^{-1}, the peak current also increases with a shift of potential toward a slightly negative value. The enhancement of the peak current with an increased scan rate is due to the thinning of the diffusion layer at a higher scan rate. For an irreversible diffusion-controlled reaction, the peak current is linearly dependent on the square root of the scan rate as per the following expression (6):

$$i_p = 2.99 \times 10^5 nAC\alpha^{1/2} {D_o}^{1/2} v^{1/2} \tag{6}$$

where C is the bulk acetic acid concentration, n is the number of heterogeneous electron transfers, α is the cathodic transfer coefficient, A is the effective surface area of the working electrode, v is the scan rate, and D_o is the diffusion coefficient of acetic acid. The peak current and square root of the scan rate showed a linear connection with a slope value of 0.48, indicating a diffusion control process, which is also compatible with the EIS results.

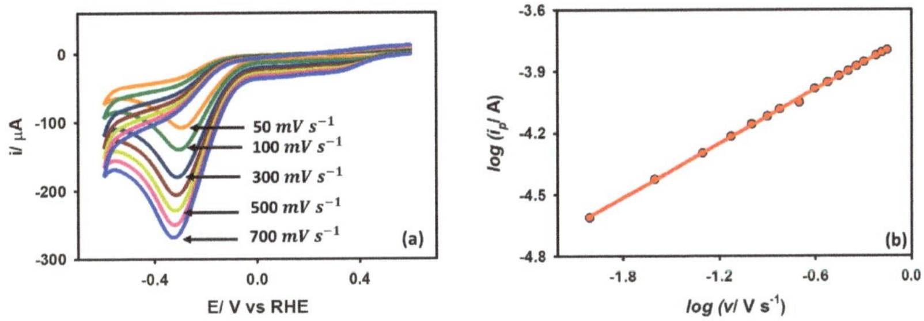

Figure 7. (**a**) CVs of Au/GCE electrode in 0.1 M KCl medium with 5 mM acetic acid at various scan rates. (**b**) Linear regression plots of log i_p vs. log v.

3.3.4. Tafel Analysis

The cathodic transfer coefficient, α, was calculated between 0.42 and 0.48 when the scan rate was raised from 0.05 V s^{-1} to 0.5 V s^{-1} according to the following formula (7):

$$\alpha = \frac{1.857RT}{F\left|\left(E_p - E_{\frac{p}{2}}\right)\right|} \tag{7}$$

The fairly constant nature of α at the current maxima indicates that the reaction follows B-V kinetics. In this regard, the Tafel slope is evaluated to understand the insight into the kinetic process of HER on the catalyst surface. Note that the Tafel slope is defined as the slope of the linear region of the plot log (current) vs. potential around the activation region of the current potential region as per Equation (8):

$$lni = lni_0 - \frac{\alpha F}{RT}\left(E - E^0\right) \tag{8}$$

Here, i is the kinetic current, i_0 is the exchange current, and $\frac{-2.303RT}{F}$ is the Tafel slope. We estimated the Tafel slope to determine the kinetic feasibility of the HER on the Au/GCE surface. Figure 8 demonstrates the Tafel plots obtained from the LSV curves with respect to HER originating from 5 mM acetic acid at 0.025, 0.05, and 0.075 V s^{-1} scan rates. The average slope at three different scan rates is found to be 178 mV dec^{-1}, which outranks many other metals reported previously for HER [43]. Note that the transfer coefficient equivalent to 0.33 obtained from the Tafel slope is smaller than that acquired at the peak region by Equation (7). This observation suggests that the electron transfer rate is slower at the activation region compared to the diffusional region, which is a highly general feature for most irreversible processes.

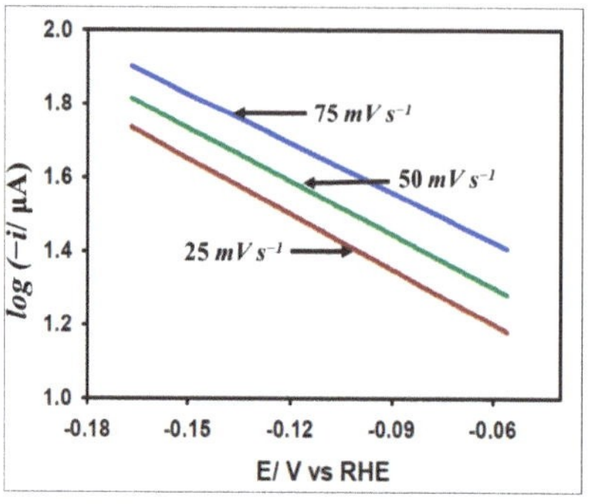

Figure 8. Variation of log(i) with the applied potential at 0.025 V s^{-1}, 0.05 V s^{-1}, and 0.075 V s^{-1} scan rates on Au/GCE electrode in 0.1 M KCl solution with 5 mM AA.

Another important parameter for understanding the HER kinetics is the exchange current density (j_o) which depends on the electrode material, electrode surface state, electrolytic composition, and experimental temperature. This current parameter can be derived by extrapolating the Tafel plots to the Y-axis as per Equation (8). It should be noted that j_o is proportional to the catalytic surface area. The large j_o value indicates a larger surface area, fast electron transfer, and favorable HER kinetics. From Figure 8, the value of j_o was calculated to be 7.90×10^{-6} A cm^{-2}, which is consistent with other metals found in the literature [17,44].

3.3.5. Diffusion Coefficient

Rotating disk electrode voltammetry (RDE) is employed to study the hydrodynamics of the HER from acetic acid. A steady-state concentration profile, and consequently a steady-state current, is produced at the electrode surface when the electrode is rotated at a fixed rate, transporting the electroactive species in the bulk solution to the electrode surface at a fixed rate. Figure 9a depicts the RDE voltammograms of the Au/GCE electrode in a 0.1 M KCl medium in the presence of 5 mM acetic acid at a 0.05 V s^{-1} scan rate. Acetic acid's diffusion coefficient is calculated using the change in steady-state current as a function of the rotation rate provided by the Levich equation. The diffusion coefficient of acetic acid according to Equation (9) is found to be 6.47×10^{-5} cm^2 s^{-1}, which is consistent with the value of the following literature [42].

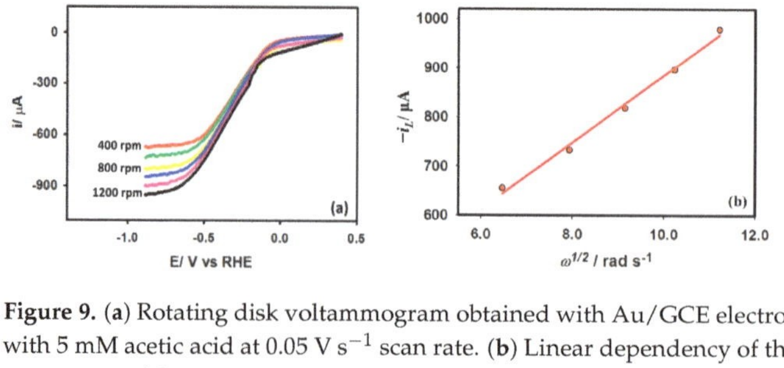

Figure 9. (**a**) Rotating disk voltammogram obtained with Au/GCE electrode in 0.1 M KCl solution with 5 mM acetic acid at 0.05 V s^{-1} scan rate. (**b**) Linear dependency of the limiting current on the square root of the rotation rate.

$$i_L = \frac{0.62nAFD^{2/3}\omega^{1/2}}{v^{1/6}} \quad (9)$$

3.3.6. Turnover Frequency

Turnover frequency (TOF) has become one of the crucial parameters in evaluating HER kinetics. It is defined by the transformation of reactive molecules per unit site at a unit time. A higher value of TOF characterizes the efficiency of a catalytic material pertaining to a specific reaction. A common method for estimating the TOF is based on the following two steps: (1) evaluation of the catalytic active sites and (2) calculation of the TOF value according to Equation (10) [45]:

$$TOF = \frac{jA}{2Fm} \quad (10)$$

Here, j is the current density at a defined overpotential, A is the geometric surface area of the electrode, 2 denotes the number of electron transfers during hydrogen formation, F is the Faraday constant, and m is the number of active sites.

The determination of the actual active surface area (A) in the catalytic surface is an extremely challenging task. The easiest approach is determining the capacitance (C_{dl}) in the double-layer region using Equation (11), which denotes the relationship between capacitive currents and scan rate. The value of C_{dl} fairly represents the active surface area.

$$i = C_{dl} \times \frac{dV}{dt} \quad (11)$$

The voltammograms concerning capacitive currents in the double-layer region as a function of scan rate are shown in Figure 10 for the Au/GCE electrode in a 0.1 M KCl solution. From the slope of Figure 10b, the C_{dl} value with respect to the Au/GCE electrode was determined to be 2.55 μF cm^{-2}. By comparing the C_{dl} value (0.65 μFcm^{-2}) obtained for a pure GCE surface in 0.1 M KCl with respect to its area of 0.0314 cm^2, the active area of the modified electrode (Au/GCE) was determined to be approximately 0.12 cm^2 using Equation (12).

$$A_{Au/GCE} = \frac{C_{dl(Au/GCE)}}{C_{dl(GCE)}} \times A_{GCE} \quad (12)$$

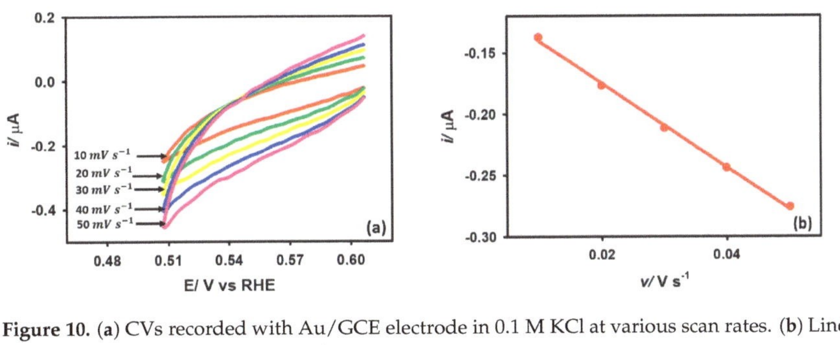

Figure 10. (**a**) CVs recorded with Au/GCE electrode in 0.1 M KCl at various scan rates. (**b**) Linear relation of the current vs. scan rate plot.

By setting the value of $A_{Au/GCE}$ and equating the slope of the i vs. v relationship (Equation (13)) with that obtained from the slope of the peak current vs. scan rate plot (w.r.t Figure 7a), the value of 'm' was determined to be 5.76 × 10^{-9}mol cm^{-2}.

$$i = \frac{n^2F^2A_{Au/GCE}m}{4RT}v \quad (13)$$

Inserting the value of 'm' in Equation (10) with respect to the peak potential of the CV obtained for 5 mM acetic acid, the TOF value was obtained to be 0.11 s^{-1}.

Thus, this work has demonstrated a green method for generating clean fuel hydrogen from acetic acid which can be obtained from agricultural waste.

4. Conclusions

A green methodology for the generation of hydrogen from acetic acid was demonstrated using a Au/GCE electrode. The electrode was prepared using a simple and straightforward electrodeposition technique and the electrokinetic investigations were carried out with CV, RDE, OCP, and EIS measurements. EIS and OCP measurements confirmed the successful modification of the GCE surface with Au and revealed the greater feasibility of the Au/GCE surface for proton reduction than that of the bare GCE surface. Cyclic voltammograms show a low overpotential (−0.2 V vs. RHE) for proton reduction in Au/GCE electrodes with around 48% higher current than that of the bare GCE surface. The Tafel slope and exchange current density were estimated to be 178 mV dec^{-1} and 7.90×10^{-6}A cm^{-2}, respectively, while the diffusion coefficient of AA in KCl was estimated to be 6.47×10^{-5} cm^2 s^{-1}. Moreover, a high TOF for proton reduction was also observed with Au/GCE electrodes in neutral media. This work not only demonstrates a green approach to clean energy production, but also proposes a recycling method for agricultural waste products.

Supplementary Materials: The following supporting information can be downloaded at: https://www.mdpi.com/article/10.3390/catal13040744/s1, Figure S1: HER observation from 5 mM acetic acid in 0.1 M KCl using Au, GCE and Au/GCE electrodes obtained scan rate of 0.1 V s^{-1}; Figure S2: Optimization of Au deposition on GCE for HER from 5 mM acetic acid in 0.1 M KCl. The electrodes having different amount of Au deposition was executed by cycling the potential in 0.01 M HAuCl4 solution from 0 V to −1.0 V with a scan rate of 0.1 V s^{-1} for different number of cycles.

Author Contributions: B.H.A.: Writing; H.B.: Experimental & draft writing; F.A.I.: editing; M.S.H.: editing, T.A.O.: Editing; N.K.: Surface analysis; M.A.H.: oveall supervision. All authors have read and agreed to the published version of the manuscript.

Funding: The authors extend their appreciation to the Deanship of Scientific Research at King Khalid University for funding this work through large group Research Project under grant number RGP2/236/44.

Data Availability Statement: Not applicable.

Conflicts of Interest: The author declares no conflict of interest.

References

1. Zhao, G.; Rui, K.; Dou, S.X.; Sun, W. Heterostructures for Electrochemical Hydrogen Evolution Reaction: A Review. *Adv. Funct. Mater.* **2018**, *28*, 1803291. [CrossRef]
2. Gyamfi, B.A.; Bein, M.A.; Bekun, F.V. Investigating the nexus between hydroelectricity energy, renewable energy, nonrenewable energy consumption on output: Evidence from E7 countries. *Environ. Sci. Pollut. Res.* **2020**, *27*, 25327–25339. [CrossRef] [PubMed]
3. Lehtola, T.; Zahedi, A. Solar energy and wind power supply supported by storage technology: A review. *Sustain. Energy Technol. Assess.* **2019**, *35*, 25–31. [CrossRef]
4. Li, Z.; Siddiqi, A.; Anadon, L.D.; Narayanamurti, V. Towards sustainability in water-energy nexus: Ocean energy for seawater desalination. *Renew. Sustain. Energy Rev.* **2018**, *82*, 3833–3847. [CrossRef]
5. Baroutaji, A.; Wilberforce, T.; Ramadan, M.; Olabi, A.G. Comprehensive investigation on hydrogen and fuel cell technology in the aviation and aerospace sectors. *Renew. Sustain. Energy Rev.* **2019**, *106*, 31–40. [CrossRef]
6. Askari, M.B.; Beheshti-Marnani, A.; Seifi, M.; Rozati, S.M.; Salarizadeh, P. *Fe_3O_4@MoS_2/RGO as an Effective Nano-Electrocatalyst Toward Electrochemical Hydrogen Evolution Reaction and Methanol Oxidation in Two Settings for Fuel Cell Application*; Elsevier Inc.: Amsterdam, The Netherlands, 2019; Volume 537. [CrossRef]
7. Duan, C.; Kee, R.; Zhu, H.; Sullivan, N.; Zhu, L.; Bian, L.; Jennings, D.; O'Hayre, R. Highly efficient reversible protonic ceramic electrochemical cells for power generation and fuel production. *Nat. Energy* **2019**, *4*, 230–240. [CrossRef]
8. Choi, S.; Davenport, T.C.; Haile, S.M. Protonic ceramic electrochemical cells for hydrogen production and electricity generation: Exceptional reversibility, stability, and demonstrated faradaic efficiency. *Energy Environ. Sci.* **2019**, *12*, 206–215. [CrossRef]

9. Chen, M.; Hu, J.; Wang, Y.; Wang, C.; Tang, Z.; Li, C.; Liang, D.; Cheng, W.; Yang, Z.; Zhang, H. Hydrogen production from acetic acid steam reforming over Ti-modified Ni/Attapulgite catalysts. *Int. J. Hydrogen Energy* **2021**, *46*, 3651–3668. [CrossRef]
10. Wang, Z.; Sun, L.; Chen, L.; Yang, S.; Xie, X.; Gao, M.; Li, T.; Zhao, B.; Si, H.; Hua, D. Steam reforming of acetic acid for hydrogen production over Ni/CaxFeyO catalysts. *Int. J. Hydrogen Energy* **2021**, *46*, 33132–33142. [CrossRef]
11. Junior, R.B.S.; Rabelo-Neto, R.C.; Gomes, R.S.; Noronha, F.B.; Fréty, R.; Brandão, S.T. Steam reforming of acetic acid over Ni-based catalysts derived from $La_{1-x}Ca_xNiO_3$ perovskite type oxides. *Fuel* **2019**, *254*, 115714. [CrossRef]
12. Zhu, K.; Liu, Z.; Zhao, X.; Yan, E.; Li, X.; Guo, J. A critical review on limitations and enhancement strategies associated with biohydrogen production. *Int. J. Hydrogen Energy* **2021**, *46*, 16565–16590. [CrossRef]
13. Xing, Z.; Zong, X.; Pan, J.; Wang, L. On the engineering part of solar hydrogen production from water splitting: Photoreactor design. *Chem. Eng. Sci.* **2013**, *104*, 125–146. [CrossRef]
14. Rajamathi, C.R.; Gupta, U.; Pal, K.; Kumar, N.; Yang, H.; Sun, Y.; Shekhar, C.; Yan, B.; Parkin, S.; Waghmare, U.V.; et al. Photochemical Water Splitting by Bismuth Chalcogenide Topological Insulators. *ChemPhysChem* **2017**, *17*, 2322–2327. [CrossRef] [PubMed]
15. Li, X.; Hao, X.; Abudula, A.; Guan, G. Nanostructured catalysts for electrochemical water splitting: Current state and prospects. *J. Mater. Chem. A* **2016**, *4*, 11973–12000. [CrossRef]
16. Bahnemann, A.A.I.D.W. Photochemical splitting of water for hydrogen production by photocatalysis: A review. *Sol. Energy Mater. Sol. Cells* **2014**, *128*, 85–101. [CrossRef]
17. Zhang, X.Z.Y. Noble metal-free hydrogen evolution catalysts for water splitting. *Chem. Soc. Rev.* **2015**, *44*, 5148–5180. [CrossRef]
18. Tiwari, J.N.; Sultan, S.; Myung, C.W.; Yoon, T.; Li, N.; Ha, M.; Harzandi, A.M.; Park, H.J.; Kim, D.Y.; Chandrasekaran, S.S.; et al. Multicomponent electrocatalyst with ultralow Pt loading and high hydrogen evolution activity. *Nat. Energy* **2018**, *3*, 773–782. [CrossRef]
19. Li, Y.; Zhang, H.; Xu, T.; Lu, Z.; Wu, X.; Wan, P.; Sun, X.; Jiang, L. Under-Water Superaerophobic Pine-Shaped Pt Nanoarray Electrode for Ultrahigh-Performance Hydrogen Evolution. *Adv. Funct. Mater.* **2015**, *25*, 1737–1744. [CrossRef]
20. Xie, C.; Chen, W.; Du, S.; Yan, D.; Zhang, Y.; Chen, J.; Liu, B.; Wang, S. In-situ phase transition of WO3 boosting electron and hydrogen transfer for enhancing hydrogen evolution on Pt. *Nano Energy* **2020**, *71*, 104653. [CrossRef]
21. Kobayashi, D.; Kobayashi, H.; Wu, D.; Okazoe, S.; Kusada, K.; Yamamoto, T.; Toriyama, T.; Matsumura, S.; Kawaguchi, S.; Kubota, Y.; et al. Significant Enhancement of Hydrogen Evolution Reaction Activity by Negatively Charged Pt through Light Doping of W. *J. Am. Chem. Soc.* **2020**, *142*, 17250–17254. [CrossRef]
22. Yu, J.; Guo, Y.; She, S.; Miao, S.; Ni, M.; Zhou, W.; Liu, M.; Shao, Z. Bigger is Surprisingly Better: Agglomerates of Larger RuP Nanoparticles Outperform Benchmark Pt Nanocatalysts for the Hydrogen Evolution Reaction. *Adv. Mater.* **2018**, *30*, e1800047. [CrossRef] [PubMed]
23. Callejas, J.F.; Read, C.G.; Roske, C.W.; Lewis, N.S.; Schaak, R.E. Synthesis, Characterization, and Properties of Metal Phosphide Catalysts for the Hydrogen-Evolution Reaction. *Chem. Mater.* **2016**, *28*, 6017–6044. [CrossRef]
24. Skúlason, Y.A.E. Hydrogen Evolution Reaction Catalyzed by Transition-Metal Nitrides. *J. Phys. Chem. C* **2017**, *121*, 24036–24045. [CrossRef]
25. Jin, H.; Gu, Q.; Chen, B.; Tang, C.; Zheng, Y.; Zhang, H.; Jaroniec, M.; Qiao, S.-Z. Molten Salt-Directed Catalytic Synthesis of 2D Layered Transition-Metal Nitrides for Efficient Hydrogen Evolution. *Chem* **2020**, *6*, 2382–2394. [CrossRef]
26. Gao, Q.; Zhang, W.; Shi, Z.; Yang, L.; Tang, Y. Structural Design and Electronic Modulation of Transition-Metal-Carbide Electrocatalysts toward Efficient Hydrogen Evolution. *Adv. Mater.* **2019**, *31*, 1802880. [CrossRef]
27. Yang, C.; Zhao, R.; Xiang, H.; Wu, J.; Zhong, W.; Li, W.; Zhang, Q.; Yang, N.; Li, X. Ni-Activated Transition Metal Carbides for Efficient Hydrogen Evolution in Acidic and Alkaline Solutions. *Adv. Energy Mater.* **2020**, *10*, 2002260. [CrossRef]
28. Chen, W.-F.; Sasaki, K.; Ma, C.; Frenkel, A.I.; Marinkovic, N.; Muckerman, J.T.; Zhu, Y.; Adzic, R.R. Hydrogen-Evolution Catalysts Based on Non-Noble Metal Nickel-Molybdenum Nitride Nanosheets. *Angew. Chemie* **2012**, *124*, 6235–6239. [CrossRef]
29. Jiang, P.; Liu, Q.; Ge, C.; Cui, W.; Pu, Z.; Asiri, A.M.; Sun, X. CoP nanostructures with different morphologies: Synthesis, characterization and a study of their electrocatalytic performance toward the hydrogen evolution reaction. *J. Mater. Chem. A* **2014**, 2, 14634–14640. [CrossRef]
30. Li, X.; Wang, X.J.; Zhu, J.Y.; Li, Y.; Zhao, J.; Li, F.T. Fabrication of two-dimensional Ni2P/ZnIn2S4 heterostructures for enhanced photocatalytic hydrogen evolution. *Chem. Eng. J.* **2018**, *353*, 15–24. [CrossRef]
31. Raoof, J.B.; Hosseini, S.R.; Ojani, R.; Mandegarzad, S. MOF-derived Cu/nanoporous carbon composite and its application for electro-catalysis of hydrogen evolution reaction. *Energy* **2015**, *90*, 1075–1081. [CrossRef]
32. Chen, Y.; Tian, G.; Ren, Z.; Pan, K.; Shi, Y.; Wang, J.; Fu, H. Hierarchical core-shell carbon Nanofiber@ZnIn2S4 composites for enhanced hydrogen evolution performance. *ACS Appl. Mater. Interfaces* **2014**, *6*, 13841–13849. [CrossRef] [PubMed]
33. Yin, J.; Jin, J.; Zhang, H.; Lu, M.; Peng, Y.; Huang, B.; Xi, P.; Yan, C. Atomic Arrangement in Metal-Doped NiS2 Boosts the Hydrogen Evolution Reaction in Alkaline Media. *Angew. Chem.-Int. Ed.* **2019**, *58*, 18676–18682. [CrossRef] [PubMed]
34. Dinh, C.-T.; Jain, A.; de Arquer, F.P.G.; De Luna, P.; Li, J.; Wang, N.; Zheng, X.; Cai, J.; Gregory, B.Z.; Voznyy, O.; et al. Multi-site electrocatalysts for hydrogen evolution in neutral media by destabilization of water molecules. *Nat. Energy* **2019**, *4*, 107–114. [CrossRef]
35. You, B.; Liu, X.; Hu, G.; Gul, S.; Yano, J.; Jiang, D.-E.; Sun, Y. Universal Surface Engineering of Transition Metals for Superior Electrocatalytic Hydrogen Evolution in Neutral Water. *J. Am. Chem. Soc.* **2017**, *139*, 12283–12290. [CrossRef] [PubMed]

36. Wang, X.; Zheng, Y.; Sheng, W.; Xu, Z.J.; Jaroniec, M.; Qiao, S.Z. Strategies for design of electrocatalysts for hydrogen evolution under alkaline conditions. *Mater. Today* **2020**, *36*, 125–138. [CrossRef]
37. Mahmood, N.; Yao, Y.; Zhang, J.W.; Pan, L.; Zhang, X.; Zou, J.J. Electrocatalysts for Hydrogen Evolution in Alkaline Electrolytes: Mechanisms, Challenges, and Prospective Solutions. *Adv. Sci.* **2018**, *5*, 1700464. [CrossRef]
38. Ibupoto, Z.H.; Tahira, A.; Tang, P.; Liu, X.; Morante, J.R.; Fahlman, M.; Arbiol, J.; Vagin, M.; Vomiero, A. MoS x @NiO Composite Nanostructures: An Advanced Nonprecious Catalyst for Hydrogen Evolution Reaction in Alkaline Media. *Adv. Funct. Mater.* **2019**, *29*, 1807562. [CrossRef]
39. Yang, F.; Chen, Y.; Cheng, G.; Chen, S.; Luo, W. Ultrathin Nitrogen-Doped Carbon Coated with CoP for Efficient Hydrogen Evolution. *ACS Catal.* **2017**, *7*, 3824–3831. [CrossRef]
40. Jiang, J.; Wang, C.; Zhang, J.; Wang, W.; Zhou, X.; Pan, B.; Tang, K.; Zuo, J.; Yang, Q. Synthesis of FeP2/C nanohybrids and their performance for hydrogen evolution reaction. *J. Mater. Chem. A* **2015**, *3*, 499–503. [CrossRef]
41. Oh, S.J.; Choi, G.G.; Kim, J.S. Production of acetic acid-rich bio-oils from the fast pyrolysis of biomass and synthesis of calcium magnesium acetate deicer. *J. Anal. Appl. Pyrolysis.* **2017**, *124*, 122–129. [CrossRef]
42. Thoi, V.S.; Karunadasa, H.I.; Surendranath, Y.; Long, J.R.H.; Chang, C.J. Electrochemical generation of hydrogen from acetic acid using a molecular molybdenum-oxo catalyst. *Energy Environ. Sci.* **2012**, *5*, 7762–7770. [CrossRef]
43. Tang, Y.; Dong, L.; Wu, H.B.; Yu, X.Y. Tungstate-modulated Ni/Ni(OH)2interface for efficient hydrogen evolution reaction in neutral media. *J. Mater. Chem. A* **2021**, *9*, 1456–1462. [CrossRef]
44. Lu, X.; Yu, T.; Wang, H.; Luo, R.; Liu, P.; Yuan, S.; Qian, L. Self-supported nanoporous gold with gradient tin oxide for sustainable and efficient hydrogen evolution in neutral media. *J. Renew. Mater.* **2020**, *8*, 133–151. [CrossRef]
45. Song, S.; Zang, J.; Zhou, S.; Gao, H.; Tian, H.; Yuan, Y.; Li, W.; Wang, Y. Self-supported amorphous nickel-iron phosphorusoxides hollow spheres on Ni-Fe foam for highly efficient overall water splitting. *Electrochim. Acta* **2021**, *392*, 138996. [CrossRef]

catalysts

Article

The Combined Impact of Ni-Based Catalysts and a Binary Carbonate Salts Mixture on the CO_2 Gasification Performance of Olive Kernel Biomass Fuel

Athanasios Lampropoulos [1,*], Stamatia A. Karakoulia [2], Georgios Varvoutis [1], Stavros Spyridakos [1], Vassilios Binas [3], Leila Zouridi [3,4], Sofia Stefa [5], Michalis Konsolakis [5] and George E. Marnellos [1,2,*]

1 Department of Mechanical Engineering, University of Western Macedonia, 50100 Kozani, Greece
2 Chemical Process and Energy Resources Institute, Centre for Research and Technology Hellas, 57001 Thessaloniki, Greece
3 Institute of Electronic Structure and Laser, Foundation for Research and Technology-Hellas, 70013 Heraklion, Greece
4 Department of Material Science and Technology, University of Crete, 71003 Heraklion, Greece
5 School of Production Engineering and Management Technical, University of Crete, 73100 Chania, Greece
* Correspondence: alabropoulos@uowm.gr (A.L.); gmarnellos@uowm.gr (G.E.M.); Tel.: +30-24610-56702 (A.L.); +30-24610-56690 (G.E.M.)

Abstract: In the present work, the individual or synergistic effect of Ni-based catalysts (Ni/CeO_2, Ni/Al_2O_3) and an eutectic carbonate salt mixture (MS) on the CO_2 gasification performance of olive kernels was investigated. It was found that the Ni/CeO_2 catalyst presented a relatively superior instant gasification reaction rate (Rco) compared to Ni/Al_2O_3, in line with the significant redox capability of CeO_2. On the other hand, the use of the binary eutectic carbonate salt mixture (MS) lowered the onset and maximum CO_2 gasification temperatures, resulting in a notably higher carbon conversion efficiency (81%) compared to the individual Ni-based catalysts and non-catalytic gasification tests (60%). Interestingly, a synergetic catalyst-carbonate salt mixture effect was revealed in the low and intermediate CO_2 gasification temperature regimes, boosting the instant gasification reaction rate (Rco). In fact, in the temperature range of 300 to 550 °C, the maximum Rco value for both MS-Ni/Al_2O_3 and MS-Ni/CeO_2 systems were four times higher (4×10^{-3} min^{-1} at 460 °C) compared to the individual counterparts. The present results demonstrated for the first time the combined effect of two different Ni-based catalysts and an eutectic carbonate salt mixture towards enhancing the CO production rate during CO_2 gasification of olive kernel biomass fuel, especially in the devolatilization and tar cracking/reforming zones. On the basis of a systematic characterization study and lab-scale gasification experiments, the beneficial role of catalysts and molten carbonate salts on the gasification process was revealed, which can be ascribed to the catalytic activity as well as the improved mass and heat transport properties offered by the molten carbonate salts.

Keywords: olive kernel; CO_2 gasification; Ni/CeO_2; Ni/Al_2O_3; molten carbonates salt (MS); instant gasification rate; carbon conversion

Citation: Lampropoulos, A.; Karakoulia, S.A.; Varvoutis, G.; Spyridakos, S.; Binas, V.; Zouridi, L.; Stefa, S.; Konsolakis, M.; Marnellos, G.E. The Combined Impact of Ni-Based Catalysts and a Binary Carbonate Salts Mixture on the CO_2 Gasification Performance of Olive Kernel Biomass Fuel. *Catalysts* **2023**, *13*, 596. https://doi.org/10.3390/catal13030596

Academic Editor: Georgios Bampos

Received: 12 February 2023
Revised: 10 March 2023
Accepted: 13 March 2023
Published: 16 March 2023

1. Introduction

The increasing demand for more energy worldwide, along with the rising concerns on climate change, are forcing humanity to speed up defossilization and prioritize the clean transition of the energy mix [1]. However, the wide deployment of intermittent renewables, such as solar and wind power, requires the development of energy and cost efficient large-scale seasonal energy storage technologies to balance electricity grids and avoid costly curtailments, which has not yet been convincingly achieved nowadays [2]. On the other hand, if logistics are resolved, the use of bioenergy potential could sufficiently meet the energy demand and decarbonize our economies. In specific, the thermochemical exploitation of biomass (combustion, pyrolysis, gasification, etc.) is expected to notably suppress greenhouse gas (GHG) emissions due to the carbon neutral nature of biomass [1,3,4]. According to Di Gulliano et al. [5], biomass-based power plants accompanied by CO_2 capture can achieve negative CO_2 emissions, leading to an estimated CO_2 sequestration potential of 2.0–12.0 Gt CO_2 by 2050.

Agricultural and agro-industrial activities are the main sources of residual biomass that could be used as renewable feedstock in thermochemical cycles, thus enhancing local economies and minimizing waste handling problems [6–10]. Greece possesses large quantities of animal manure and agricultural and agro-industrial residues that are equal to 50 Mt/year [11]. Olive oil industry residues (tree pruning, leaves, pits, kernel and pomace) represent 20% of the total annual residual biomass in Greece, which is the third largest olive oil industry worldwide (350 ktons/year) [12,13]. Specifically, olive kernel accounts for 15% of the total residues resulting as by-products from the olive oil extraction process [14,15]. Among other residues, olive kernel exhibits the lowest moisture content (5–15%), and is thus considered a saleable fuel without further treatment [14–16]. Assuming an average lower heating value (LHV) of olive-based biomass residues equal to 16–18 MJ/kg [17–21] and an electrical efficiency of 25–35% [22–24], the thermochemical exploitation of the total annual quantity of olive oil-industry residues in Greece (10 Mt) could potentially produce around 14–15 GWh/year of energy. However, uncertainties associated with the Greek institutional framework towards biomass-based power plants are slowing down practical applications at larger scales [25–27].

Converting biomass to heat, electrical power, biofuels, and chemicals via the gasification process is of paramount importance in achieving energy independence and increasing the share of renewables in the energy mix. In principle, gasification process includes two main consecutive steps: (a) a relatively fast step ($t < 60$ s @ 300–700 °C), involving the thermal decomposition of volatile components to gaseous products, tars (condensable hydrocarbon compounds), and a solid char residue, and (b) a slower step ($t > 3300$ s @ $T > 700$ °C), including the gasification of the pyrolyzed char (gas-solid reactions) along with other gas-phase reforming reactions [28]. Suitable gasifying agents (air, steam, CO_2, or their mixture) can benefit the gasification reaction of the remaining char [4,29,30]. Apart from the gasifying agent, from a process system perspective, the main factors affecting syngas quality and quantity are the employed operating temperature and pressure, gasifier design and heating mode, the addition of catalysts, and the feedstock composition [31,32].

Despite the advantages of biomass exploitation through the gasification process, there are specific challenges that seriously affect the gasification performance in terms of the as-produced syngas quantity and quality. Syngas clean-up from impurities such as tars and water is of high priority towards producing a final product suitable for several downstream processes, i.e., in high-temperature fuel cells for the co-generation of heat and power or in catalytic reactors for the production of synthetic fuels and/or chemicals [33–35]. Mechanical separation methods for the removal of particulate matter (wet scrubber, cyclone, filter, and electrostatic precipitator), along with tar thermal cracking, have been widely used for syngas clean-up and conditioning [36–38]. However, the relatively high capital cost and the formation of small, complex tar structures restrict the commercialization of gasification technology [38]. In this regard, the use of highly active catalysts, in situ or outside the gasifier, has been proposed to reform tars, generating additional syngas with negligible impurities. Moreover, the use of catalysts in gasification leads to faster kinetics at decreased temperatures, further enhancing its efficiency [33,34,39]. Numerous studies [40–43] highlighted the main criteria for a good catalyst selection, which include the decrease of energy activation for tar cracking, gas-solid and gas-phase reactions, the reduced supply of gasifying agents, and the generation of high-value chemicals.

In general, catalysts used in biomass gasification can be divided into three major categories: (a) natural mineral catalysts (dolomite, olivine, zeolites, alumina), (b) alkali and alkaline Earth metal catalysts, AAEMs (Li, Na, K, Mg, Ca, and Rb), and (c) transition metal catalysts [44,45]. Among all of the transition metals (group VIII), Ni-based catalysts have been widely reported in the literature [28,44,46–50]. This can be attributed to their remarkable role on tar removal, steam reforming, and water gas shift (WGS) reactions, for H_2-rich syngas production [44,45]. For instance, Baker et al. [51] and Li et al. [52] investigated the performance of commercial nickel catalysts in fluidized bed gasifiers in the temperature range of 700–800 °C during sawdust and pine flakes steam gasification. They both noticed the pronounced role of Ni catalysts during steam gasification in terms of gaseous product yield and tar elimination. However, in both studies, the catalysts suffered from rapid deactivation (<5 h) due to carbon deposition on the catalyst surface and nickel particles sintering. In this regard, numerous studies investigated and proposed alternative routes to modify Ni based catalysts, including the use of different supporting materials and/or aliovalent doping (Fe, Co, La Ce, Mg, etc.) [46–50,53–62]. Several types of supporting materials have been examined in the literature, involving metal oxides, among others [44,45]. Metal oxides (CeO_2, Al_2O_3, ZrO_2), natural minerals (olivine, dolomite), and zeolites can provide mechanical strength, improved textural properties, and adsorption ability if they are employed as supports in Ni-based catalysts [63,64].

In addition to the use of catalysts, molten alkali carbonate (MAC) salts have been extensively studied in biomass gasification for tar elimination and enhanced syngas production [65–67]. In general, MAC salts (Li_2CO_3, Na_2CO_3, K_2CO_3, etc.) in eutectic compositions can be utilized as excellent industrial fluids in a wide range of high temperature operating technologies (solar power applications, waste oxidation, direct carbon fuel cells, and catalytic biomass gasification) [66]. A well-established application of MAC salts is as heating carriers, since they exhibit a dual function as solar heat storage and heat transfer media [65]. Unlike the use of solar-heated MACs as an indirect heating medium for pyrolysis and/or gasification processes, the direct contact of molten salts and biomass particles can lead to an enhanced gasification rate and syngas production. This enhancement can be attributed to the facilitation of mass and heat transfer phenomena induced by MACs in conjunction with their alkali-based catalytic nature [66].

Various studies have been devoted to interpreting the gasification mechanism of solid fuels in the presence of MAC salts [68–74]. Notably, the majority of works are dealing with pure carbon and coal/biomass chars. For instance, Mckee and co-investigators [68] concluded that the catalytic gasification of coal char over K_2CO_3 involves the following steps: (a) melting of K_2CO_3 salt, (b) its precipitation at the char surface area (pores), and (c) the formation of a thin layer favoring the carbon-catalyst interaction.

Gasifying agents notably affect the overall reaction network and, in turn, the syngas formation reaction rate. Strong gasifying agents (pure oxygen and steam) can affect the thermal stability of MAC salts [75], whereas the formation of hydroxide intermediates (MOH) and salt oxides (M_2O_2) may lead to fouling and plugging of the reactor and pipeline system [28,76]. However, CO_2 is expected to be a suitable gasifying agent for biomass gasification in the presence of MAC catalysts since it can maintain MACs' stability at high temperatures [28]. Taking into account the efforts towards CO_2 mitigation, CO_2 biomass gasification over MAC salts can be a promising route towards producing biofuels and high value products with negative CO_2 emissions [77–80]. In the light of the above aspects, the majority of the studies regarding biomass gasification over MAC salts are restricted to the use of CO_2 as a gasifying agent [28,81–87].

As mentioned above, the char-CO_2 gasification reaction requires more energy compared to other gasifying agents (air, oxygen, and steam) and can be considered the rate-determining step of the overall reaction network [88]. Interestingly, catalyst addition to MACs could synergistically lead to enhanced tar-free syngas production, sufficiently simplify gas clean-up steps, and render the whole process less complex and economically feasible for large-scale applications [28]. As extensively mentioned above, commercial nickel and modified Ni-based catalysts can facilitate biomass gasification, tar cracking, and reforming reactions. In fact, Ni-based catalysts are both selective for hydrogen production and highly efficient towards carbon conversion. Very few works have reported on the combination of Ni-based catalysts and molten alkali carbonate salts during biomass gasification [28,76]. Specifically, Sakhon Ratchahat and co-workers [28,76] investigated the catalytic effect of a combined Ni/Al_2O_3 and ternary carbonate salt (Li_2-K_2-Na_2-CO_3) catalytic system on the pyrolysis and CO_2 gasification of different biomass wastes.

In this context, the present work aims for the first time to systematically explore the combined impact of different Ni-based catalysts (Ni/CeO_2 and Ni/Al_2O_3) and an eutectic binary carbonate salt mixture (Li_2CO_3-K_2CO_3 molten salt, MS) on the CO_2 gasification performance of olive kernels under non-isothermal operating conditions in a lab-scale fixed-bed reactor. Various characterization techniques (H_2-TPR, SEM, HR/SEM_EDS, TEM, HR/TEM, and XRD) were applied to investigate the morphological and structural properties of these catalysts. It was disclosed that the type of commercial metal oxide used as support (Al_2O_3 or CeO_2) does not significantly affect the activity of Ni catalysts towards CO_2 gasification. However, the addition of the binary eutectic carbonate mixture (MS) resulted in a notably favorable performance during olive kernel-CO_2 gasification tests, leading to a high CO yield and carbon conversion efficiency. More interestingly, a synergistic MS-Ni medium effect in the low-temperature CO_2 gasification zone was revealed, boosting carbon monoxide production yields and carbon conversion efficiency.

2. Results and Discussion

2.1. Structural Characterization (XRD Analysis)

Figure 1 shows the XRD patterns of Ni/CeO_2, Ni/Al_2O_3 and their blends with the binary carbonate salt mixture. All catalysts exhibited the characteristic peaks of cubic NiO at 2θ angles of ca. 37, 43, 63, 76, and 79.5° (PDF #73-1519). No characteristic peaks of precursor $Ni(NO_3)_3$ or of other nitrates were detected, indicating that nickel nitrate was totally utilized and decomposed during the catalysts' preparation procedure. Furthermore, the d-spacing (d) values of as-prepared catalysts, calculated by Bragg's law, along with the corresponding lattice parameters (α), are summarized in Table 1. In addition, by applying the Scherrer equation [89], the primary crystallite sizes (D_{XRD}) of ceria and alumina were found to be 54.92 and 73.22 nm, respectively (Table 1). Moreover, the crystallite size (D_{XRD}) of the NiO particles on both the alumina and ceria supports was calculated at ca. 40 nm (37.22 and 42.83, respectively, in Table 1), in accordance with HR-SEM observations (Section 2.2). Both MS-Ni catalysts exhibited characteristic peaks at 2θ angles of 28–35, 40, and 47°, which are attributed to Li_2CO_3-K_2CO_3 carbonates (Figure 1).

Notably, the characteristic peaks attributed to NiO and carbonates phases present lower intensities in the case of CeO_2-containing samples, ascribed to the higher crystallinity of ceria compared to that of alumina. However, in the case of Ni/Al and MS-Ni/Al catalysts, the high-intensity characteristic peaks at 2θ = 37 and 43° can be ascribed to the co-existence of NiO, Al_2O_3, $NiAl_2O_4$ and carbonates phases. Interestingly, it should be noticed that small shoulder peaks standing for $NiAl_2O_4$ spinel oxide (#77-1877) can be found at 2θ = 37, 45, and 65°, indicating its formation, which has also been confirmed by H_2-TPR (see in Section 2.3). In particular, under high calcination temperatures (above 700 °C), nickel ions are migrating into the alumina lattice, forming the spinel structure ($NiAl_2O_4$), through the solid-solid interaction of NiO with Al_2O_3.

2.2. Morphological Characterization (SEM, HR-SEM, EDS Mapping, and TEM)

Figure 2 illustrates the SEM images of the fresh samples prior to gasification experiments. It can be observed from Figure 2a,b that both Ni/Al_2O_3 and Ni/CeO_2 catalysts display an irregular morphology consisting of nano-sized particles. Additionally, micro-scaled agglomerates are present on both catalysts, attributed to the high calcination temperature, which presumably leads to the sintering of NiO particles and/or supporting materials [44,45].

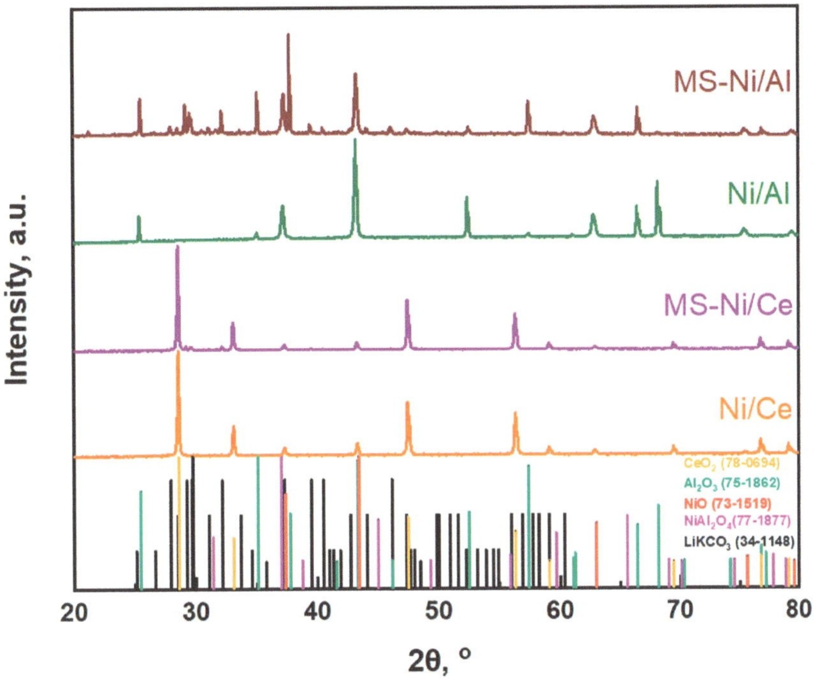

Figure 1. XRD patterns of Ni-based and MS-Ni-based catalysts.

Table 1. XRD and TEM results for the as-prepared Ni/CeO_2 and Ni/Al_2O_3 catalysts.

Sample	**XRD**						**TEM**
Ni/CeO_2	CeO_2 (111)			NiO (200)			NiO mean particle size
	D_{XRD} (nm) [1]	d (nm) [2]	a (nm)	D_{XRD} (nm) [1]	d (nm) [2]	a (nm)	38.1 ± 8.5
	54.92	0.31	0.54	42.83	0.21	0.42	
Ni/Al_2O_3	Al_2O_3 (113)			NiO (200)			NiO mean particle size
	D_{XRD} (nm) [1]	d (nm) [2]	a (nm)	D_{XRD} (nm) [1]	d (nm) [2]	a (nm)	37.7 ± 9.0
	73.22	0.21	n/a *	37.22	0.21	0.42	

* Aluminum forms a rhombohedral crystal lattice, not a cubic lattice such as CeO_2 and NiO. [1] Calculated applying Scherrer's equation. [2] Calculated by Bragg's law.

Figure 2c illustrates a uniform surface for the binary molten salt carbonate mixture (MS) with particle sizes less than 10 μm, confirming that the preparation protocol resulted in a relatively homogeneous blend. In the case of MS/catalyst mixtures (Figure 2d,e), the catalytic particles retained their irregular shape together with the presence of certain agglomerates due to the calcination pretreatment. At the same time, EDS analysis (not shown for the shake of brevity) demonstrates that Ni loading is practically equal to the nominal one (ca. 15 wt.%) in the case of Ni/CeO_2, whereas a slightly lower value was revealed for Ni/Al_2O_3 (13 wt.%).

Figure 2. SEM images of the as-prepared samples: (**a**) Ni/CeO_2; (**b**) Ni/Al_2O_3; (**c**) MS (**d**) MS-Ni/CeO_2; and (**e**) MS-Ni/Al_2O_3.

TEM analysis was applied to gain further insight into the morphological features of the samples (Figure 3). The bare support materials are presented in Figure 3a,b, while Ni/CeO_2 and Ni/Al_2O_3 are shown in Figure 3c,d. In the case of both Ni/CeO_2 and Ni/Al_2O_3, the mean particle size of NiO was ca. 40 nm, verifying the XRD findings, as shown in Table 1. It is worth noting that the wet-impregnation method and calcination at 850 °C resulted in similar NiO particle sizes. Given the inherent difficulties in distinguishing between commercial CeO_2 and NiO particles, the lattice fringes of NiO and CeO_2 were identified, which are reflective of the crystallinity of the samples. In particular, for the Ni/CeO_2 sample, the distinction was established by the calculation of the d-spacing values in an indicative HR-TEM image (Figure 3e). The lattice spacing for the phases of NiO and CeO_2 were 0.2 and 0.3 nm, respectively, which confirms the presence of NiO particles by exposing (200) planes that are in contact with CeO_2 (111), further confirming the results from XRD. In addition, it can be clearly seen that NiO cluster particles are semi-submerged upon the CeO_2 support and are characterized by a quasi-spherical shape.

2.3. Redox Behavior (H_2-TPR Analysis)

The reduction profiles (H_2-TPR) of Ni/CeO_2 and Ni/Al_2O_3 samples are compared in Figure 4. Four major H_2 consumption peaks were observed for both Ni-based catalysts. Notably, for Ni/CeO_2, the onset reduction temperature is around 180 °C, whereas for Ni/Al_2O_3, this temperature is significantly higher (ca. 235 °C), implying the beneficial effect of ceria in the low-temperature reducibility [58,90–92]. Specifically, a low-temperature peak at 295 °C (peak α) and a prominent medium-temperature peak at 376 °C (peak β) can be distinguished in the low-temperature region (below 500 °C) for the Ni/CeO_2 catalyst. These peaks can be associated with the reduction of NiO species strongly interacting with the CeO_2 support [46,50,91,93–96]. Notably, a shoulder at 242 °C can be in addition observed, which can be ascribed to the reduction of the loosely bound NiO species and/or to surface-absorbed O^{2-}/O^- species (oxygen vacancies associated with the formation of the Ni-O-Ce structure) [97,98]. The peak at 502 °C (peak γ) corresponds to the reduction of bulk NiO species. Lastly, the low intensity reduction peak above 750 °C (peak δ) can be attributed to the reduction of the CeO_2 bulk [46,50,91,93]. For the Ni/Al_2O_3 catalyst, one smaller and two major reduction peaks were observed at 310, 436, and 507 °C, respectively. The first peak (peak α) is assigned to the reduction of the loosely bound NiO species, whereas the two peaks at 436 and 507 °C (peaks β, γ) are attributed to the NiO species interacting with the Al_2O_3 support [91–93]. Finally, the last reduction feature (peak δ) at the high temperature zone (above 900 °C) corresponds to the reduction of $NiAl_2O_4$ spinel phases, suggesting a strong interaction between NiO and Al_2O_3 species, in good agreement with the XRD findings and relevant literature studies [50,91,93,97,98].

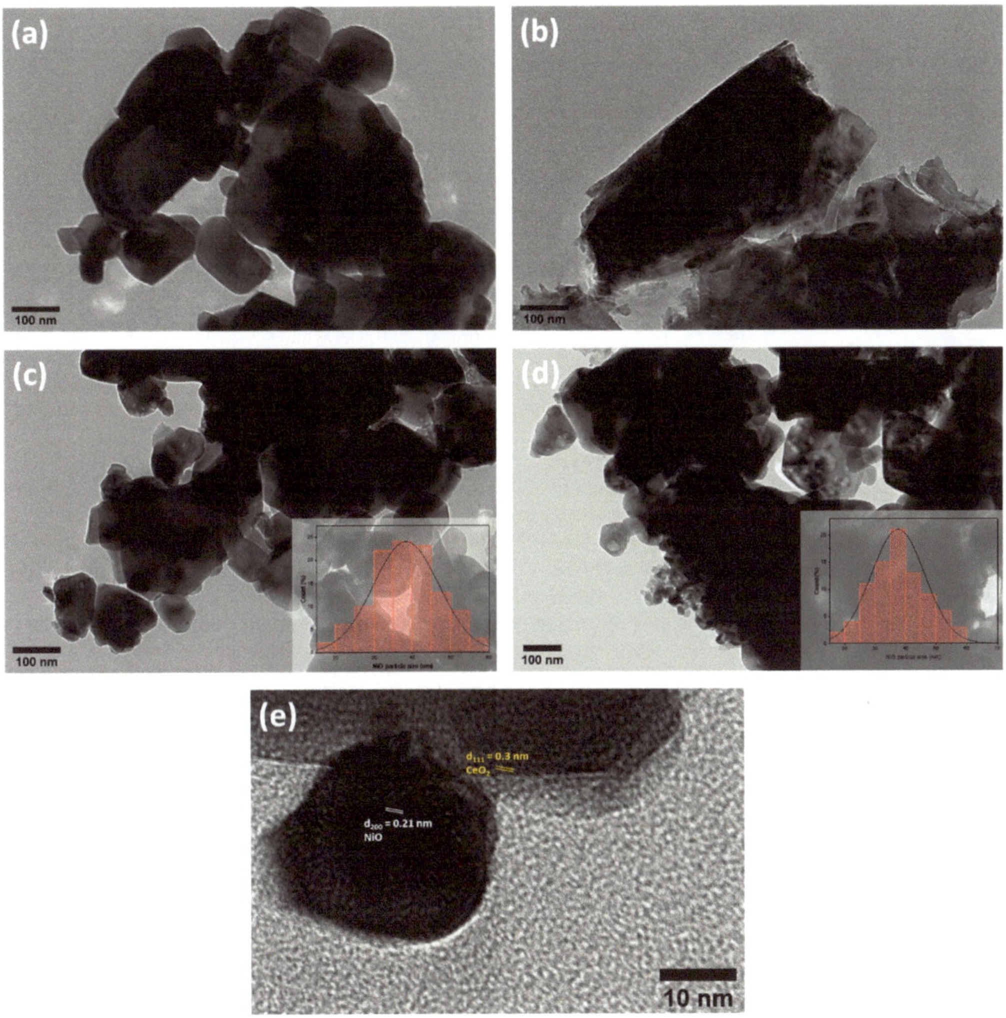

Figure 3. (**a**) TEM image of CeO_2; (**b**) TEM image of Al_2O_3; (**c**,**d**) TEM images and NiO particle size distribution (PSD histograms) of the as-prepared Ni/CeO_2 and Ni/Al_2O_3, respectively; (**e**) HRTEM image of Ni/CeO_2.

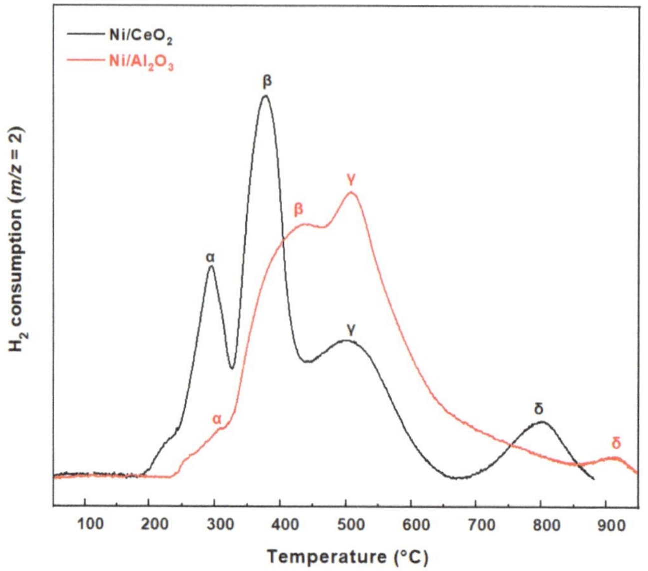

Figure 4. H_2-TPR profiles of the as-prepared Ni/CeO_2 and Ni/Al_2O_3 catalysts.

Quantitative analysis of the H_2-TPR profiles was also carried out by calculating the H_2 uptake (mmol H_2/g) of each peak (Table 2). As expected, regarding the reducibility of the samples in the low-temperature region (T < 500 °C), the calculated H_2 consumption values were almost identical for both Ni/CeO_2 and Ni/Al_2O_3 catalysts (2.5 mmol/g_{cat}). However, H_2 is consumed at relatively lower temperatures in the case of the Ni/CeO_2 catalyst, indicative of the superior reducibility of ceria-based oxides, facilitated by metal-support interactions [46,90,97,99]. Moreover, the total hydrogen consumption in the case of Ni/Al_2O_3 catalyst (2.65 mmol H_2/g_{cat}), which almost coincides with the theoretical amount of H_2 required for the complete reduction of NiO to metallic Ni° (~2.5 mmol H_2/g_{cat}), implies a complete reduction of NiO species. On the other hand, in the case of the Ni/CeO_2 catalyst, the relatively higher H_2 consumption (compared to the theoretical one) can be presumably attributed to the reduction of surface oxygen species of CeO_2, which is facilitated by nickel-ceria interactions. To summarize, the results show that both Ni catalysts present a redox ability at the lower (250–400 °C) and intermediate (400–600 °C) temperature regimes. This can be an indication that more labile oxygen species could be involved in the primary and secondary tar cracking and/or reforming reactions, effectively reducing the tar content in the as-produced syngas [100]. Moreover, the peak observed at 800 °C and attributed to the reduction of bulk CeO_2 species may lead to additional syngas production by the oxidation of the remaining char and/or formatted carbon on the catalyst's surface (carbon deposition) [99].

Table 2. Redox properties of the Ni/CeO_2 and Ni/Al_2O_3 samples.

	Total H_2 Uptake (mmol H_2/g)	α	β	γ	δ
Ni/CeO_2	2.85	0.69 (295 °C)	1.10 (376 °C)	0.70 (502 °C)	0.35 (802 °C)
Ni/Al_2O_3	2.65	0.04 (310 °C)	1.33 (436 °C)	1.21 (507 °C)	0.06 (916 °C)

2.4. Catalyst-Aided CO_2 Gasification Experiments under Batch Mode of Operation

Figures 5 and 6 depict the instant CO production rate, R_{CO} (min^{-1}), and carbon to CO conversion efficiency, Xco, as a function of temperature during the batch non-isothermal

CO_2 catalytic gasification tests over the different catalytic systems examined (Ni/CeO_2, Ni/Al_2O_3, MS, and MS-Ni systems). For comparison purposes, the catalyst-free gasification experiments are also included. The CO_2 gasification process includes two consecutive steps (Table 3): the volatile decomposition and tar cracking/reforming reactions (R1-R6,) in the temperature range of 300–600 °C (step 1), and the reaction of the remaining char with CO_2 (R9) along with other solid and/or gas-phase reactions (R7, R8), denoted as step 2 [36].

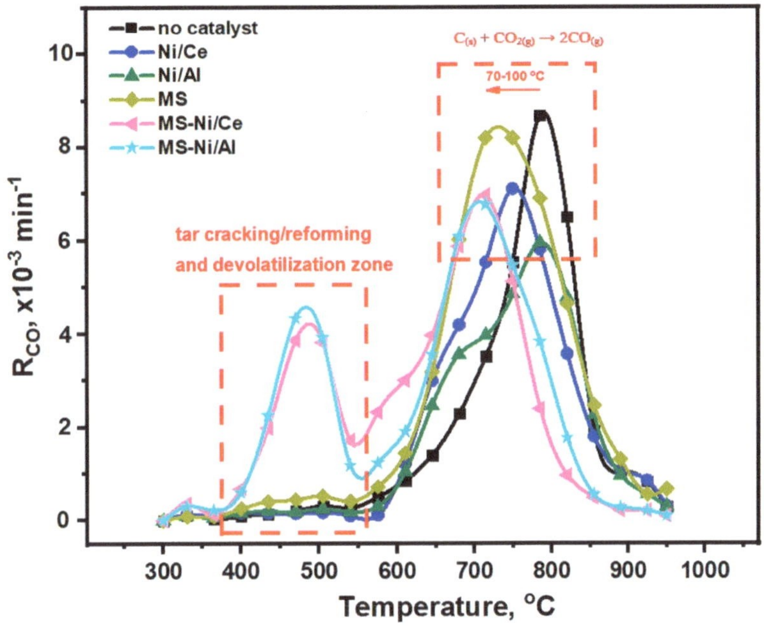

Figure 5. Effect of temperature and catalyst addition on the instant CO production rate (R_{CO}) during non-isothermal CO_2 gasification experiments. Gasifying agent flowrate = 30 cm^3/min of 20 vol.% CO_2 in Helium.

In this regard, Xco (Equation (1) in Section 3.4) involves a first term standing for the CO production during the tar cracking and devolatilization stage and a second term corresponding to the CO formation associated with the Boudouard reaction (R9) during the gasification stage [81,101]. In all cases, carbon monoxide was the principal product of gasification, appearing at different temperature regimes depending on the presence and the of catalyst. CO production, which is favored with temperature due to the endothermic reaction R9 [77], reaches its maximum and then gradually decreases until fuel depletion. During the olive kernel non-catalytic CO_2 gasification, a maximum R_{CO} value (8.5×10^{-3} min^{-1}) was observed at a temperature of 785 °C. Both Ni/CeO_2 and Ni/Al_2O_3 catalysts did not reveal any significant effect on the CO production rate. However, a slight reduction in the onset and peak gasification temperature was noticed in the case of the Ni/CeO_2 catalyst. Indeed, the maximum Rco value for Ni/CeO_2 sample (ca. 7×10^{-3} min^{-1}) was attained at 750 °C, which is 35 °C lower compared to the non-catalytic gasification. On the other hand, Ni/Al_2O_3 exhibited a maximum Rco value equal to 6×10^{-3} min^{-1} at 780 °C, a slightly lower temperature compared to the un-catalyzed reaction (795 °C). The lower Rco values obtained during the Ni-based catalytic CO_2 gasification reaction as compared to the non-catalytic process (8.5×10^{-3} min^{-1}) can possibly be attributed to the faster fuel depletion in conjunction with the favorable conditions for the endothermic Boudouard reaction (R9), since the gasification process takes place at lower temperatures. However, in the temperature regime of 600–750 °C, the Rco values for the Ni-based catalysts were higher compared to those obtained during the uncatalyzed gasification reaction. Apart from the

CO generated via the endothermic R9 reaction, Sakhon Ratchahat and co-investigators [28] proposed a different pathway for CO production. In brief, solid carbon atoms react with the adsorbed oxygen on a support surface, resulting from the CO_2 dissociation on Ni active sites, leading to additional CO production (carbon cleavage reaction on Ni active sites). Notably, for the Ni/Al_2O_3 catalytic gasification, at temperatures above 650–700 °C, the Rco values present a pronounced descending trend compared to the Ni/CeO_2 catalyst. This can presumably be attributed to two different phenomena that take place during the high temperature char-gasification stage: (a) the inhibiting role of $NiAl_2O_4$ spinel species observed in XRD and H_2-TPR tests on catalytic gasification performance [102–104], and (b) the coke/carbon formation through the thermal and catalytic decomposition of carbonaceous intermediates and/or undesired gas phase side reactions. Indeed, since $NiAl_2O_4$ cannot be easily reduced to Ni^0, its formation probably lowers the amount of active Ni sites, thus leading to a reduced catalytic activity [102]. On the other hand, a superior gasification rate is observed for Ni/CeO_2 sample up to 800 °C, which can be related to the high catalytic activity of Ni/CeO_2 catalyst and its tolerance toward carbon formation. CeO_2-based catalysts present a significant redox ability by releasing and up-taking oxygen through the reversible reaction, $CeO_2 \leftrightarrow CeO_{2-x} + O_x$, where O_x stands for the lattice oxygen at CeO_2 [33,34]. This lattice oxygen possibly enhances the catalytic gasification performance via two different pathways. The first one is by the reaction of CO with CeO_2 lattice oxygen, resulting in CO_2 formation ($CO + O_x \rightarrow CO_2 + O_{x-1}$). The second one is related to the oxidation of the remaining solid char/carbon which could be facilitated by CeO_2. Indeed, as a supplier of lattice oxygen, CeO_2 may oxidize solid carbon ($C_{(s)} + O_x \rightarrow CO + O_{x-1}$) towards additional CO formation. These induced effects are in line with the enhanced Ni/CeO_2 reducibility results (Figure 4, Table 2); the peak observed at 800 °C and attributed to the reduction of bulk CeO_2 species confirms the presence of lattice oxygen at such high temperatures, which can further oxidize the remaining carbon species [99].

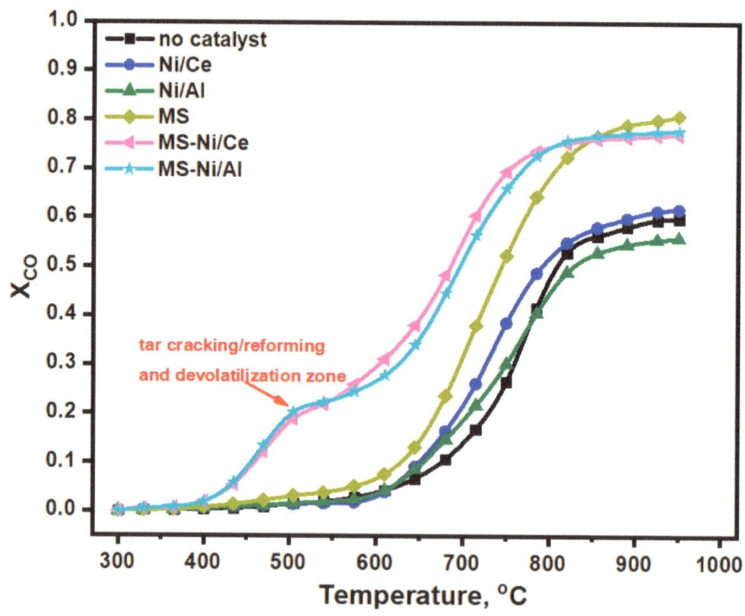

Figure 6. Effect of temperature and catalyst addition on the conversion of carbon to CO during non-isothermal CO_2 gasification experiments. Gasifying agent flowrate = 30 cm^3/min of 20 vol.% CO_2 in Helium.

Table 3. The main reaction network during biomass CO_2 gasification.

	Reaction	No
Tar thermal cracking	Tar (C_mH_n) → Smaller Tar (C_qH_q) + $H_{2(g)}$	R1
Tar dry reforming	Tar (C_mH_n) + $mCO_{2(g)}$ → $2mCO_{2(g)}$ + $(n/2)H_{2(g)}$	R2
Tar dry reforming	Tar (C_mH_n) + $(m + n/2)CO_{2(g)}$ → $(2m + n/2)CO_{(g)}$ + $(n/2)H_{2(g)}$	R3
Tar steam cracking	Tar (C_mH_n) + $mH_2O_{(g)}$ → $mCO_{(g)}$ + $(m + n/2)H_{2(g)}$	R4
Tar steam cracking	Tar (C_mH_n) + $2mH_2O_{(g)}$ → $mCO_{2(g)}$ + $(2m + n/2)H_{2(g)}$	R5
Tar partial oxidation	Tar (C_mH_n) + $(n/2)O_{2(g)}$ → $mCO_{(g)}$ + $(n/2)H_{2(g)}$	R6
Carbon-steam reforming	$C_{(s)} + H_2O_{(g)} \leftrightarrow CO_{(g)} + H_{2(g)}$	R7
Water gas shift	$CO_{(g)} + H_2O_{(g)} \leftrightarrow CO_{2(g)} + H_{2(g)}$	R8
Boudouard reaction	$C_{(s)} + CO_{2(g)} \leftrightarrow 2CO_{(g)}$	R9

In the case of the molten carbonate salt mixture (MS), the instant gasification rate approached its maximum value (8.3×10^{-3} min^{-1}) at ca. 55 °C lower compared to non-catalytic CO_2 gasification experiments. Moreover, the Rco values obtained during MS-assisted gasification were significantly higher, compared to non-catalytic and Ni-assisted catalytic gasification, in the temperature range of 600–800 °C. This remarkable result is in agreement with relevant studies on the MS-assisted CO_2 gasification of various biomass fuels [28,71,81,86,105]. Obviously, MS favors CO production at lower gasification temperatures due to its ability to improve mass and heat transfer phenomena in conjunction with the catalytic role of contained alkali metals in the three-cyclic-step Boudouard reaction (Table 4) [65–67].

Table 4. Mechanism of the alkali carbonate-assisted carbon CO_2 gasification.

Reaction	No
$M_2CO_3 + 2C \rightarrow 2M^* + 3CO$	R11
$2M^* + CO_2 \rightarrow M_2O + CO$	R12
$M_2O + CO_2 \rightarrow M_2CO_3$	R13
$C_{(s)} + CO_{2(g)} \rightarrow 2CO(g)$	R14

* M stands for the alkali metal.

Notably, the CO production rate maxima for the combined MS-Ni/X (X: CeO_2, Al_2O_3) medium was shifted to 15, 40, and 70 °C lower compared to MS and Ni catalysts and non-catalytic gasification tests, respectively. Most importantly, the mixed MS-Ni/CeO_2 and MS-Ni/Al_2O_3 composites revealed an excellent synergy regarding their beneficial role in the devolatilization and tar cracking/reforming zones. Indeed, in the temperature range of 300 to 550 °C, the maximum Rco value for both MS-Ni systems was four times higher (4×10^{-3} min^{-1}, 460 °C) compared to their counterparts. This can be attributed to the combined catalytic effect of both Ni catalysts and MS on tar/volatiles' thermal and/or dry cracking/reforming reactions, as well as on the oxidation of the dissociated carbon on nickel active sites [44,45]. Interestingly, the additional CO formation at this temperature regime coincides with the melting point of the MS (Li_2CO_3-K_2CO_3) which is ca. 490 °C [65]. As the salt mixture melts, it forms a thin layer that favors the carbon-MS contact [68,70]. This in turn facilitates the tar oxidation reaction (R6) at the carbon-catalyst interface through the labile oxygen species in the low temperature region (see H_2-TPR analysis above).

To further shed light on the CO_2 gasification performance, the carbon to CO conversion efficiency, Xco (Figure 6), was calculated via Equation 1. Notably, there are three distinct temperature regimes. At temperatures up to 600 °C, where tar and volatiles' cracking/reforming reactions prevail, the combined MS-Ni systems lead to a carbon to CO conversion of ca. 0.3, in complete contrast to the bare constituents (Ni catalysts or MS), which are almost inactive. This further verifies the synergy between the Ni-based catalysts and MS towards tar and volatile dry cracking/reforming reactions. At the temperature range between 600 and 800 °C, the CO production is mainly ascribed to the endothermic Boudouard reaction (R9). In this temperature regime, the Xco values for the MS-Ni systems are higher compared to their individual counterparts, highlighting the synergetic combination of the MS-induced mass/heat transfer phenomena and catalytic activity towards the Boudouard reaction (R9), along with the Ni ability to favor the C-C bond cleavage reaction into an additional CO formation. At 800 °C, the X_{CO} follows the order: MS-Ni/Ce ≈ MS-Ni/Al (0.75) > MS (0.68) > Ni/Ce (0.51) > Ni/Al ≈ no catalyst (ca. 0.45). At temperatures higher than 800 °C, the carbon to CO conversion for the combined MS-Ni samples reached a plateau (Xco = 0.78) due to fuel depletion above this temperature. The Xco for the MS-assisted gasification tests reached the highest value (0.81 at 950 °C), highlighting the pronounced effect of MS at higher temperatures.

To further explore the synergetic role between NiO and CeO_2 in combination with molten carbonate salt (MS) in the low/intermediate tar cracking/reforming temperature regime, the effect of MS-NiO and MS-CeO_2 systems on the CO_2 gasification reaction rate and carbon to CO conversion was examined (Figure 7). It is evident that the presence of all the catalyst's components is necessary to achieve high gasification performance at temperatures below 700 °C. This coincides with the well-established hydrocarbon (tar) dissociation and dehydrogenation on Ni active sites, the enhanced CeO_2 redox properties in the low/intermediate temperature regime, and the key role of MS on the decomposition of volatiles during steam and/or dry reforming reactions [44,45].

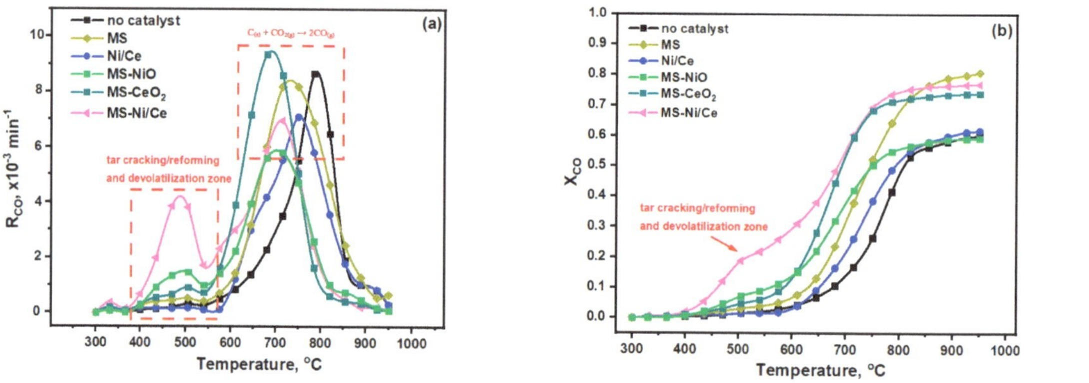

Figure 7. Effect of temperature and catalyst components on the instant CO production rate, R_{CO} (**a**), and carbon to CO conversion (**b**) during non-isothermal CO_2 gasification experiments. Gasifying agent flowrate = 30 cm^3/min of 20 vol.% CO_2 in Helium.

3. Materials and Methods

3.1. Fuel Feedstock

Olive kernel (OK) from Cretan olive tree cultures was selected as raw fuel. OK was initially crushed to a particle size between 1–3 mm using a jaw crusher. Then, part of this fraction was milled to 100–200 μm using an agate mortar (Retsch RM200) producing a large amount (ca. 70 wt.%) of fines (<100 μm). The elemental and proximate analysis of olive kernel fuel has been presented in our previous work [31], following a typical pattern for lignocellulosic residual biomass with negligible nitrogen, sulfur, and chlorine contents. Moreover, the proximate analysis revealed the high volatile matter content (75.8 wt.%) and low ash concentration (2.9 wt.%) of olive kernel biomass.

3.2. Catalyst Preparation

Ni-based catalysts were prepared using the wet impregnation method. Commercial aluminum oxide (α-Al_2O_3, ~100 mesh, 99.9% purity, Aldrich) and cerium oxide (CeO_2, 99.9% purity) were used as catalyst supports. In brief, Ni/Al_2O_3 and Ni/CeO_2 catalysts were prepared by the aqueous impregnation of nickel nitrate precursor salt ($Ni(NO_3)_2 \cdot 6H_2O$) in commercial α-Al_2O_3 and CeO_2 by vigorously agitating at ca. 80 °C for about 3 h until a green slurry was obtained. The slurry was dried at 90 °C for 12 h and then calcined in an air flow at 850 °C for 4 h. The nominal loading of Ni over Al_2O_3 and CeO_2 was set at 15 wt.% based on relevant literature studies and reviews [33,39]. For sake of brevity, the Ni-based catalysts will be noted as Ni/Ce and Ni/Al, with the procedure followed to synthesize the former sample being illustrated in Figure 8.

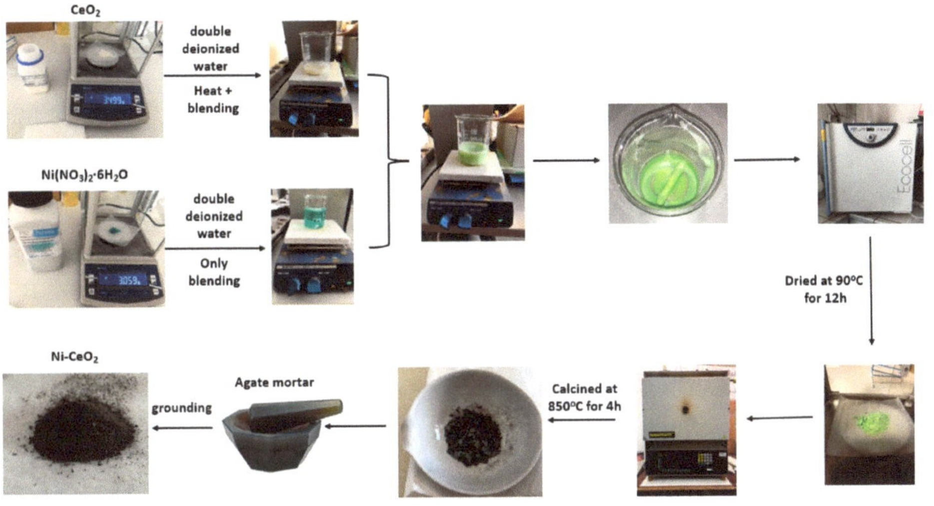

Figure 8. Illustrated procedure for the synthesis of the 15 wt.% Ni/CeO_2 sample.

The molten salts used for preparing the eutectic carbonates mixture were Li_2CO_3 and K_2CO_3 (Sigma Aldrich, Munich, Germany) with a molar ratio of 62/38 mol.%, respectively. The salts were firstly rolled and milled at ca. 200 rpm for approximately 12 h. The melting point of the binary carbonates eutectic mixture (MS) is approximately 498 °C, thus it could be utilized as a mass and heat transfer medium for the gasification of solid fuels [65]. Then the Li_2-K_2/CO_3 mixture (MS) was placed on a hot plate at 60 °C so as to evaporate the liquid carrier (ethanol) used during the roll milling process [65]. Finally, the molten salts were cooled to room temperature, crushed, and finely ground in an agate mortar.

The combined Ni/M_xO_y (M: Al, Ce) and MS blends were prepared by homogeneously mixing the individual counterparts in an agate mortar.

3.3. Experimental Apparatus for the Olive Kernel Catalytic Gasification Studies

Figure 9 schematically illustrates the experimental setup for the olive kernel catalytic gasification tests, which were performed under a batch mode of operation. The experimental apparatus and conditions are described in detail in our previous work [31]. In brief, a mixture of fuel feedstock (0.1 g) and MS or MS-Ni catalyst was loaded into an Inox, U-tube, fixed bed reactor. The employed catalyst-to-biomass weight ratio was fixed at 0.5:1 for the MS-OK experiments, 2.5:1 for the Ni-aided gasification tests, and 0.5:2:1 for the ternary MS-Ni catalyst/OK mixtures. The mass ratio of the employed catalysts to OK was fixed at 2.5, while the MS to OK weight ratio was set at 0.5. The feedstock mixtures were prepared by physically mixing the olive kernel with the examined catalyst composites in an agate mortar. The flowrate of 20 vol.% CO_2/He, used as gasifying agent mixture, was set at 30 cm^3/min. After passing through a cooling trap (ice bath) to collect tars, the gaseous products were driven to a gas chromatograph (Shimadzu GC-14B) to monitor the effluent's gas composition in the temperature range of 300–950 °C, by increasing the temperature at a constant rate of 2 °C/min. In order to obtain reliable and accurate results, all gasification tests were repeated at least twice.

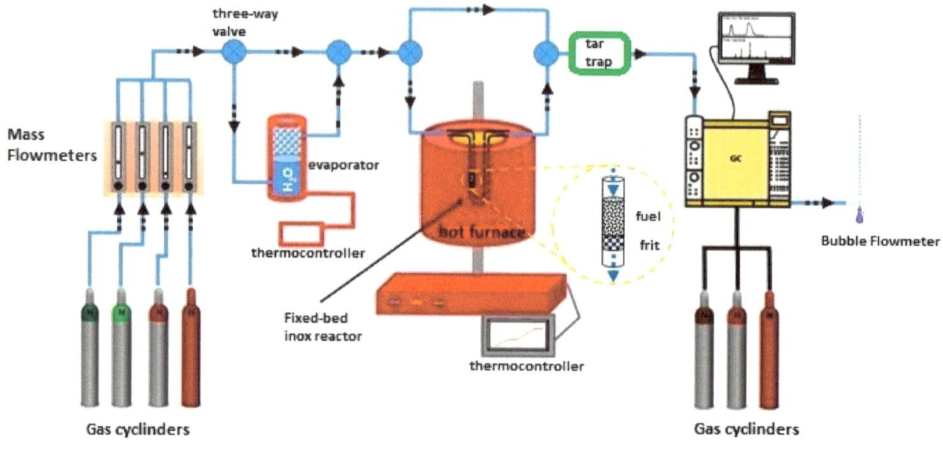

Figure 9. Experimental apparatus of the lab-scale CO_2-gasification experiments.

Here, it should be noted that non-isothermal gasification studies provide a more reliable overview on the gasification performance in the whole operational range since the gasification temperature often varies along the gasifier in real applications. Moreover, lower heating rates result in more accurate studies since higher ones may lead to the formation of larger temperature gradients within the sample, leading to higher temperature differences inside the fuel bed [106].

3.4. Gasification Performance Indicators

In the case of the non-isothermal catalytic CO_2-gasification experiments, since CO was practically the main product during gasification, the catalysts' activity was evaluated through the calculation of two performance parameters: X_{CO} expressing the carbon to CO conversion; and R_{CO} (min^{-1}) standing for the instant CO production rate, as described in our previous work [107]:

$$X_{CO} = \frac{\sum_{t=0}^{t_{step\ 1}} y_{CO,t} \times 12}{m \times w_c} + \frac{\sum_{t_{step\ 1}}^{t_f} y_{CO,t} \times 12}{2 \times m \times w_c} \quad (1)$$

$$R_{CO} = \frac{dX_{CO}}{dt} \quad (2)$$

where $y_{CO,t}$ corresponds to the cumulative CO production moles in time t, m is the initial mass of fuel sample (gr), and w_c is the elemental carbon content of the olive kernel biomass fuel. It should be noted that the first term in Equation (1) stands for the CO production during the devolatilization stage, whereas the second term corresponds to the CO formation associated with the Boudouard reaction ($C + CO_2 \rightarrow 2\ CO$) during the gasification stage. The same expression has also been employed in relevant studies in the literature to estimate the CO_2 gasification conversion of various coal fuels [81,101].

3.5. Physicochemical Characterizations

X-ray diffraction patterns of powdered samples were obtained by a Bruker AXS D8 Advance copper anode diffractometer (CuKα radiation) equipped with a nickel foil monochromator, operating at 40 kV and 40 mA over the 2θ collection range of 10–80° at a scan rate of $0.05° s^{-1}$.

The d-spacing was calculated by applying the Bragg's law:

$$2d \sin\theta_B = n\ \lambda \quad (3)$$

where λ is the X-ray wavelength, d is the spacing of the diffraction planes, and θ_B is the angle between the incident rays and the diffracting planes, otherwise known as the Bragg angle. For CeO_2 and NiO, which are both crystallized in cubic form (a = b = c, $\alpha = \beta = \gamma = 90°$), the lattice parameters can be calculated by the following equation:

$$1/d^2 = (h^2 + k^2 + l^2)/a^2 \quad (4)$$

The morphological analysis of the samples was performed by scanning electron microscopy (SEM, JEOL JSM-6390LV, JEOL Ltd., Akishima, Tokyo, Japan) operating at 20 keV. Elemental analysis/mapping and morphological observation were obtained by high-resolution scanning electron microscopy (HR-SEM) using a JS-IT700HR (JEOL, Tokyo, Japan) instrument at 20 kV. Transmission electron microscopy (TEM) was performed using a JEM-2100 instrument (JEOL, Tokyo, Japan) equipped with a LaB6 filament, working at 200 kV. Specimens for TEM were prepared by the deposition of dispersed powder samples after ultrasonication.

The reducibility of the Ni/Ce and Ni/Al catalysts was examined by temperature-programmed reduction with H_2 (TPR-H_2). In a typical experiment, 0.1 g of the catalyst sample was loaded in a fixed-bed quartz reactor and pretreated at 500 °C for one hour under helium flow (50 mL/min). The catalyst was then cooled down to 35–40 °C, and TPR-H_2 analysis was carried out from 40 to 920 °C at a heating rate of 10 °C/min in a 5 *v*/*v*% H_2/He flow (50 mL/min). The composition of the exit gas was monitored online with a quadrupole mass analyzer (Omnistar, Balzer). The *m*/*z* fragments registered were as follows: H_2 = 2, H_2O = 18, and He = 4. The quantitative analysis of the consumed/desorbed H_2 was based on *m*/*z* = 2.

4. Conclusions

The main objective of this work was to explore the impact of catalyst addition (Ni/CeO_2, Ni/Al_2O_3) and/or carbonates (eutectic binary mixture of carbonate salts, MS) on the CO_2 gasification of olive kernel biomass fuel. It was found that Ni/CeO_2 and Ni/Al_2O_3 catalysts demonstrated a beneficial role on CO_2 gasification performance by slightly reducing the onset and peak CO_2 gasification temperatures. Interestingly, the Ni/CeO_2 catalyst presented an enhanced instant gasification reaction rate for temperatures up to 800 °C, in line with the significant redox capability of CeO_2. In the presence of MS, the instant gasification rate approached its maximum value (8.3×10^{-3} min^{-1}) at ca. 55 °C lower compared to the non-catalytic gasification tests. The pronounced effect of MS at higher temperatures was also reflected in the high carbon to CO conversion value (Xco = 0.81 at 950 °C). More importantly, the combined use of MS with Ni catalysts notably enhances the CO_2 gasification performance, specifically on the devolatilization and tar cracking/reforming zones. Indeed, in the temperature range of 300 to 550 °C, the maximum Rco and Xco values for both MS-Ni systems were notably higher (4×10^{-3} min^{-1} and 0.25 at 460 °C) compared to the insignificant corresponding performance (Rco < 0.5×10^{-3} min^{-1}) obtained by the individual catalyst counterparts. Collectively, the synergistic combination of molten salt and nickel-based catalysts can provide an innovative approach to enhance biomass CO_2 gasification and, in parallel, improve syngas quality for downstream processes, substantially reducing syngas cleaning requirements and system costs.

Author Contributions: A.L., V.B., L.Z., S.S. (Sofia Stefa), S.A.K. and G.V. contributed to samples preparation and characterization; A.L. and S.S. (Stavros Spyridakos) contributed to gasification studies; A.L., G.E.M. and M.K. contributed to the conception, design, and results interpretation; G.E.M. and M.K. validated the results and administered the project. A.L. wrote the original draft. Finally, A.L., G.E.M. and M.K. reviewed, edited, and submitted the manuscript in the final form. All authors have read and agreed to the published version of the manuscript.

Funding: This research has been co-financed by the project "Development of New Innovative Low Carbon Footprint Energy Technologies to Enhance Excellence in the Region of Western Macedonia"(MIS 5047197), which is implemented under the action "Reinforcement of the Research and Innovation Infrastructure ", funded by the Operational Programme "Competitiveness, Entrepreneurship and Innovation" (NSRF 2014–2020) and co-financed by Greece and the European Union (European Regional Development Fund).

Conflicts of Interest: The authors declare no conflict of interest.

References

1. Brown, R.C. The Role of Pyrolysis and Gasification in a Carbon Negative Economy. *Processes* **2021**, *9*, 882. [CrossRef]
2. Varvoutis, G.; Lampropoulos, A.; Mandela, E.; Konsolakis, M. Recent Advances on CO_2 Mitigation Technologies: On the Role of Hydrogenation Route via Green H_2. *Energies* **2022**, *15*, 4790. [CrossRef]
3. Zhang, X. Essential Scientific Mapping of the Value Chain of Thermochemically Converted Second-Generation Bio-Fuels. *Green Chem.* **2016**, *18*, 5086–5117. [CrossRef]
4. Mishra, S.; Upadhyay, R.K. Review on Biomass Gasification: Gasifiers, Gasifying Mediums, and Operational Parameters. *Mater. Sci. Energy Technol.* **2021**, *4*, 329–340. [CrossRef]
5. Di Giuliano, A.; Funcia, I.; Pérez-Vega, R.; Gil, J.; Gallucci, K. Novel Application of Pretreatment and Diagnostic Method Using Dynamic Pressure Fluctuations to Resolve and Detect Issues Related to Biogenic Residue Ash in Chemical Looping Gasification. *Processes* **2020**, *8*, 1137. [CrossRef]
6. Narnaware, S.L.; Panwar, N.L. Biomass Gasification for Climate Change Mitigation and Policy Framework in India: A Review. *Bioresour. Technol. Reports* **2022**, *17*, 100892. [CrossRef]
7. Allesina, G.; Pedrazzi, S. Barriers to Success: A Technical Review on the Limits and Possible Future Roles of Small Scale Gasifiers. *Energies* **2021**, *14*, 6711. [CrossRef]
8. Ongen, A.; Ozgu, C.; Ayol, A. ScienceDirect Biomass Gasification for Sustainable Energy Production: A Review. *Int. J. Hydrog. Energy* **2022**, *47*, 15419–15433. [CrossRef]
9. Pio, D.T.; Tarelho, L.A.C. Industrial Gasification Systems (>3 MWth) for Bioenergy in Europe: Current Status and Future Perspectives. *Renew. Sustain. Energy Rev.* **2021**, *145*, 111108. [CrossRef]
10. Situmorang, Y.A.; Zhao, Z.; Yoshida, A.; Abudula, A.; Guan, G. Small-Scale Biomass Gasification Systems for Power Generation (<200 kW Class): A Review. *Renew. Sustain. Energy Rev.* **2020**, *117*, 109486. [CrossRef]
11. Aravani, V.P.; Sun, H.; Yang, Z.; Liu, G.; Wang, W.; Anagnostopoulos, G.; Syriopoulos, G.; Charisiou, N.D.; Goula, M.A.; Kornaros, M.; et al. Agricultural and Livestock Sector's Residues in Greece & China: Comparative Qualitative and Quantitative Characterization for Assessing Their Potential for Biogas Production. *Renew. Sustain. Energy Rev.* **2022**, *154*, 111821. [CrossRef]
12. Marks, E.A.N.; Kinigopoulou, V.; Akrout, H.; Azzaz, A.A.; Doulgeris, C.; Jellali, S.; Rad, C.; Zulueta, P.S.; Tziritis, E.; El-Bassi, L.; et al. Potential for Production of Biochar-Based Fertilizers from Olive Millwaste in Mediterranean Basin Countries: An Initial Assessment for Spain, Tunisia, and Greece. *Sustainability* **2020**, *12*, 6081. [CrossRef]
13. Diez, C.M.; Trujillo, I.; Martinez-Urdiroz, N.; Barranco, D.; Rallo, L.; Marfil, P.; Gaut, B.S. Olive Domestication and Diversification in the Mediterranean Basin. *New Phytol.* **2015**, *206*, 436–447. [CrossRef]
14. del Mar Contreras, M.; Romero, I.; Moya, M.; Castro, E. Olive-Derived Biomass as a Renewable Source of Value-Added Products. *Process Biochem.* **2020**, *97*, 43–56. [CrossRef]
15. Garcìa Martìn, J.F.; Cuevas, M.; Feng, C.; Mateos, P.Á.; Torres, M. Energetic Valorisation of Olive Biomass: Olive-Tree Pruning, Olive Stones and Pomaces. *Processes* **2020**, *8*, 511. [CrossRef]
16. Christoforou, E.; Fokaides, P.A. A Review of Olive Mill Solid Wastes to Energy Utilization Techniques. *Waste Manag.* **2016**, *49*, 346–363. [CrossRef]
17. Damartzis, T.; Michailos, S.; Zabaniotou, A. Energetic Assessment of a Combined Heat and Power Integrated Biomass Gasification-Internal Combustion Engine System by Using Aspen Plus®. *Fuel Process. Technol.* **2012**, *95*, 37–44. [CrossRef]
18. Vera, D.; Jurado, F.; Margaritis, N.K.; Grammelis, P. Experimental and Economic Study of a Gasification Plant Fuelled with Olive Industry Wastes. *Energy Sustain. Dev.* **2014**, *23*, 247–257. [CrossRef]
19. Vera, D.; De Mena, B.; Jurado, F.; Schories, G. Study of a Downdraft Gasifier and Gas Engine Fueled with Olive Oil Industry Wastes. *Appl. Therm. Eng.* **2013**, *51*, 119–129. [CrossRef]
20. Fryda, L.; Panopoulos, K.D.; Kakaras, E. Integrated CHP with Autothermal Biomass Gasification and SOFC-MGT. *Energy Convers. Manag.* **2008**, *49*, 281–290. [CrossRef]
21. Manara, P.; Zabaniotou, A. Indicator-Based Economic, Environmental, and Social Sustainability Assessment of a Small Gasification Bioenergy System Fuelled with Food Processing Residues from the Mediterranean Agro-Industrial Sector. *Sustain. Energy Technol. Assessments* **2014**, *8*, 159–171. [CrossRef]
22. Vera, D.; Jurado, F.; De Mena, B.; Hernández, J.C. A Distributed Generation Hybrid System for Electric Energy Boosting Fueled with Olive Industry Wastes. *Energies* **2019**, *12*, 500. [CrossRef]

23. Kougioumtzis, M.A.; Kanaveli, I.P.; Karampinis, E.; Grammelis, P.; Kakaras, E. Combustion of Olive Tree Pruning Pellets versus Sunflower Husk Pellets at Industrial Boiler. Monitoring of Emissions and Combustion Efficiency. *Renew. Energy* **2021**, *171*, 516–525. [CrossRef]
24. Colantoni, A.; Villarini, M.; Marcantonio, V.; Gallucci, F.; Cecchini, M. Performance Analysis of a Small-Scale ORC Trigeneration System Powered by the Combustion of Olive Pomace. *Energies* **2019**, *12*, 2279. [CrossRef]
25. Alatzas, S.; Moustakas, K.; Malamis, D.; Vakalis, S. Biomass Potential from Agricultural Waste for Energetic Utilization in Greece. *Energies* **2019**, *12*, 1095. [CrossRef]
26. Vourdoubas, J. Use of Renewable Energy Sources for Energy Generation in Rural Areas in the Island of Crete, Greece. *Eur. J. Environ. Earth Sci.* **2020**, *1*, 1–7. [CrossRef]
27. Vourdoubas, J. Description and Assessment of a Small Renewable Energy Community in the Island of Crete, Greece. *Open J. Energy Effic.* **2017**, *06*, 97–111. [CrossRef]
28. Ratchahat, S.; Kodama, S.; Tanthapanichakoon, W.; Sekiguchi, H. CO_2 Gasification of Biomass Wastes Enhanced by Ni/Al_2O_3 Catalyst in Molten Eutectic Carbonate Salt. *Int. J. Hydrogen Energy* **2015**, *40*, 11809–11822. [CrossRef]
29. Teh, J.S.; Teoh, Y.H.; How, H.G.; Sher, F. Thermal Analysis Technologies for Biomass Feedstocks: A State-of-the-Art Review. *Processes* **2021**, *9*, 1610. [CrossRef]
30. Suárez-Ruiz, I.; Diez, M.A.; Rubiera, F. *New Trends in Coal Conversion: Combustion, Gasification, Emissions, and Coking*; Elsevier: Amsterdam, The Netherlands, 2018; ISBN 9780081022016.
31. Lampropoulos, A.; Kaklidis, N.; Athanasiou, C.; Montes-Morán, M.A.; Arenillas, A.; Menéndez, J.A.; Binas, V.D.; Konsolakis, M.; Marnellos, G.E. Effect of Olive Kernel Thermal Treatment (Torrefaction vs. Slow Pyrolysis) on the Physicochemical Characteristics and the CO_2 or H_2O Gasification Performance of as-Prepared Biochars. *Int. J. Hydrogen Energy* **2020**, *46*, 29126–29141. [CrossRef]
32. Lampropoulos, A.; Binas, V.; Konsolakis, M.; Marnellos, G.E. Steam Gasification of Greek Lignite and Its Chars by Co-Feeding CO_2 toward Syngas Production with an Adjustable H_2/CO Ratio. *Int. J. Hydrogen Energy* **2021**, *46*, 28486–28500. [CrossRef]
33. Galadima, A.; Masudi, A.; Muraza, O. Catalyst Development for Tar Reduction in Biomass Gasification: Recent Progress and the Way Forward. *J. Environ. Manage.* **2022**, *305*, 114274. [CrossRef] [PubMed]
34. Lee, D.; Nam, H.; Won Seo, M.; Hoon Lee, S.; Tokmurzin, D.; Wang, S.; Park, Y.K. Recent Progress in the Catalytic Thermochemical Conversion Process of Biomass for Biofuels. *Chem. Eng. J.* **2022**, *447*, 137501. [CrossRef]
35. Straczewski, G.; Mai, R.; Gerhards, U.; Garbev, K.; Leibold, H. Catalytic Tar Conversion in Two Different Hot Syngas Cleaning Systems. *Catalysts* **2021**, *11*, 1231. [CrossRef]
36. Guan, G.; Kaewpanha, M.; Hao, X.; Abudula, A. Catalytic Steam Reforming of Biomass Tar: Prospects and Challenges. *Renew. Sustain. Energy Rev.* **2016**, *58*, 450–461. [CrossRef]
37. Mun, T.Y.; Seon, P.G.; Kim, J.S. Production of a Producer Gas from Woody Waste via Air Gasification Using Activated Carbon and a Two-Stage Gasifier and Characterization of Tar. *Fuel* **2010**, *89*, 3226–3234. [CrossRef]
38. Guo, F.; Jia, X.; Liang, S.; Zhou, N.; Chen, P.; Ruan, R. Development of Biochar-Based Nanocatalysts for Tar Cracking/Reforming during Biomass Pyrolysis and Gasification. *Bioresour. Technol.* **2020**, *298*, 122263. [CrossRef]
39. Ren, J.; Cao, J.P.; Zhao, X.Y.; Yang, F.L.; Wei, X.Y. Recent Advances in Syngas Production from Biomass Catalytic Gasification: A Critical Review on Reactors, Catalysts, Catalytic Mechanisms and Mathematical Models. *Renew. Sustain. Energy Rev.* **2019**, *116*, 109426. [CrossRef]
40. Binte Mohamed, D.K.; Veksha, A.; Ha, Q.L.M.; Chan, W.P.; Lim, T.T.; Lisak, G. Advanced Ni Tar Reforming Catalysts Resistant to Syngas Impurities: Current Knowledge, Research Gaps and Future Prospects. *Fuel* **2022**, *318*, 123602. [CrossRef]
41. Ren, J.; Cao, J.P.; Zhao, X.Y.; Liu, Y.L. Recent Progress and Perspectives of Catalyst Design and Downstream Integration in Biomass Tar Reforming. *Chem. Eng. J.* **2022**, *429*, 132316. [CrossRef]
42. Wu, Y.; Wang, H.; Li, H.; Han, X.; Zhang, M.; Sun, Y.; Fan, X.; Tu, R.; Zeng, Y.; Xu, C.C.; et al. Applications of Catalysts in Thermochemical Conversion of Biomass (Pyrolysis, Hydrothermal Liquefaction and Gasification): A Critical Review. *Renew. Energy* **2022**, *196*, 462–481. [CrossRef]
43. Qin, T.; Yuan, S. Research Progress of Catalysts for Catalytic Steam Reforming of High Temperature Tar: A Review. *Fuel* **2023**, *331*, 125790. [CrossRef]
44. Gao, N.; Salisu, J.; Quan, C.; Williams, P. Modified Nickel-Based Catalysts for Improved Steam Reforming of Biomass Tar: A Critical Review. *Renew. Sustain. Energy Rev.* **2021**, *145*, 111023. [CrossRef]
45. Chan, F.L.; Tanksale, A. Review of Recent Developments in Ni-Based Catalysts for Biomass Gasification. *Renew. Sustain. Energy Rev.* **2014**, *38*, 428–438. [CrossRef]
46. Abedi, A.; Dalai, A.K. Steam Gasification of Oat Hull Pellets over Ni-Based Catalysts: Syngas Yield and Tar Reduction. *Fuel* **2019**, *254*, 115585. [CrossRef]
47. Hu, J.; Jia, Z.; Zhao, S.; Wang, W.; Zhang, Q.; Liu, R.; Huang, Z. Activated Char Supported Fe-Ni Catalyst for Syngas Production from Catalytic Gasification of Pine Wood. *Bioresour. Technol.* **2021**, *340*, 125600. [CrossRef]
48. Khalifa, O.; Xu, M.; Zhang, R.; Iqbal, T.; Li, M.; Lu, Q. Steam Reforming of Toluene as a Tar Model Compound with Modified Nickel-Based Catalyst. *Front. Energy* **2021**, *16*, 492–501. [CrossRef]
49. Lu, Y.; Jin, H.; Zhang, R. Evaluation of Stability and Catalytic Activity of Ni Catalysts for Hydrogen Production by Biomass Gasification in Supercritical Water. *Carbon Resour. Convers.* **2019**, *2*, 95–101. [CrossRef]

50. Granados-Fernández, R.; Cortés-Reyes, M.; Poggio-Fraccari, E.; Herrera, C.; Larrubia, M.; Alemany, L.J. Biomass Catalytic Gasification Performance over Unsupported Ni-Ce Catalyst for High-Yield Hydrogen Production. *Biofuels Bioprod. Biorefining* **2020**, *14*, 20–29. [CrossRef]
51. Baker, E.G.; Mudge, L.K.; Brown, M.D. Steam Gasification of Biomass with Nickel Secondary Catalysts. *Ind. Eng. Chem. Res.* **1987**, *26*, 1335–1339. [CrossRef]
52. Li, X.T.; Grace, J.R.; Lim, C.J.; Watkinson, A.P.; Chen, H.P.; Kim, J.R. Biomass Gasification in a Circulating Fluidized Bed. *Biomass Bioenergy* **2004**, *26*, 171–193. [CrossRef]
53. Ashok, J.; Kawi, S. Steam Reforming of Toluene as a Biomass Tar Model Compound over CeO_2 Promoted Ni/$CaOeAl_2O_3$ Catalytic Systems. *Int. J. Hydrogen Energy* **2013**, *38*, 13938–13949. [CrossRef]
54. Furusawa, T.; Saito, K.; Kori, Y.; Miura, Y.; Sato, M.; Suzuki, N. Steam Reforming of Naphthalene/Benzene with Various Types of Pt- and Ni-Based Catalysts for Hydrogen Production. *Fuel* **2013**, *103*, 111–121. [CrossRef]
55. Artetxe, M.; Nahil, M.A.; Olazar, M.; Williams, P.T. Steam Reforming of Phenol as Biomass Tar Model Compound over Ni/Al2O3 Catalyst. *Fuel* **2016**, *184*, 629–636. [CrossRef]
56. Artetxe, M.; Alvarez, J.; Nahil, M.A.; Olazar, M.; Williams, P.T. Steam Reforming of Different Biomass Tar Model Compounds over Ni/Al2O3 Catalysts. *Energy Convers. Manag.* **2017**, *136*, 119–126. [CrossRef]
57. Kihlman, J.; Simell, P. Carbon Formation in the Reforming of Simulated Biomass Gasification Gas on Nickel and Rhodium Catalysts. *Catalysts* **2022**, *12*, 410. [CrossRef]
58. Tomishige, K.; Kimura, T.; Nishikawa, J.; Miyazawa, T.; Kunimori, K. Promoting Effect of the Interaction between Ni and CeO_2 on Steam Gasification of Biomass. *Catal. Commun.* **2007**, *8*, 1074–1079. [CrossRef]
59. Park, H.J.; Park, S.H.; Sohn, J.M.; Park, J.; Jeon, J.K.; Kim, S.S.; Park, Y.K. Steam Reforming of Biomass Gasification Tar Using Benzene as a Model Compound over Various Ni Supported Metal Oxide Catalysts. *Bioresour. Technol.* **2010**, *101*, S101–S103. [CrossRef] [PubMed]
60. Abou Rached, J.; El Hayek, C.; Dahdah, E.; Gennequin, C.; Aouad, S.; Tidahy, H.L.; Estephane, J.; Nsouli, B.; Aboukaïs, A.; Abi-Aad, E. Ni Based Catalysts Promoted with Cerium Used in the Steam Reforming of Toluene for Hydrogen Production. *Int. J. Hydrogen Energy* **2017**, *42*, 12829–12840. [CrossRef]
61. Lu, Y.; Li, S.; Guo, L.; Zhang, X. Hydrogen Production by Biomass Gasification in Supercritical Water over Ni/γAl_2O_3 and Ni/CeO_2-γAl_2O_3 Catalysts. *Int. J. Hydrogen Energy* **2010**, *35*, 7161–7168. [CrossRef]
62. Nishikawa, J.; Nakamura, K.; Asadullah, M.; Miyazawa, T.; Kunimori, K.; Tomishige, K. Catalytic Performance of Ni/CeO_2/Al_2O_3 Modified with Noble Metals in Steam Gasification of Biomass. *Catal. Today* **2008**, *131*, 146–155. [CrossRef]
63. Łamacz, A. Toluene Steam Reforming over Ni/$CeZrO_2$—The Influence of Performance and Carbon Deposition. *Catalysts* **2022**, *12*, 219. [CrossRef]
64. Hernandez, A.R.C.; Castañeda, D.G.G.; Enriquez, A.S.; de Lasa, H.; Rosales, B.S. Ru-Promoted Ni/γAl_2O_3 Fluidized Catalyst for Biomass Gasification. *Catalysts* **2020**, *10*, 316. [CrossRef]
65. Frangini, S.; Masi, A. Molten Carbonates for Advanced and Sustainable Energy Applications: Part I. Revisiting Molten Carbonate Properties from a Sustainable Viewpoint. *Int. J. Hydrogen Energy* **2016**, *41*, 18739–18746. [CrossRef]
66. Abdalazeez, A.; Li, T.; Wang, W.; Abuelgasim, S. A Brief Review of CO_2 utilization for Alkali Carbonate Gasification and Biomass/Coal Co-Gasification: Reactivity, Products and Process. *J. CO_2 Util.* **2021**, *43*, 101370. [CrossRef]
67. Yu, J.; Guo, Q.; Gong, Y.; Ding, L.; Wang, J.; Yu, G. A Review of the Effects of Alkali and Alkaline Earth Metal Species on Biomass Gasification. *Fuel Process. Technol.* **2021**, *214*, 106723. [CrossRef]
68. Mckee, D.W. Gasification of Graphite in CO_2 and Water Vapor- the Catalytic Effect of Metal Salts. *Carbon N. Y.* **1982**, *20*, 59–66. [CrossRef]
69. Wen, W.Y. Mechanisms of Alkali Metal Catalysis in the Gasification of Coal, Char, or Graphite. *Catal. Rev.* **1980**, *22*, 1–28. [CrossRef]
70. Mallick, D.; Mahanta, P.; Moholkar, V.S. Co-Gasification of Coal and Biomass Blends: Chemistry and Engineering. *Fuel* **2017**, *204*, 106–128. [CrossRef]
71. Jia, S.; Ning, S.; Ying, H.; Sun, Y.; Xu, W.; Yin, H. High Quality Syngas Production from Catalytic Gasification of Woodchip Char. *Energy Convers. Manag.* **2017**, *151*, 457–464. [CrossRef]
72. Nzihou, A.; Stanmore, B.; Sharrock, P. A Review of Catalysts for the Gasification of Biomass Char, with Some Reference to Coal. *Energy* **2013**, *58*, 305–317. [CrossRef]
73. Freriks, I.L.C.; van Wechem, H.M.H.; Stuiver, J.C.M.; Bouwman, R. Potassium-Catalysed Gasification of Carbon with Steam: A Temperature-Programmed Desorption and Fourier Transform Infrared Study. *Fuel* **1981**, *60*, 463–470. [CrossRef]
74. Di Blasi, C. Combustion and Gasification Rates of Lignocellulosic Chars. *Prog. Energy Combust. Sci.* **2009**, *35*, 121–140. [CrossRef]
75. Olivares, R.I.; Chen, C.; Wright, S. The Thermal Stability of Molten Lithium-Sodium-Potassium Carbonate and the Influence of Additives on the Melting Point. *J. Sol. Energy Eng. Trans. ASME* **2012**, *134*, 041002. [CrossRef]
76. Ratchahat, S.; Kodama, S.; Tanthapanichakoon, W.; Sekiguchi, H. Combined Molten Salt-Ni/Al2O3 as Synergistic Medium for High-Quality Syngas Production. *Chem. Eng. J.* **2015**, *278*, 224–233. [CrossRef]
77. Parvez, A.M.; Afzal, M.T.; Victor Hebb, T.G.; Schmid, M. Utilization of CO_2 in Thermochemical Conversion of Biomass for Enhanced Product Properties: A Review. *J. CO_2 Util.* **2020**, *40*, 101217. [CrossRef]

78. Lahijani, P.; Zainal, Z.A.; Mohammadi, M.; Mohamed, A.R. Conversion of the Greenhouse Gas CO_2 to the Fuel Gas CO via the Boudouard Reaction: A Review. *Renew. Sustain. Energy Rev.* **2015**, *41*, 615–632. [CrossRef]
79. Roncancio, R.; Gore, J.P. CO_2 Char Gasification: A Systematic Review from 2014 to 2020. *Energy Convers. Manag. X* **2021**, *10*, 100060. [CrossRef]
80. Renganathan, T.; Yadav, M.V.; Pushpavanam, S.; Voolapalli, R.K.; Cho, Y.S. CO_2 Utilization for Gasification of Carbonaceous Feedstocks: A Thermodynamic Analysis. *Chem. Eng. Sci.* **2012**, *83*, 159–170. [CrossRef]
81. Xie, Y.; Yang, H.; Zeng, K.; Zhu, Y.; Hu, J.; Mao, Q.; Liu, Q.; Chen, H. Study on CO_2 Gasification of Biochar in Molten Salts: Reactivity and Structure Evolution. *Fuel* **2019**, *254*, 115614. [CrossRef]
82. Iwaki, H.; Ye, S.; Katagiri, H.; Kitagawa, K. Wastepaper Gasification with CO_2 or Steam Using Catalysts of Molten Carbonates. *Appl. Catal. A Gen.* **2004**, *270*, 237–243. [CrossRef]
83. Kirtania, K.; Axelsson, J.; Matsakas, L.; Christakopoulos, P.; Umeki, K.; Furusjö, E. Kinetic Study of Catalytic Gasification of Wood Char Impregnated with Different Alkali Salts. *Energy* **2017**, *118*, 1055–1065. [CrossRef]
84. Jin, G.; Iwaki, H.; Arai, N.; Kitagawa, K. Study on the Gasification of Wastepaper/Carbon Dioxide Catalyzed by Molten Carbonate Salts. *Energy* **2005**, *30*, 1192–1203. [CrossRef]
85. Sadhwani, N.; Adhikari, S.; Eden, M.R.; Wang, Z.; Baker, R. Southern Pines Char Gasification with CO_2—Kinetics and Effect of Alkali and Alkaline Earth Metals. *Fuel Process. Technol.* **2016**, *150*, 64–70. [CrossRef]
86. Kramb, J.; Gómez-Barea, A.; DeMartini, N.; Romar, H.; Doddapaneni, T.R.K.C.; Konttinen, J. The Effects of Calcium and Potassium on CO_2 Gasification of Birch Wood in a Fluidized Bed. *Fuel* **2017**, *196*, 398–407. [CrossRef]
87. Vamvuka, D.; Karouki, E.; Sfakiotakis, S.; Salatino, P. Gasification of Waste Biomass Chars by Carbon Dioxide via Thermogravimetry-Effect of Catalysts. *Combust. Sci. Technol.* **2012**, *184*, 64–77. [CrossRef]
88. Song, Q.; Wang, X.; Gu, C.; Wang, N.; Li, H.; Su, H.; Huo, J.; Qiao, Y. A Comprehensive Model of Biomass Char-CO_2 Gasification Reactivity with Inorganic Element Catalysis in the Kinetic Control Zone Based on TGA Analysis. *Chem. Eng. J.* **2020**, *398*, 125624. [CrossRef]
89. Fatimah, S.; Ragadhita, R.; Al Husaeni, D.F.; Nandiyanto, A.B.D. How to Calculate Crystallite Size from X-Ray Diffraction (XRD) Using Scherrer Method. *ASEAN J. Sci. Eng.* **2021**, *2*, 65–76. [CrossRef]
90. Bampenrat, A.; Meeyoo, V.; Kitiyanan, B.; Rangsunvigit, P.; Rirksomboon, T. Naphthalene Steam Reforming over Mn-Doped CeO_2-ZrO_2 Supported Nickel Catalysts. *Appl. Catal. A Gen.* **2010**, *373*, 154–159. [CrossRef]
91. Lee, Y.H.; Ahn, J.Y.; Nguyen, D.D.; Chang, S.W.; Kim, S.S.; Lee, S.M. Role of Oxide Support in Ni Based Catalysts for CO_2 Methanation. *RSC Adv.* **2021**, *11*, 17648–17657. [CrossRef]
92. Ding, M.Y.; Tu, J.Y.; Wang, T.J.; Ma, L.L.; Wang, C.G.; Chen, L.G. Bio-Syngas Methanation towards Synthetic Natural Gas (SNG) over Highly Active Al_2O_3-CeO_2 Supported Ni Catalyst. *Fuel Process. Technol.* **2015**, *134*, 480–486. [CrossRef]
93. Sun, D.; Du, Y.; Wang, Z.; Zhang, J.; Li, Y.; Li, J.; Kou, L.; Li, C.; Li, J.; Feng, H.; et al. Effects of CaO Addition on Ni/CeO_2–ZrO_2–Al_2O_3 Coated Monolith Catalysts for Steam Reforming of N-Decane. *Int. J. Hydrogen Energy* **2020**, *45*, 16421–16431. [CrossRef]
94. Jayaprakash, S.; Dewangan, N.; Jangam, A.; Kawi, S. H_2S-Resistant CeO_2-NiO-MgO-Al_2O_3 LDH-Derived Catalysts for Steam Reforming of Toluene. *Fuel Process. Technol.* **2021**, *219*, 106871. [CrossRef]
95. Ashok, J.; Kawi, S. Low-Temperature Biomass Tar Model Reforming over Perovskite Materials with DBD Plasma: Role of Surface Oxygen Mobility. *Energy Convers. Manag.* **2021**, *248*, 114802. [CrossRef]
96. Chen, X.; Ma, X.; Peng, X. Effect of Lattice Oxygen in Ni-Fe/Bio-Char on Filamentous Coke Resistance during CO_2 Reforming of Tar. *Fuel* **2022**, *307*, 121878. [CrossRef]
97. Varvoutis, G.; Lykaki, M.; Stefa, S.; Binas, V.; Marnellos, G.E.; Konsolakis, M. Deciphering the Role of Ni Particle Size and Nickel-Ceria Interfacial Perimeter in the Low-Temperature CO_2 Methanation Reaction over Remarkably Active Ni/CeO_2 Nanorods. *Appl. Catal. B Environ.* **2021**, *297*, 120401. [CrossRef]
98. Varvoutis, G.; Karakoulia, S.A.; Lykaki, M.; Stefa, S.; Binas, V.; Marnellos, G.E.; Konsolakis, M. Support-Induced Modifications on the CO_2 Hydrogenation Performance of Ni/CeO_2: The Effect of ZnO Doping on CeO_2 Nanorods. *J. CO_2 Util.* **2022**, *61*, 102057. [CrossRef]
99. Lu, Y.; Li, S.; Guo, L. Hydrogen Production by Supercritical Water Gasification of Glucose with Ni/CeO_2/Al_2O_3: Effect of Ce Loading. *Fuel* **2013**, *103*, 193–199. [CrossRef]
100. de Caprariis, B.; Bracciale, M.P.; De Filippis, P.; Hernandez, A.D.; Petrullo, A.; Scarsella, M. Steam Reforming of Tar Model Compounds over Ni Supported on CeO_2 and Mayenite. *Can. J. Chem. Eng.* **2017**, *95*, 1745–1751. [CrossRef]
101. Xu, M.; Hu, H.; Yang, Y.; Huang, Y.; Xie, K.; Liu, H.; Li, X.; Yao, H.; Naruse, I. A Deep Insight into Carbon Conversion during Zhundong Coal Molten Salt Gasification. *Fuel* **2018**, *220*, 890–897. [CrossRef]
102. Zhou, L.; Li, L.; Wei, N.; Li, J.; Basset, J.M. Effect of $NiAl_2O_4$ Formation on Ni/Al_2O_3 Stability during Dry Reforming of Methane. *Chem. Cat. Chem.* **2015**, *7*, 2508–2516. [CrossRef]
103. Qiu, H.; Ran, J.; Huang, X.; Ou, Z.; Niu, J. Unrevealing the Influence That Preparation and Reaction Parameters Have on Ni/Al_2O_3 Catalysts for Dry Reforming of Methane. *Int. J. Hydrogen Energy* **2022**, *47*, 34066–34074. [CrossRef]
104. Norouzi, O.; Safari, F.; Jafarian, S.; Tavasoli, A.; Karimi, A. Hydrothermal Gasification Performance of Enteromorpha Intestinalis as an Algal Biomass for Hydrogen-Rich Gas Production Using Ru Promoted Fe–Ni/γ-Al_2O_3 Nanocatalysts. *Energy Convers. Manag.* **2017**, *141*, 63–71. [CrossRef]

105. Furusjö, E.; Ma, C.; Ji, X.; Carvalho, L.; Lundgren, J.; Wetterlund, E. Alkali Enhanced Biomass Gasification with in Situ S Capture and Novel Syngas Cleaning. Part 1: Gasifier Performance. *Energy* **2018**, *157*, 96–105. [CrossRef]
106. Chan, Y.H.; Syed Abdul Rahman, S.N.F.; Lahuri, H.M.; Khalid, A. Recent Progress on CO-Rich Syngas Production via CO_2 Gasification of Various Wastes: A Critical Review on Efficiency, Challenges and Outlook. *Environ. Pollut.* **2021**, *278*, 116843. [CrossRef]
107. Lampropoulos, A.; Binas, V.D.; Zouridi, L.; Athanasiou, C.; Montes-Morán, M.A.; Menéndez, J.A.; Konsolakis, M.; Marnellos, G.E. CO_2 Gasification Reactivity and Syngas Production of Greek Lignite Coal and Ex-Situ Produced Chars under Non-Isothermal and Isothermal Conditions: Structure-Performance Relationships. *Energies* **2022**, *15*, 679. [CrossRef]

catalysts

MDPI

Article

Transition Metal (Fe_2O_3, Co_3O_4 and NiO)-Promoted CuO-Based α-MnO_2 Nanowire Catalysts for Low-Temperature CO Oxidation

Haiou Zhang [1,†], Yixin Zhang [1,†], Huikang Song [1], Yan Cui [1], Yingying Xue [1], Cai-e Wu [2], Chao Pan [3,*], Jingxin Xu [3], Jian Qiu [4], Leilei Xu [1,*] and Mindong Chen [1,*]

1 Collaborative Innovation Centre of the Atmospheric Environment and Equipment Technology, School of Environmental Science and Engineering, Nanjing University of Information Science & Technology, Jiangsu Key Laboratory of Atmospheric Environment Monitoring and Pollution Control, Nanjing 210044, China

2 College of Light Industry and Food Engineering, Nanjing Forestry University, Nanjing 210037, China

3 State Environmental Protection Key Laboratory of Atmospheric Physical Modeling and Pollution Control, China Energy Science and Technology Research Institute Co., Ltd., Nanjing 210023, China

4 Jiangsu Shuang Liang Environmental Technology Co., Ltd., Jiangyin 214400, China

* Correspondence: pcxyz@139.com (C.P.); leileixu88@gmail.com (L.X.); chenmdnuist@163.com (M.C.)

† These authors contributed equally to this work.

Abstract: As a toxic pollutant, carbon monoxide (CO) usually causes harmful effects on human health. Therefore, the thermally catalytic oxidation of CO has received extensive attention in recent years. The CuO-based catalysts have been widely investigated due to their availability. In this study, a series of transition metal oxides (Fe_2O_3, Co_3O_4 and NiO) promoted CuO-based catalysts supported on the α-MnO_2 nanowire catalysts were prepared by the deposition precipitation method for catalytic CO oxidation reactions. The effects of the loaded transition metal type, the loading amount, and the calcination temperature on the catalytic performances were systematically investigated. Further catalyst characterization showed that the CuO/α-MnO_2 catalyst modified with 3 wt% Co_3O_4 and calcined at 400 °C performed the highest CO catalytic activity (T_{90} = 75 °C) among the investigated catalysts. It was supposed that the loading of the Co_3O_4 dopant not only increased the content of oxygen vacancies in the catalyst but also increased the specific surface area and pore volume of the CuO/α-MnO_2 nanowire catalyst, which would further enhance the catalytic activity. The CuO/α-MnO_2 catalyst modified with 3 wt% NiO and calcined at 400 °C exhibited the highest surface adsorbed oxygen content and the best normalized reaction rate, but the specific surface area limited its activity. Therefore, the appropriate loading of the Co_3O_4 modifier could greatly enhance the activity of CuO/α-MnO_2. This research could provide a reference method for constructing efficient low-temperature CO oxidation catalysts.

Keywords: transition metal oxides; CuO-based catalysts; Co_3O_4 modification; α-MnO_2 nanowire; catalytic oxidation of CO

Citation: Zhang, H.; Zhang, Y.; Song, H.; Cui, Y.; Xue, Y.; Wu, C.-e.; Pan, C.; Xu, J.; Qiu, J.; Xu, L.; et al. Transition Metal (Fe_2O_3, Co_3O_4 and NiO)-Promoted CuO-Based α-MnO_2 Nanowire Catalysts for Low-Temperature CO Oxidation. *Catalysts* **2023**, *13*, 588. https://doi.org/10.3390/catal13030588

Academic Editors: Georgios Bampos, Athanasia Petala and Zacharias Frontistis

Received: 3 February 2023
Revised: 10 March 2023
Accepted: 10 March 2023
Published: 15 March 2023

1. Introduction

As a toxic pollutant in the atmosphere, carbon monoxide (CO) has widely received public attention in recent years [1]. CO is not only mainly produced during the incomplete combustion of fossil fuels and motor vehicle emissions [2], but it is also a precursor of the ozone pollution [3], which can be harmful even at low concentrations. In recent years, scientists around the world have employed different methods for the removal of CO, such as photocatalysis [4], thermocatalysis oxidation [5], etc. Thermocatalysis oxidation is the most commonly used method for the oxidation of CO. The noble metal-based catalysts (such as Au [6,7], Pt [8–10], Pd [11], Ru [12,13], Rh [14] and Ag [15]), transition metal oxide based catalysts (such as MnO_2 [16], CeO_2 [17], ZrO_2 [18], Co_3O_4 [16]), and metal hybrid catalysts (such as $MnCo_2O_4$ spinel [19], Ce-Zr solid solution [20]) can be used for the thermocatalytic oxidation of CO at low temperatures without generating the secondary contaminants [16]. Transition metal-based catalysts have good potential for application due to their low cost, high stability, and good activity [21,22]. Among the transition metal oxides, manganese dioxide (MnO_2) has many advantages, such as a low price, being environmentally friendly, and abundance in nature [2,23]. As is well known, MnO_2 usually exists in different crystalline phases with different structures, such as the α-, β-, and γ-types of the one-dimensional pore structure, the δ-type of the two-dimensional pore structure, and the λ-type of the three-dimensional network structure [24], which greatly depend on the different connectivity (corner-edge sharing) exhibited by the [MnO_6] octahedra [25–28]. Liang et al. [28] prepared MnO_2 nanorods with four different crystalline phases for the oxidation of CO. It was found that the order of oxidation activity of different crystalline phases with the same nanorod morphology was greatly different, in decreasing order from α- ≈ δ- > γ- > β-MnO_2 (The temperature at which the conversion of CO on α-MnO_2 reached 100% was approximately 130 °C). This indicated that the oxidation activity of CO significantly depended on the phase structure and channel structure of the MnO_2. Tian et al. [29] investigated the effects of the crystalline phases of MnO_2 (α-, β-, and ε-MnO_2) on the performances of the oxidation of CO and toluene. It was found that the β-MnO_2 performed the highest activity for CO oxidation (T_{90} = 75 °C) among the three crystalline phases of MnO_2, and the α-MnO_2 behaved with the lowest activity for CO oxidation (T_{90} = 118 °C). The content of oxygen vacancy in the catalyst was also determined by in situ EPR spectra and the results showed that the catalytic activity of the catalyst was proportional to the concentration of oxygen vacancies, which were regarded as the active sites for the adsorption and dissociation of oxygen molecules. It was believed that the catalytic activity of MnO_2 was greatly related to the oxygen vacancy activity [29]. As is well known, the α-MnO2 is provided with the 1D (1 × 1) (2 × 2) tunnel structures, which are attributed to the tetragonal crystal system [30].

CuO has been widely used as the active site of the low-temperature catalytic oxidation of CO due to its excellent activity and abundant availability [31]. For example, Raziyeh Jokar et al. [31] used CuO/α-MnO_2 as the catalyst of the preferential oxidation of CO in the hydrogen-rich gas stream and investigated the interaction between the MnO_2 and CuO (T_{97} = 130 °C), the superior activity of the catalyst due to the beneficial synergistic interaction between CuO and α-MnO_2. Meanwhile, the catalytic activity was also influenced by several factors, such as specific surface area, crystallinity, oxygen vacancies, and redox properties. Qian et al. [32] prepared a series of CuO/MnO_2 catalysts with different CuO loading amounts by the incipient wetness impregnation method for the oxidation of CO. The catalyst activity was almost the same for the CuO loadings, from 1% to 40%. Sun et al. [33] prepared a CuO/$Cu_{1.5}Mn_{1.5}O_4$ spinel-type composite oxide for synergistic catalysis of CO oxidation. It was found that the synergistic effect between $Cu_{1.5}Mn_{1.5}O_4$ spinel and CuO can promote the oxidation of CO, and CuO-$Cu_{1.5}Mn_{1.5}O_4$ had the best oxidation activity for CO (T_{100} = 177 °C).

Nowadays, for the modification of transition metal oxide-based catalysts, in addition to the carrier and active site, the promoter also plays an active role in the improvement of catalytic performance [34]. Gao et al. [35] doped transition metals (Fe, Co, Ni, and Cu) with a 1:10 molar ratio on α-MnO_2 nanowires by a one-step hydrothermal method to oxidize CO. Among the four transition metals, $Cu_{0.1}MnOx$ had the best oxidation activity for CO (T_{100}~120 °C). Krasimir et al. [36] investigated the effects of different molar ratios of chemical compositions on the γ-Al_2O_3-supported CuO/MnO_2/Cr_2O_3 catalysts for the oxidation of CO, dimethyl ether (DME), and methanol. The results showed that the Cu-Mn-Cr/γ-Al_2O_3 catalyst, which Cu/(Mn + Cr) has a molar ratio of 2:1 and a Mn/Cr molar ratio of 0.25, can achieve the complete oxidation of CO at 200 °C.

In order to further investigate the contribution of promoters to the catalytic performance of CuO/MnO_2 catalysts in the oxidation of CO, in this work, the α-MnO_2 nanowire was successfully prepared by the hydrothermal method and used as the support for the

CuO-based catalysts. A series of the transition metal oxides (Fe_2O_3, Co_3O_4, and NiO) promoted CuO-based α-MnO_2 nanowire catalysts were prepared by the deposition precipitation method. The effects of the type, the loading amount, and the calcination temperature of three transition metal oxides on the performance of the catalytic oxidation of CO were systematically studied. The obtained catalysts were carefully characterized by X-ray diffraction (XRD), scanning electron microscopy (SEM), nitrogen physisorption, X-ray photoelectron spectroscopy (XPS), and H_2 temperature-programmed reduction (H_2-TPR). These catalysts were evaluated for their catalytic performances in the oxidation of CO. The results show that the Co_3O_4 (3 wt.%) promoted 10 wt.% CuO/α-MnO_2 catalyst calcined at 400 °C performed the greatest CO reactivity (T_{90} = 75 °C).

2. Results and Discussion

2.1. XRD Analysis

The crystalline phase structures of the catalysts were obtained using X-ray diffraction (XRD) analysis. The results of XRD are shown in the Figure 1. Figure 1a shows the XRD patterns of the transition metal oxide (MOx = Fe_2O_3, Co_3O_4, NiO) doped catalysts and the pristine 10CuO/α-MnO_2-400 catalyst. It can be observed in Figure 1a that all the catalysts show wide and clear diffraction peaks at 2θ = 12.78°, 18.11°, 25.71°, 37.52°, 41.97°, 49.86°, 56.37°, 65.11°, and 69.71°, which could be conformed to the characteristic peaks of the α-MnO_2 (PDF#44-0141). It could be observed that the intensity of the diffraction peaks of MnO_2 increased after the loading of the second transition metal oxides, especially over the 10CuO-3Co_3O_4/α-MnO_2 catalyst. This phenomenon suggests that the crystallinity of the catalysts increased after loading the transition metal oxides, especially for the catalyst loading Co_3O_4. Meanwhile, two weak diffraction peaks could be detected at 2θ = 35.5° and 38.8°, which corresponded to the characteristic peaks of CuO, according to the standard card of PDF#05-0661. The diffraction peak intensity of CuO became weaker after the addition of the transition metal to the catalyst, indicating that the addition of the transition metal promoted the dispersion of CuO.

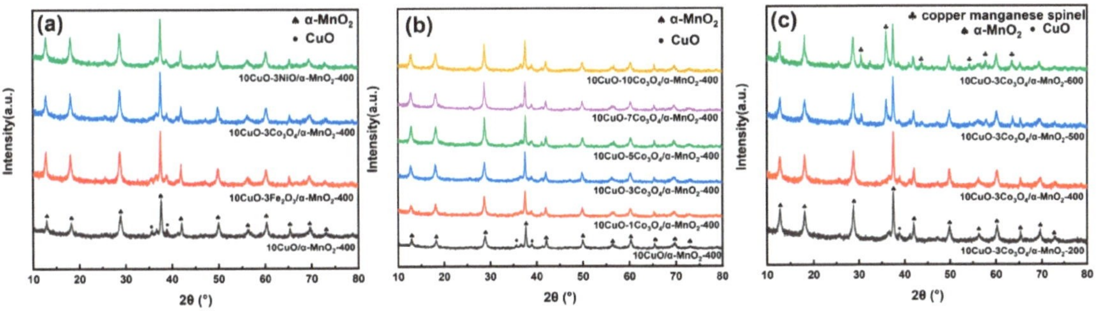

Figure 1. Powder XRD patterns of the (**a**) 10CuO-3MOx/α-MnO_2-400 catalysts doped with different transition metals (MOx = Fe_2O_3, Co_3O_4, NiO); (**b**) 10CuO-yCo_3O_4/α-MnO_2-400 catalysts doped with different contents of Co_3O_4 (y = 1, 3, 5, 7, 10); (**c**) 10CuO-3Co_3O_4/α-MnO_2-T catalysts calcined at different temperatures (T = 200, 400, 500, 600 °C).

The XRD patterns of the 10CuO-yCo_3O_4/α-MnO_2-400 catalysts with different contents of Co_3O_4 are shown in the Figure 1b. It can be observed that the diffraction peaks of CuO are very clear over the pristine 10CuO/α-MnO_2-400 catalyst without Co_3O_4 loading. However, it can be observed that the diffraction peaks of CuO became blurry after loading Co_3O_4 from 1 wt.% to 10 wt.%. The possible reason accounting for this phenomenon was that the appropriate loading amount of Co_3O_4 could greatly enhance the dispersion of CuO on the surface of α-MnO_2 nanowire catalyst. Nevertheless, the characteristic peaks of Co_3O_4 could be not observed in the catalysts due to the good dispersion of Co_3O_4.

The XRD patterns of 10CuO-3Co_3O_4/α-MnO_2 catalysis calcined at different temperatures are shown in the Figure 1c. It can be recognized that the intensity of the diffraction peaks of α-MnO_2 became weak at high calcination temperatures. The reason might be due to the formation of a copper manganese spinel (JCPDS No.70-0260) [37]. Only a diffraction peak of CuO was observed at 38.8°. Meanwhile, the intensity of the diffraction peaks of CuO became sharp with the increase of the calcination temperature (only from 200 °C to 400 °C). This indicated that the dispersion of CuO on the catalyst surface gradually deteriorated and the crystal size of CuO nanoparticles grew due to the finite surface area of the MnO_2 nanowire from 200 °C to 400 °C. As the calcination temperature rose from 400 °C to 600 °C, the diffraction peak strength of copper manganese spinel increased. The diffraction peak strength of CuO decreased, indicating that the dispersion of CuO increased.

2.2. SEM Observation

The morphologies of the 10CuO-3MOx/α-MnO_2-400 catalysts loaded with different transition metal oxides (MOx = Fe_2O_3, Co_3O_4, NiO) and the pristine 10CuO/α-MnO_2-400 catalyst were characterized by SEM (Figure 2). It can be observed that the morphology of the α-MnO_2 nanowire support greatly changed after loading the CuO active sites and the transition metal oxides. Specifically, the length to diameter ratio of the α-MnO_2 nanowire significantly decreased compared to the pristine α-MnO_2 nanowire. This was caused by the high loading contents of the CuO and transition metal oxides. In addition, the α-MnO_2 nanowire support might experience thermal sintering and self-assemble at high calcination temperatures. The spatial distribution of the elements over the one-dimensional 10CuO-3MOx/α-MnO_2-400 (MOx = Fe_2O_3, Co_3O_4, NiO) nanowire catalyst was studied by energy dispersive X-ray spectroscopy mapping (EDS-mapping). As shown in Figure 2, the Cu and the doped Fe/Co/Ni elements were homogenously distributed over these investigated catalysts. This indicated that the great dispersion of the CuO and the doped transition metal oxides on the surface of the α-MnO_2 nanowire supports could be facilely achieved by the precipitation deposition method.

2.3. BET Analysis

In order to further investigate the structural properties of the catalysts, the specific surface areas, pore volumes, and pore size distributions of the catalysts were measured by nitrogen physisorption measurements. As shown in Figure 3a, all catalysts show IV isotherms with H3-shaped hysteresis loops. These results prove the presence of mesopores with a narrow slit-shape in the catalyst [38]. It is also interesting to find that the 10CuO-3MOx/α-MnO_2 catalysts still possess the similar mesoporous structure to the 10CuO/α-MnO_2 nanowire catalysts after loading different transition metal oxides. This demonstrates that the 10CuO/α-MnO_2 nanowire catalyst was provided with good thermal stability. The pore size distribution curves of the corresponding catalysts are shown in Figure 3b. The pore diameter of the catalysts is in the range of 2–15 nm after loading with transition metals. Moreover, the specific parameters of the structural properties of these catalysts are shown in Table 1. It can be observed that the specific surface areas of both the 10CuO-3Fe_2O_3/α-MnO_2 and 10CuO-3NiO/α-MnO_2 catalysts decreased after the loading of transition metals. On the contrary, the specific surface area of the 10CuO-3Co_3O_4/α-MnO_2 catalysts increased substantially. This indicates that the Co_3O_4 on the α-MnO_2 nanowire support surface was highly dispersed. Meanwhile, the average pore sizes of all the 10CuO-3MOx/α-MnO_2 (MOx = Fe_2O_3, Co_3O_4, NiO) catalysts are very similar to the pristine 10CuO/α-MnO_2 nanowire. Specifically, the 10CuO-3Fe_2O_3/α-MnO_2 catalyst has a similar pore size distribution as the 10CuO/α-MnO_2 catalyst. This indicates that their mesoporous structures are not sharply impaired by the loaded metal oxides. In addition, the pore capacities of all 10CuO-3MOx/α-MnO_2 (MOx = Fe_2O_3, Co_3O_4, NiO) catalysts are enhanced compared to the pristine 10CuO/α-MnO_2. As for the 10CuO-3Co_3O_4/α-MnO_2 catalyst, its surface area was twice as large as the pristine 10CuO/α-MnO_2 catalyst. The higher specific surface area is beneficial for the catalyst to expose more active sites, and the larger pore volume helps the reactant accelerate the reactant mass diffusion and has a better adsorption ability to the reactant [39].

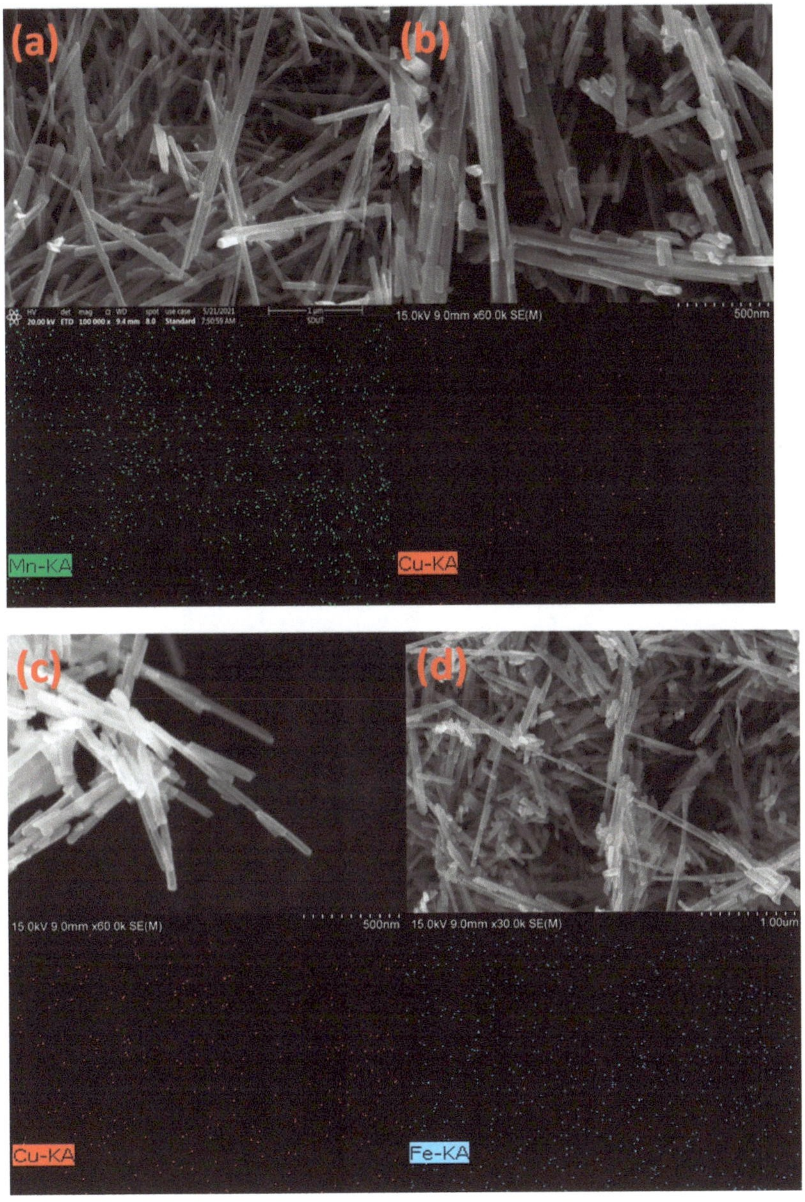

Figure 2. *Cont.*

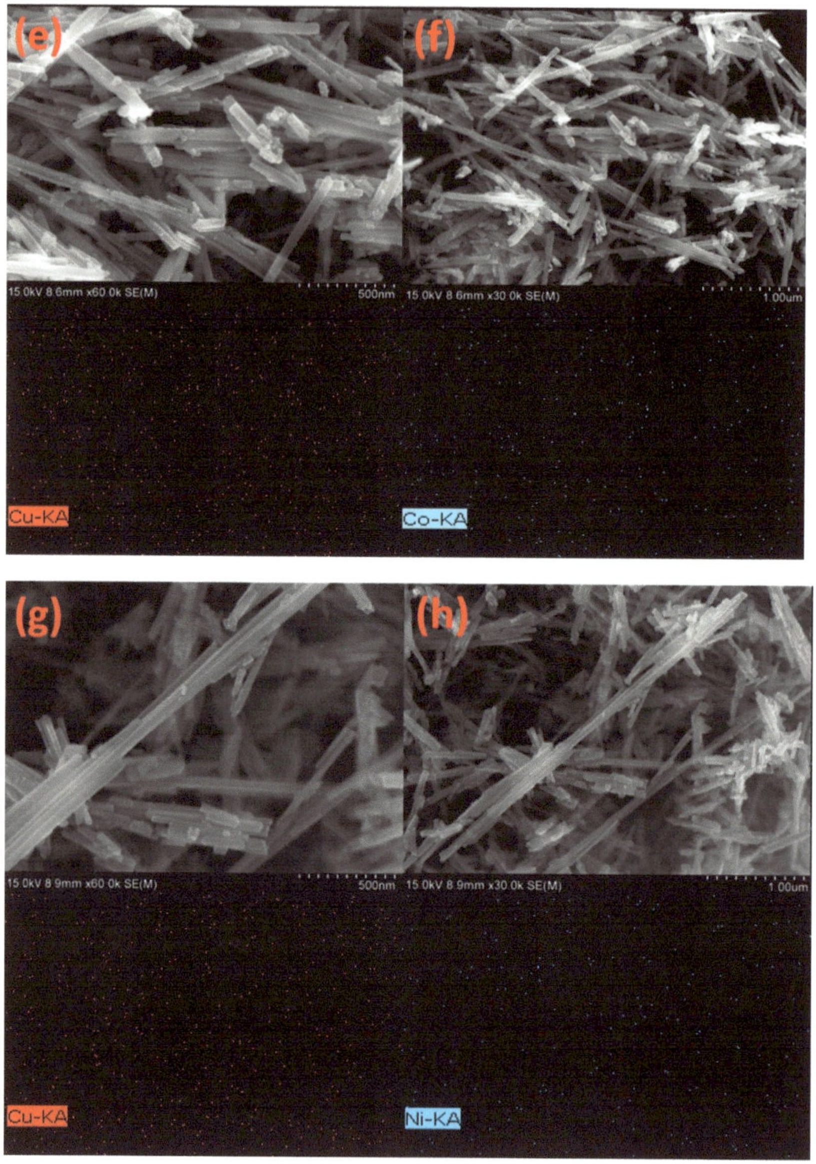

Figure 2. SEM-EDS images of the (**a**) α-MnO_2 nanowire, (**b**) 10CuO/α-MnO_2-400, (**c**,**d**) 10CuO-3Fe_2O_3/α-MnO_2-400, (**e**,**f**) 10CuO-3Co_3O_4/α-MnO_2-400, and (**g**,**h**) 10CuO-3NiO/α-MnO_2-400.

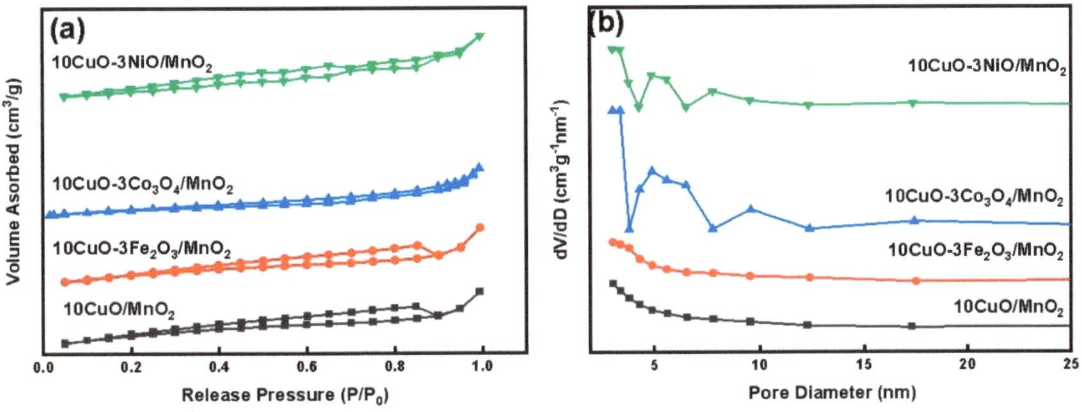

Figure 3. (**a**) Nitrogen adsorption-desorption isotherms and (**b**) BJH pore size distribution of the $10CuO/\alpha\text{-}MnO_2$-400 and $10CuO\text{-}3MOx/\alpha\text{-}MnO_2$-400 (MOx = Fe_2O_3, Co_3O_4, NiO) catalysts.

Table 1. Structural properties of the $10CuO/\alpha\text{-}MnO_2$-400 and $10CuO\text{-}3MOx/\alpha\text{-}MnO_2$-400 (MOx = Fe_2O_3, Co_3O_4, NiO) catalysts.

Catalyst	Specific Surface Area (m^2/g)	Pore Volume (cm^3/g)	Average Pore Diameter (nm)	Isotherm Type
$10CuO/MnO_2$	72	0.09	3.1	IV H3
$10CuO\text{-}3Fe_2O_3/MnO_2$	64	0.10	3.1	IV H3
$10CuO\text{-}3Co_3O_4/MnO_2$	84	0.18	3.1	IV H3
$10CuO\text{-}3NiO/MnO_2$	52	0.10	3.1	IV H3

2.4. XPS Analysis

The coordination, composition, and valence state of the elements over the catalyst surface were investigated by X-ray photoelectron spectroscopy (XPS). The XPS of Mn 2p, O 1s, and Cu 2p of the $10CuO/\alpha\text{-}MnO_2$-400 and the $10CuO\text{-}3MOx/\alpha\text{-}MnO_2$-400 (MOx = Fe_2O_3, Co_3O_4, NiO) are shown in Figure 4. It can be observed in Figure 4a that $10CuO/\alpha\text{-}MnO_2$-400 had two main peaks around 654.0 eV and 642.0 eV, which could be attributed to the binding energies of the Mn $2p_{1/2}$ at BE = 654.0 eV and Mn $2p_{3/2}$ at BE = 642.0 eV, respectively. The two main peaks were divided into four peaks after the peak fitting. The fitted peaks of Mn $2p_{3/2}$ at 642.0 eV and 643.5 eV indicate the existence of Mn^{3+} and Mn^{4+} in the $10CuO\text{-}3MOx/\alpha\text{-}MnO_2$-400 catalyst [40–44]. The ratios of the $Mn^{3+}/(Mn^{3+} + Mn^{4+})$ over different catalysts followed the below order: $10CuO\text{-}3Co_3O_4/\alpha\text{-}MnO_2$ (0.606) > $10CuO\text{-}3NiO/\alpha\text{-}MnO_2$ (0.541) > $10CuO/\alpha\text{-}MnO_2$ (0.414) > $10CuO\text{-}3Fe_2O_3/\alpha\text{-}MnO_2$ (0.406). The redox electron pair in Cu-Mn oxide was the $\text{-}Cu^{2+}\text{-}O^{2-}\text{-}Mn^{4+}\text{-}\rightarrow\text{-}Cu^{+}\text{-}\square\text{-}Mn^{3+}\text{-} + 1/2O_2$ (□ indicates the oxygen vacancy) [45]. The content of Mn^{3+} and oxygen vacancies are proportional, or indirectly proportional, to the oxidation capacity of the catalyst [25,29,46,47]. The Mn^{3+} may cause the Jahn-Teller effect, which could prolong the Mn-O bond in $[MnO_6]$ [48,49], thereby prolonging the distance between the oxygen pairs and causing the stretching of the Mn-O bond length [48]. As a result, the Mn-O bond was easier to break, and the mobility of oxygen became higher. Therefore, the released surface oxygen atoms are more likely to participate in the reaction and thus promote the catalytic performance. The XPS of O 1s of $10CuO\text{-}3MOx/\alpha\text{-}MnO_2$-400 was measured to elucidate the nature of the oxygen species over the $10CuO\text{-}3MOx/\alpha\text{-}MnO_2$-400 catalysts. As shown in Figure 4b, all the samples show double peaks of different oxygen species. Specifically, the BEs at around 529.8 eV and 531.4 eV could be ascribable to the surface lattice oxygen (O_{latt}) and surface adsorbed oxygen (O_{ads}) species [50,51], respectively. The ratios of the

O_{ads}/O_{latt} are also summarized in Table 2. It can be observed that the binding energy of surface lattice oxygen (O_{latt}) shifted to a higher binding energy with the addition of transition metals. The surface oxidation states of the Cu species were also investigated to show the redox properties of the as-prepared catalysts. As shown in Figure 4c, all the catalysts displayed two main peaks of Cu $2p_{1/2}$ (953.6 eV) and Cu $2p_{3/2}$ (933.7 eV) [43]. The Cu $2p_{3/2}$, orbitals with binding energy in the range of 930.0–935.0 eV, could be divided into two peaks. Specifically, the binding energy peak at 933.4 eV was attributed to Cu^+, and the peak at 934.4 eV was attributed to Cu^{2+} [43]. Furthermore, it is worth noting that the Cu $2p_{3/2}$ peak is accompanied by the vibrating satellite peaks in the range of 940.0–944.0 eV [52]. Combined with the results of the Mn 2p spectrum, the catalysts formed redox pairs of Cu^+/Cu^{2+} and Mn^{3+}/Mn^{4+}, which would promote the charge transference to generate more oxygen defects [37,53].

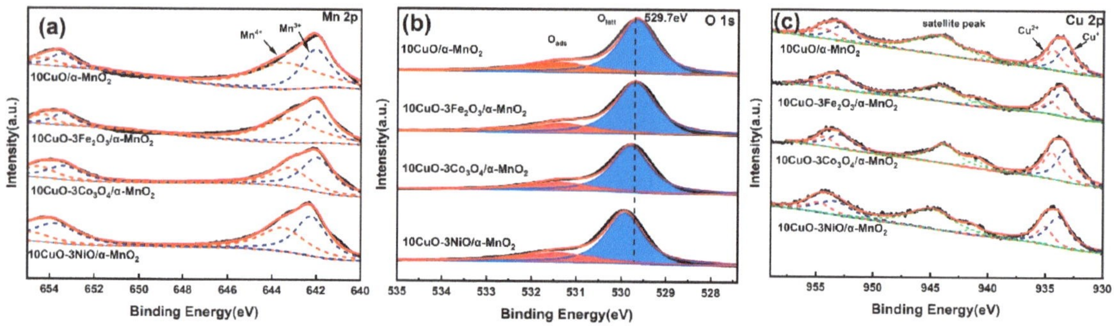

Figure 4. (**a**) Mn 2p spectra, (**b**) O 1s spectra, and (**c**) Cu 2p spectra of the $10CuO/\alpha\text{-}MnO_2$, $10CuO\text{-}3Fe_2O_3/\alpha\text{-}MnO_2$, $10CuO\text{-}3Co_3O_4/\alpha\text{-}MnO_2$, and $10CuO\text{-}3NiO/\alpha\text{-}MnO_2$ (T = 400 °C) catalysts.

Table 2. The ratio of $Mn^{3+}/(Mn^{3+} + Mn^{4+})$, O_{ads}/O_{latt}, and $Cu^+/(Cu^{2+} + Cu^+)$ of the $10CuO\text{-}3MOx/\alpha\text{-}MnO_2$(MOx = Fe_2O_3, Co_3O_4, NiO) and the $10CuO/\alpha\text{-}MnO_2$ catalyst.

Sample	$Mn^{3+}/(Mn^{3+} + Mn^{4+})$	O_{ads}/O_{latt}	$Cu^+/(Cu^{2+} + Cu^+)$
$10CuO/\alpha\text{-}MnO_2$	0.414	0.324	0.551
$10CuO\text{-}3Fe_2O_3/\alpha\text{-}MnO_2$	0.406	0.283	0.644
$10CuO\text{-}3Co_3O_4/\alpha\text{-}MnO_2$	0.606	0.300	0.563
$10CuO\text{-}3NiO/\alpha\text{-}MnO_2$	0.541	0.352	0.688

To determine the valence state of the transition metals loaded on the catalyst, the XPS spectra of Fe 2p, Co 2p, and Ni 2p are determined. The XPS profile of Fe 2p over the $10CuO\text{-}3Fe_2O_3/\alpha\text{-}MnO_2$-400 catalyst is shown in Figure 5a. The binding energies at 710.4 eV and 725.1 eV are ascribed to Fe $2p_{3/2}$ and Fe $2p_{1/2}$, respectively [54,55]. The peak of Fe $2p_{3/2}$ can be divided into two peaks (710.3 eV and 712.5 eV) [56]. In addition, a satellite peak was observed at about 718.3 eV. This indicates that the iron species existed in the form of Fe^{3+} on the surface of the $10CuO\text{-}3Fe_2O_3/\alpha\text{-}MnO_2$-400 [54,56]. The XPS of Co 2p over the $10CuO\text{-}3Co_3O_4/\alpha\text{-}MnO_2$-400 catalyst is shown in Figure 5b. The binding energies at 780.0 eV are ascribed to the Co $2p_{3/2}$ [57]. Meanwhile, there was no significant satellite shake-up intensity at 786 eV, indicating the dominance of Co^{3+} on the surface of Co_3O_4 [57]. The XPS of Ni 2p over the $10CuO\text{-}3NiO/\alpha\text{-}MnO_2$-400 catalyst is shown in Figure 5c. The binding energies at 855.1eV are ascribed to the Ni $2p_{3/2}$ [58]. The peak of Ni $2p_{3/2}$ of metallic Ni was basically at 852.6 eV, and the binding energy of NiO is about 1 eV higher than that of Ni^0 [58]. The peak of NiO $2p_{3/2}$ is lower than that of Ni $2p_{3/2}$ in this work (854.8 eV). Therefore, the oxidation state of the indicated Ni element is mainly in the form of Ni^{2+}. The higher binding energy compared to pure NiO binding energy indicates that NiO did not exist in the free form. The strong interaction was formed between the

Ni and the support. The results indicate that Fe_2O_3, Co_3O_4, and NiO were successfully loaded on the catalyst. The XPS electronic binding energies of the surface elements of the 10CuO-3MOx/α-MnO_2-400 (MOx = Fe_2O_3, Co_3O_4, NiO) are summarized in Table 3.

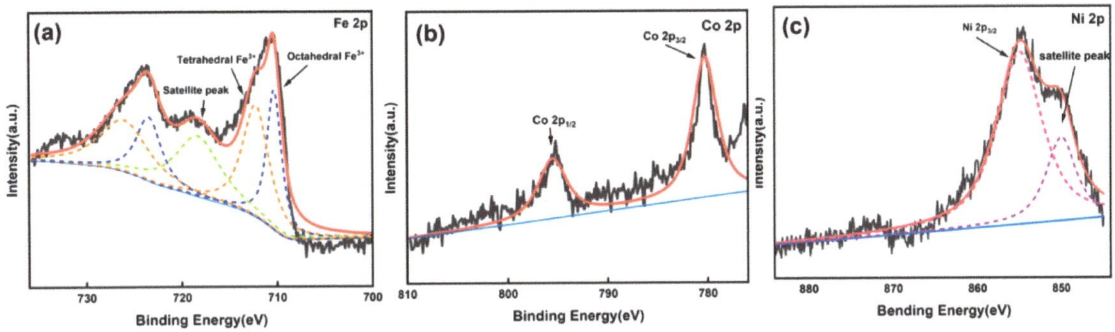

Figure 5. (**a**) Fe 2p, (**b**) Co 2p, and (**c**) Ni 2p spectra of the 10CuO-3MOx/α-MnO_2-400 (MOx = Fe_2O_3, Co_3O_4, NiO) catalysts.

Table 3. XPS electronic binding energies of surface elements of the 10CuO-3MOx/α-MnO_2-400 (MOx = Fe_2O_3, Co_3O_4, NiO) catalysts.

Sample	Cu $2p_{3/2}$	O 1s	Mn $2p_{3/2}$
10CuO/α-MnO_2	933.6	529.7	642.1
10CuO-3Fe_2O_3/α-MnO_2	933.7	529.7	642.1
10CuO-3Co_3O_4/α-MnO_2	933.7	529.8	642.3
10CuO-3NiO/α-MnO_2	934.2	529.9	642.4

2.5. H_2-TPR

The H_2-TPR profiles of 10CuO-3MOx/α-MnO_2-400 (MOx = Fe_2O_3, Co_3O_4, NiO), 10CuO-yCo_3O_4/α-MnO_2-400 (y = 0, 1, 3, 5, 7, 10), and 10CuO-3Co_3O_4/α-MnO_2-T (T = 200, 400, 500, 600 °C) are shown in Figure 6. It can be observed that both the 10CuO-3MOx/α-MnO_2-400 catalyst and the 10CuO/α-MnO_2-400 catalyst show similar hydrogen consumption peaks in strength and shape in Figure 6a. Specifically, there were two sets of peaks in the range of 297–342 °C and 462–472 °C, which might be attributed to the hydrogen consumption derived from the gradual reduction of the α-MnO_2 nanowire ($MnO_2 \rightarrow Mn_2O_3 \rightarrow Mn_3O_4$), according to the pioneer report [31]. Meanwhile, it is worth noting that the loading of the transition metal oxides on the 10CuO/α-MnO_2 support changed the reduction behavior of the 10CuO/α-MnO_2 catalyst. Specifically, the reduction of 10CuO/α-MnO_2 nanowires shifted to higher temperatures after loading the transition metal oxides. In general, the reducibility of the 10CuO/α-MnO_2 catalyst decreases with the addition of transition metal oxides. The H_2-TPR profiles of the 10CuO-yCo_3O_4/α-MnO_2-400 catalysts with different Co_3O_4 loading amounts are shown in Figure 6b. It is of great interest to find that the positions of the reduction peaks gradually shifted to the high-temperature region with the Co_3O_4 loading amount increasing from 1 wt% to 10 wt%. This illustrated that the reducibility of the catalysts also gradually decreases. Therefore, the reduction temperatures of the 10CuO-yCo_3O_4/α-MnO_2-400 catalyst were significantly higher than that of the 10CuO/α-MnO_2 catalyst, except for the 10CuO-7Co_3O_4/α-MnO_2-400 catalyst. The H_2-TPR profiles of the 10CuO-3MOx/α-MnO_2-T catalyst with different calcination temperatures is shown in Figure 6c. It can be observed that the reduction temperature of the 10CuO-3Co_3O_4/α-MnO_2-T catalyst gradually shifted to a higher reduction temperature with the increase of the calcination temperature from 200 °C to 600 °C. The positions of the two reduction peaks became closer. This suggests that the interaction between the CuO and the α-MnO_2 nanowire support became stronger at higher temperatures.

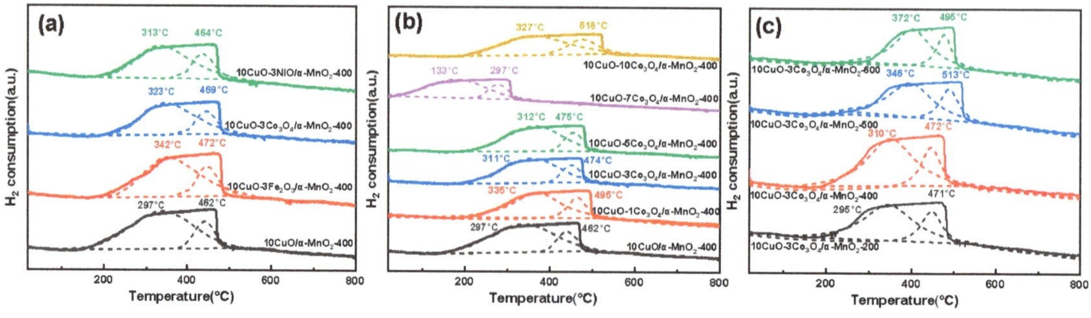

Figure 6. H_2-TPR profiles of the (**a**) 10CuO-3MOx/α-MnO_2-400 (MOx = Fe_2O_3, Co_3O_4, NiO), (**b**) 10CuO-yCo_3O_4/α-MnO_2-400 (y = 0, 1, 3, 5, 7, 10), and (**c**) 10CuO-3Co_3O_4/α-MnO_2-T (T = 200, 400, 500, 600 °C) catalysts.

2.6. Catalytic Performance of the CO Oxidation

2.6.1. The Effect of the Transition Metal Oxides (MOx) on the Activities

The catalytic CO oxidation activities of the 10CuO-3MOx/α-MnO_2-400 (MOx = Fe_2O_3, Co_3O_4, NiO) catalysts were studied to investigate the effect of the dual loading of transition metals and CuO on the catalytic activity. As shown in Figure 7a, it can be observed that the CO conversion gradually increases with the increase of the reaction temperature until it reached 100%. As can be observed, the temperature of the 90% CO conversion over the 10CuO-3Co_3O_4/α-MnO_2-400 catalyst was 75 °C. The activity of the 10CuO-3NiO/α-MnO_2-400 catalyst basically shows a similar catalytic activity to that of the pristine 10CuO/α-MnO_2-400 catalyst without modification (T_{90} = 77 °C). However, the CO oxidation activity of the 10CuO-3Fe_2O_3/α-MnO_2-400 catalyst (T_{90} = 80 °C) was even worse than that of the pristine 10CuO/α-MnO_2-400 catalyst. Therefore, only the catalytic activity of the 10CuO-3Co_3O_4/α-MnO_2-400 was significantly improved compared with the pristine 10CuO/α-MnO_2-400 catalyst. The reason for this might be that the loading of Cu and Co could generate more oxygen vacancies and active sites to the α-MnO_2 nanowire. From the order of catalyst activity, it can be observed that the catalytic activity of the catalyst increases with the increase in of Mn^{3+} content. Meanwhile, the loading of Co_3O_4 increased the specific surface area and pore volume of the catalyst, providing more active sites for the reaction. The ratio of O_{ads}/O_{latt} was not exactly the same as the catalytic activity of the catalyst. The reasons for this phenomenon were stated in the discussion. The CO oxidation activity was significantly improved over the 10CuO-3Co_3O_4/α-MnO_2-400 catalyst. This result was well consistent with the Mn 2p XPS analysis. The results of the normalized reaction rates of the four catalysts are shown in Figure 7b. It can be observed that the normalized reaction rates gradually increase with the increase of the reaction temperature. The order of reaction rates of the catalysts per surface area was 10CuO-3NiO/MnO_2 > 10CuO/MnO_2 > 10CuO-3Co_3O_4/MnO_2 > 10CuO-3Fe_2O_3/MnO_2. The normalized reaction rates excluded the effect of specific surface area on catalytic activity and expressed the intrinsic catalytic activity of the catalysts. The order of the O_{ads}/O_{latt} ratios was consistent with the order of the normalized reaction rates.

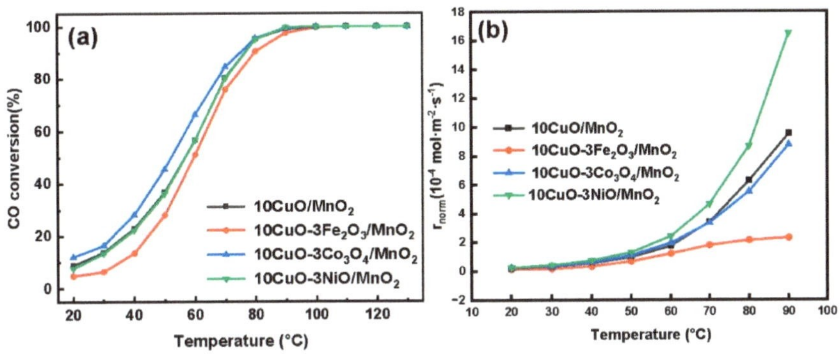

Figure 7. (**a**) CO conversions and (**b**) normalized reaction rates over the 10CuO-3MOx/α-MnO_2-400 (MOx = Fe_2O_3, Co_3O_4, NiO) catalysts under the reaction conditions: $CO/O_2/N_2$ = 1/20/79, GHSV = 12,000 mL$\cdot g^{-1}\cdot h^{-1}$, 1 atm.

2.6.2. The Effect of the Co_3O_4 Loading Amount on the Activities

The effect of Co_3O_4 loading on the catalytic activity of the CO oxidation was investigated, and the profiles of the CO conversion are shown in Figure 8a. It could be noticed that the Co_3O_4 loading amount in the range of 0 wt.%–10 wt.% had little effect on the CO oxidation activity of the 10CuO/α-MnO_2 catalyst. The order of CO catalytic activity of the catalysts is 10CuO-3Co_3O_4/α-MnO_2-400(T_{90} = 75 °C) > 10CuO-10Co_3O_4/α-MnO_2-400 (T_{90} = 77 °C) $\approx$ 10CuO/α-MnO_2-400 > 10CuO-1Co_3O_4/α-MnO_2-400 (T_{90} = 79 °C) $\approx$ 10CuO-5Co_3O_4/α-MnO_2-400 > 10CuO-7Co_3O_4/α-MnO_2-400 (T_{90} = 84 °C). The 10CuO-3Co_3O_4/α-MnO_2-400 catalyst performed the highest activity in the low temperature region. It is shown that a certain increase in the loading of Co_3O_4 was beneficial to the catalytic activity of the catalyst.

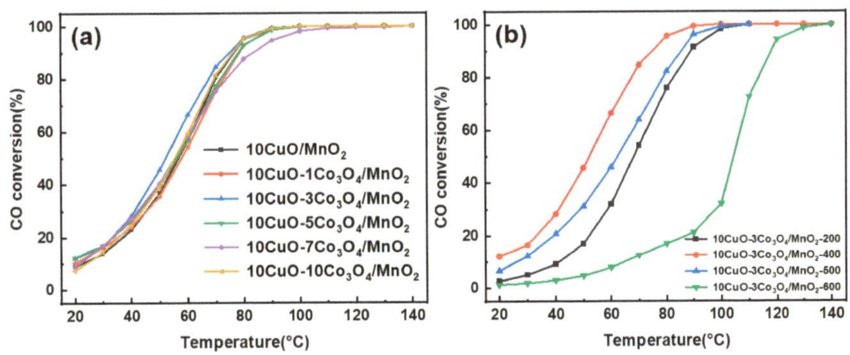

Figure 8. CO conversions over the (**a**) 10CuO-yCo_3O_4/α-MnO_2-400 (y = 0, 1, 3, 5, 7, 10), and (**b**) 10CuO-3Co_3O_4/α-MnO_2-T (T = 200, 400, 500, 600 °C) catalysts under the reaction conditions: $CO/O_2/N_2$ = 1/20/79, GHSV = 12,000 mL$\cdot g^{-1}\cdot h^{-1}$, 1 atm.

2.6.3. The Effect of the Calcination Temperature on the Activities

The catalytic CO oxidation of the 10CuO-3Co_3O_4/α-MnO_2-T (T = 200, 400, 500, 600 °C) catalyst was conducted to study the influence of the calcination temperature on catalytic activity, and the CO conversion profiles are shown in Figure 8b. The order of CO catalytic activity of catalysts is 10CuO-3Co_3O_4/α-MnO_2-400 (T_{90} = 75 °C) > 10CuO-3Co_3O_4/α-MnO_2-500 (T_{90} = 86 °C) > 10CuO-3Co_3O_4/α-MnO_2-200 (T_{90} = 89 °C) > 10CuO-3Co_3O_4/α-MnO_2-600 (T_{90} = 118 °C). It can be observed that catalytic activity has significantly decreased with the increase of the calcination temperature from 400 °C to 600 °C, especially over the 10CuO-3Co_3O_4/α-MnO_2-600 catalyst. Specifically, the 10CuO-3Co_3O_4/α-MnO_2-600 catalyst had low activity of CO oxidation in the low temperature region. The possible reason is that the formation of a copper manganese spinel in the catalyst after calcination at high temperatures led to the significant decrease in the CO adsorption and oxidation capacities of the catalysts. This result is consistent with the XRD analysis. The 10CuO-3Co_3O_4/α-MnO_2-200 catalyst was also prepared for comparison. The results show that the catalytic activity of the 10CuO-3Co_3O_4/α-MnO_2-200 catalyst was significantly lower than that of the 10CuO-3Co_3O_4/α-MnO_2-400 catalyst. The possible reason is that the precursor of the Co_3O_4 could not be completely decomposed at 200 °C.

2.6.4. Stability Tests

The 12 h stability measurement was conducted over the 10CuO-3Co_3O_4/α-MnO_2-400 catalyst under the specific conditions (80 °C, $CO/O_2/N_2$ = 1/20/79, GHSV = 12,000 $mL \cdot g^{-1} \cdot h^{-1}$, 1 atm), and the result is shown in Figure 9a. It can be observed that the initial activity of the 10CuO-3Co_3O_4/α-MnO_2-400 catalyst can achieve 100% CO conversion in the first 2 h. Then, the CO conversion gradually decreased from 100% to 80% in the subsequent 2 h test. This indicates that the catalyst began to deactivate. However, the CO conversion could remain stable in the subsequent 8 h. This suggests that the 10CuO-3Co_3O_4/α-MnO_2-400 catalyst was provided with relatively good stability to some degree.

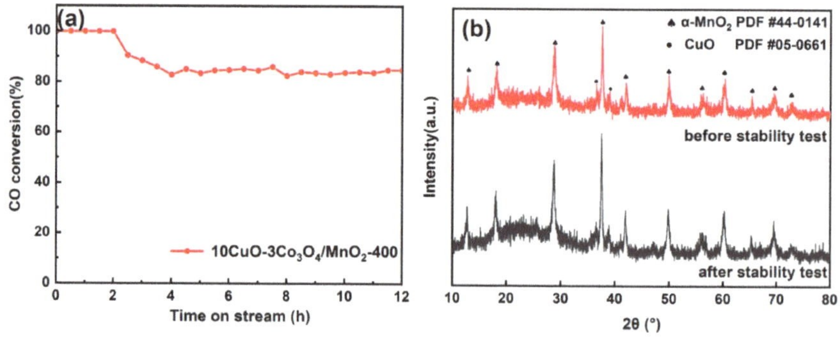

Figure 9. (**a**) The 12 h stability test of CO oxidation over the 10CuO-3Co_3O_4/α-MnO_2-400 catalyst under the conditions: 80 °C, $CO/O_2/N_2$ = 1/20/79, GHSV = 12,000 $mL \cdot g^{-1} \cdot h^{-1}$, 1 atm; (**b**) The XRD patterns of the fresh and the spent 10CuO-3Co_3O_4/α-MnO_2-400 catalysts before and after the stability test.

The XRD pattern of the 10CuO-3Co_3O_4/α-MnO_2-400 catalyst after the 12 h stability test is shown in Figure 9b. It can be observed that the diffraction peaks of the spent 10CuO-3Co_3O_4/α-MnO_2-400 catalyst after the 12 h stability test were a bit narrower and sharper than the fresh catalyst before the stability test. The possible reason is that the 10CuO-3Co_3O_4/α-MnO_2-400 catalyst underwent a bit of thermal agglomeration of the CuO active sites and the α-MnO_2 nanowire support during the catalytic process due to the hot spots of the catalyst bed, which could partly reduce the stability of the catalyst.

3. Discussion

The oxidation of CO on Cu-doped MnO_2 follows the Mars–van Krevelen (MvK) mechanism [35]. The reaction is divided into two parts: CO is first adsorbed on the catalyst surface and then reacts with surface-active oxygen on the catalyst surface to produce CO_2, which then generates oxygen vacancies on the catalyst surface. O_2 replenishes the depleted surface-active oxygen. After these two processes, the reaction completes a cycle [59]. There are redox electron pairs in the Cu-doped MnO_2 catalyst: -Cu^{2+}-O^{2-}-Mn^{4+}-→-Cu^{+}-□-Mn^{3+}- + $1/2O_2$ (□ indicates the oxygen vacancy) [45]. The content of Mn^{3+} on the MnO_2 catalyst is higher, presumably with more oxygen vacancies on the MnO_2 [60].

The order of the oxygen vacancy content of the catalysts is 10CuO-3Co_3O_4/α-MnO_2 > 10CuO-3NiO/α-MnO_2 > 10CuO/α-MnO_2 > 10CuO-3Fe_2O_3/α-MnO_2. The oxygen vacancy content is consistent with the Mn^{3+} content and catalyst activity. O_2 is commonly activated near the oxygen vacancy, producing surface active oxygen species (O_{sur}) [61]. It is well known that the higher the surface oxygen vacancy, the more easily O_2 is activated to reactive oxygen species [50]. However, the XPS spectra of O 1s showed that the order of the O_{ads}/O_{latt} ratios was not consistent with the oxygen vacancy content and catalyst activity. To evaluate the intrinsic activity of the catalysts, the surface area normalized reaction rates are determined. The results of the surface area normalized reaction rates show that the loading of NiO has the greatest effect on the intrinsic activity of the catalyst. The 10CuO-3NiO/α-MnO_2 catalyst did not exhibit superior catalytic activity because the specific surface area of the catalyst after NiO loading was reduced, and the effect of specific surface area on the activity of the CO oxidation reaction could not be ignored. The 10CuO-3NiO/α-MnO_2 catalyst has the most surface adsorbed oxygen and reaction rates per unit surface area, but the small specific surface area results in the catalytic activity being similar to that of the pristine 10CuO/α-MnO_2-400 catalyst. The order of the intrinsic activity of the catalyst is consistent with the order of the O_{ads}/O_{latt} ratio (10CuO-3NiO/α-MnO_2 > 10CuO/α-MnO_2 > 10CuO-3Co_3O_4/α-MnO_2 > 10CuO-3Fe_2O_3/α-MnO_2). This suggests that the surface-adsorbed oxygen is the reactive oxygen species involved in the oxidation of CO. CO_2 was produced by CO reacting with surface-adsorbed oxygen species [62,63]. After loading different transition metals, the catalysts form different types of oxygen vacancies, which have different electron densities and affect the production of reactive oxygen species [39]. The oxygen vacancies of 10CuO-3Co_3O_4/α-MnO_2 did not activate O_2 as well, and the 10CuO-3Co_3O_4/α-MnO_2 catalyst did not form more surface adsorbed oxygen. This might be the cause of the 10CuO-3Co_3O_4/α-MnO_2 catalyst providing the highest oxygen vacancy but poor surface-adsorbed oxygen. Although the normalized reaction rate of the 10CuO-3Co_3O_4/α-MnO_2 catalyst is not the highest, its high specific surface area allows for a greater number of oxygen vacancies. The large number of oxygen vacancies of the 10CuO-3Co_3O_4/α-MnO_2 catalyst counteracted the low activity of the oxygen vacancies and performed the high catalytic activity of CO oxidation. Therefore, the 10CuO-3Co_3O_4/α-MnO_2 catalyst exhibited the highest activity owing to its maximum specific surface area. The activity of the 10CuO-3Co_3O_4/α-MnO_2 catalyst was slightly higher than that of the 10CuO-3NiO/α-MnO_2 catalyst.

4. Materials and Method

4.1. Synthesis of α-MnO_2 Nanowire Supports

The α-MnO_2 nanowire support was facilely prepared via the typical hydrothermal method. Typically, 2 mmol $KMnO_4$ (Sinopharm Chemical Reagent Co., Ltd., Shanghai, China, AR, >99.5%) and 3 mmol $MnSO_4 \cdot H_2O$ (Shanghai Aladdin Biochemical Technology Co., Ltd., Shanghai, China, AR, 99%) were absolutely dissolved in 40 mL deionized water and vigorously stirred for 5 min, respectively. Then, these two solutions were mixed together by adding the $KMnO_4$ solution into the $MnSO_4$ solution. After stirring for 30 min, the obtained suspension was transferred to the 100 mL Teflon-lined stainless-steel autoclave. The autoclave was kept at 160 °C for 12 h. The obtained precipitate after the hydrothermal reaction was separated by the centrifuge and washed with ethanol (Sinopharm Chemical Reagent Co., Ltd., Shanghai, China, AR) and deionized water for three times. Then, the final powder was dried in the 100 °C oven overnight after the centrifugation. The obtained α-MnO_2 was used as a support for subsequent experiments.

4.2. The Preparation of the Transition Metal Oxide (Fe_2O_3, Co_3O_4, NiO)-Doped CuO-Based Catalysts Supported on the α-MnO_2 Nanowire

The transition metal oxides promoted CuO-based α-MnO_2 nanowire supported catalysts were prepared by the deposition precipitation method as reported in our previous work [64]. The weight percentages of the CuO and transition metal oxides were controlled at x% and y% (x% = $m_{CuO}/(m_{CuO} + m_{MOx} + m_{support}) \times 100\%$; y% = $m_{MOx}/(m_{CuO} + m_{MOx} + m_{support}) \times 100\%$) by using the $Cu(NO_3)_2 \cdot 6H_2O$ (Shanghai Xinbao Fine Chemical Industry Factory, Shanghai, China, AR, >99.5%), $Fe(NO_3)_3 \cdot 9H_2O$ (Shanghai Macklin Bio-Chem Co., Ltd., Shanghai, China, AR, 98.5%), $Co(NO_3)_2 \cdot 6H_2O$ (Shanghai Macklin Bio-Chem Co., Ltd., Shanghai, China, AR, 99%), $Ni(NO_3)_2 \cdot 6H_2O$ (Sinopharm Chemical Reagent Co., Ltd., Shanghai, China, AR, 99%) as the precursors. For the specific procedure, the α-MnO_2 nanowire was firstly dispersed in a $Cu(NO_3)_2 \cdot 6H_2O$ and $Fe(NO_3)_3 \cdot 9H_2O/Co(NO_3)_2 \cdot 6H_2O/Ni(NO_3)_2 \cdot 6H_2O$ aqueous solution Then, the Na_2CO_3 (0.01M, Shanghai Ling Feng Chemical Reagent Co., Ltd., Shanghai, China, AR, >99.8%) solution was added by droplet into the above solution to adjust the pH to 8–9 under vigorously stirring. The resultant solution was stirred for 30 min and then kept still for 1 h. The solid powder was separated by filtration and washed by the deionized water. The obtained catalyst was dried at 120 °C in an oven overnight and then calcinated at 400 °C for 5 h with a heating rate of 2 °C/min. The α-MnO_2 nanowire-supported catalysts with 10 wt.% CuO and 3 wt.% MOx (MOx = Fe_2O_3, Co_3O_4, NiO) were denoted as the 10CuO-3MOx/α-MnO_2 (MOx = Fe_2O_3, Co_3O_4, NiO). Meanwhile, the loading amount of the Co_3O_4 (wt.%) was subsequently changed in the same way. The obtained catalysts with loading amounts of 10 wt.% CuO and y wt.% Co_3O_4 were denoted as the 10CuO-yCo_3O_4/α-MnO_2 (y = 0, 1, 3, 5, 7, 10). As for the influence of the calcination temperature, the catalysts with loading of the 10 wt.% CuO and 3 wt.% Co_3O_4 were calcinated at different temperatures and named 10CuO-3Co_3O_4/α-MnO_2-T (T = 200, 400, 500, 600 °C).

4.3. Catalyst Characterizations

The X-ray diffraction (XRD) patterns of the catalysts were recorded on an X-ray power diffractometer (XRD-6100) from the Shimadzu Corporation by using the Cu Kα rays (0.15046 nm), 40 KV tube voltage, and 40 mA tube current. The 2θ scanning range was from 10° to 80°, and the scanning speed was controlled at 3°/min. The scanning electron microscope (SEM) images were taken on an Apreo S Hivac instrument (Thermo Fisher Science, Waltham, MA, USA) with the accelerating voltage of 5 kV. The nitrogen physisorption was performed on an Autosorb-iQ-AG-MP instrument (Quantachrome, Boynton Beach, FL, USA) at liquid nitrogen temperature (−196 °C). The samples were degassed at 300 °C for 3 h to remove the surface-adsorbed water and impurities before the regular test. The specific surface areas of the catalysts were calculated by the Brunauer-Emmett-Teller (BET) method, and the pore size distribution and pore volume were calculated from the adsorption branch of the isotherm by the Barrett-Joyner-Halenda (BJH) method in the range of 0–1.0 P/P0. The X-ray photoelectron spectroscopy (XPS) measurements were performed on a Thermo Science K-Alpha + spectrometer (Thermo Fisher Science, Waltham, MA, USA). For the XPS measurement, the penetration depth of each catalyst was about 1–2 nm. The binding energy (BE) was calibrated by using C 1s = 284.8 eV as the standard. A fixed-bed reactor was used to conduct the hydrogen temperature-programmed reduction (H_2-TPR) experiment. The hydrogen consumption profile was recorded and analyzed with the online LC-D200 mass spectrometer (TILON GRP TECHNOLOGY LIMITED, Shanghai, China). The mixture of H_2 (1.2 mL/min) and Ar (23.7 mL/min) was introduced into the reactor. For each test, 100 mg of catalyst was loaded. After the hydrogen signal baseline line (*m/z* = 2) was stabilized, the H_2-TPR test was performed with a heating rate of 20 °C/min from room temperature to 800 °C.

4.4. Catalyst Evaluation

The activity of the catalysts was evaluated in a fixed-bed reactor equipped with a quartz tube (i.d. = 8.00 mm). For each test, 100 mg catalyst was placed in the center of the quartz tube. The reactant gases, with a composition of 1% CO, 21% O_2, 79% N_2 (20 mL/min), were introduced into the reactor with the gas hourly space velocity (GHSV) of 12,000 $mL \cdot g^{-1} \cdot h^{-1}$. The catalytic activities of CO oxidation over different catalysts were tested in the specified temperature range. Each catalyst was tested three times at each temperature. The outlet gases were analyzed online by using the GC-680 Perkin Elmer gas chromatograph equipped with the thermal conductivity detector (TCD). The 24 h stability tests of catalysts were carried out at 80 °C with the GHSV of 12,000 $mL\cdot g^{-1}\cdot h^{-1}$.

The catalytic performances of CO oxidation over these catalysts were stated in the form of the CO conversion. The calculated formula is listed below:

$$X_{CO} = (C_{CO,\,Inlet} - C_{CO,\,Outlet})/C_{CO,\,Inlet} \times 100\% \quad (1)$$

where X_{CO} is the conversion rate of CO; and $C_{CO,\,Inlet}$ (ppm), and $C_{CO,\,Outlet}$ (ppm) are CO flowing into and out of the reactor, respectively.

In order to evaluate the intrinsic rate of CO oxidation on these catalysts, the calculated formula of the specific surface area normalization reaction rate is listed below [25]:

$$r_{norm}\left(mol\cdot m^{-2}\cdot s^{-1}\right) = \frac{C_{inlet}\cdot F}{m_{cat}\cdot S_{BET}}\cdot \ln\left(\frac{1}{1-X_{CO}}\right) \quad (2)$$

where r_{norm} ($mol\cdot m^{-2}\cdot s^{-1}$) is the normalized reaction rate, F ($mol\cdot s^{-1}$) is the CO flow rate, m_{cat} (g) is the mass of catalyst, and S_{BET} ($m^2\cdot g^{-1}$) is the BET surface area.

5. Conclusions

In this work, the novel α-MnO_2 nanowire was prepared by the one-step hydrothermal method. A series of transition metal oxides (Fe_2O_3, Co_3O_4, NiO) promoted the CuO-based catalyst supported on the α-MnO_2 nanowire and were synthesized by the co-precipitation method for the CO oxidation reaction. The effects of the transition metal oxide type, the loading amount, and the calcination temperature on the CO oxidation reaction had been systematically investigated. It was found that the 10CuO-3Co_3O_4/α-MnO_2-400 catalyst showed the highest reactivity with T_{90} = 75 °C. It was found that the 10CuO-3Co_3O_4/α-MnO_2-400 catalyst possessed the largest specific surface area and exposed more active sites, which could further enhance the catalytic activity. The 10CuO-3NiO/α-MnO_2-400 catalyst had the highest surface-adsorbed oxygen content and normalized reaction rate. This indicated that the surface-adsorbed oxygen was the surface-active oxygen involved in the oxidation of CO. It was found that the 10CuO-3Co_3O_4/α-MnO_2-400 catalyst suffered from some deactivation during the 12 h stability test, which might be caused by the thermal sintering and agglomeration of the CuO active sites and α-MnO_2 nanowire support. This should be the key consideration when designing the CuO-based CO oxidation catalyst in the future.

Author Contributions: Investigation, H.Z., H.S., Y.Z., Y.C., C.-e.W. and J.Q.; formal analysis, H.Z., H.S., Y.Z., Y.C., Y.X., C.-e.W. and J.X.; conceptualization, H.Z., Y.Z., H.S. and J.Q.; writing—original draft preparation, Y.Z. and L.X.; writing—review and editing, C.P., L.X. and M.C.; supervision, C.P., L.X. and M.C.; funding acquisition, L.X. and M.C.; project administration, L.X. and M.C. All authors have read and agreed to the published version of the manuscript.

Funding: This research was funded by the National Natural Science Foundation of China (Grant Nos. 22276098, 21976094, and 22176100), the National Key Research and Development Project (Grant No.2018YFC0213802), the Jiangsu Province "Carbon Peak and Carbon Neutrality Science and Technology Innovation Special Fund (The Third Batch)—Industry Foresight and Key Core Technology Research (Grant No. BE2022033-2), and the Postgraduate Research and Practice Innovation Program of Jiangsu Province (SJCX22_0367 and KYCX22_1216).

Data Availability Statement: The data presented in this study are available on request from the corresponding author. The data are not publicly available due to privacy.

Conflicts of Interest: The authors declare no conflict of interest.

References

1. Schnitzhofer, R.; Beauchamp, J.; Dunkl, J.; Wisthaler, A.; Weber, A.; Hansel, A. Long-term measurements of CO, NO, NO_2, benzene, toluene and PM10 at a motorway location in an Austrian valley. *Atmos. Environ.* **2008**, *42*, 1012–1024. [CrossRef]
2. Kunkalekar, R.K.; Salker, A.V. Low temperature carbon monoxide oxidation over nanosized silver doped manganese dioxide catalysts. *Catal. Commun.* **2010**, *12*, 193–196. [CrossRef]
3. Zheng, B.; de Beurs, K.M.; Owsley, B.C.; Henebry, G.M. Scaling relationship between CO pollution and population size over major US metropolitan statistical areas. *Landsc. Urban Plan.* **2019**, *187*, 191–198. [CrossRef]
4. Wu, X.; Lang, J.; Sun, Z.; Jin, F.; Hu, Y.H. Photocatalytic conversion of carbon monoxide: From pollutant removal to fuel production. *Appl. Catal. B Environ.* **2021**, *295*, 120312. [CrossRef]
5. Yoo, S.; Lee, E.; Jang, G.H.; Kim, D.H. Effect of Pd precursors on the catalytic properties of Pd/CeO2 catalysts for CH4 and CO oxidation. *Mol. Catal.* **2022**, *533*, 112791. [CrossRef]
6. Schubert, M.M.; Hackenberg, S.; van Veen, A.C.; Muhler, M.; Plzak, V.; Behm, R.J. CO Oxidation over Supported Gold Catalysts—"Inert" and "Active" Support Materials and Their Role for the Oxygen Supply during Reaction. *J. Catal.* **2001**, *197*, 113–122. [CrossRef]
7. Min, B.K.; Friend, C.M. Heterogeneous Gold-Based Catalysis for Green Chemistry: Low-Temperature CO Oxidation and Propene Oxidation. *ChemInform* **2007**, *107*, 2709–2724. [CrossRef]
8. Qiao, B.; Wang, A.; Yang, X.; Allard, L.F.; Jiang, Z.; Cui, Y.; Liu, J.; Li, G.; Zhang, T. Single-atom catalysis of CO oxidation using Pt_1/FeOx. *Nat. Chem.* **2011**, *3*, 634–641. [CrossRef] [PubMed]
9. Chen, G.; Zhao, Y.; Fu, G.; Duchesne, P.; Gu, L.; Zheng, Y.; Weng, X.; Chen, M.; Pao, C. Interfacial effects in iron-nickel hydroxide-platinum nanoparticles enhance catalytic oxidation. *Science* **2014**, *344*, 495–499. [CrossRef]
10. Liu, K.; Wang, A.; Zhang, T. Recent Advances in Preferential Oxidation of CO Reaction over Platinum Group Metal Catalysts. *ACS Catal.* **2012**, *2*, 1165–1178. [CrossRef]
11. Zhang, L.; Liu, H.; Huang, X.; Sun, X.; Jiang, Z.; Schlögl, R.; Su, D. Stabilization of Palladium Nanoparticles on Nanodiamond-Graphene Core-Shell Supports for CO Oxidation. *Angew. Chem.* **2015**, *127*, 16049–16052. [CrossRef]
12. Gonzalez-A, E.; Rangel, R.; Solís-Garcia, A.; Venezia, A.M.; Zepeda, T.A. FTIR investigation under reaction conditions during CO oxidation over Ru(x)-CeO_2 catalysts. *Mol. Catal.* **2020**, *493*, 111086. [CrossRef]
13. Yadav, P.K.; Kumari, S.; Naveena, U.; Deshpande, P.A.; Sharma, S. Insights into the substitutional chemistry of $La_{1-x}Sr_xCo_{1-y}MyO_3$ (M = Pd, Ru, Rh, and Pt) probed by in situ DRIFTS and DFT analysis of CO oxidation. *Appl. Catal. A Gen.* **2022**, *643*, 118768. [CrossRef]
14. Bunluesin, T.; Cordatos, H.; Gorte, R.J. Study of CO Oxidation Kinetics on Rh/Ceria. *J. Catal.* **1995**, *157*, 222–226. [CrossRef]
15. Dey, S.; Dhal, G.C. Applications of silver nanocatalysts for low-temperature oxidation of carbon monoxide. *Inorg. Chem. Commun.* **2019**, *110*, 107614. [CrossRef]
16. Du, Y.; Meng, Q.; Wang, J.; Yan, J.; Fan, H.; Liu, Y.; Dai, H. Three-dimensional mesoporous manganese oxides and cobalt oxides: High-efficiency catalysts for the removal of toluene and carbon monoxide. *Micropor. Mesopor. Mat.* **2012**, *162*, 199–206. [CrossRef]
17. Xiao, M.; Zhang, X.; Yang, Y.; Cui, X.; Chen, T.; Wang, Y. M (M = Mn, Co, Cu)-CeO_2 catalysts to enhance their CO catalytic oxidation at a low temperature: Synergistic effects of the interaction between Ce^{3+}-M^{x+}-$Ce^{4+}$and the oxygen vacancy defects. *Fuel* **2022**, *323*, 124379. [CrossRef]
18. Bi, F.; Zhang, X.; Xiang, S.; Wang, Y. Effect of Pd loading on ZrO_2 support resulting from pyrolysis of UiO-66: Application to CO oxidation. *J. Colloid Interf. Sci.* **2020**, *573*, 11–20. [CrossRef] [PubMed]
19. Dey, S.; Dhal, G.C. Catalytic Conversion of Carbon Monoxide into Carbon Dioxide over Spinel Catalysts: An Overview. *Mater. Sci. Energy Technol.* **2019**, *2*, 575–588. [CrossRef]
20. Kaplin, I.Y.; Lokteva, E.S.; Golubina, E.V.; Shishova, V.V.; Maslakov, K.I.; Fionov, A.V.; Isaikina, O.Y.; Lunin, V.V. Efficiency of manganese modified CTAB-templated ceria-zirconia catalysts in total CO oxidation. *Appl. Surf. Sci.* **2019**, *485*, 432–440. [CrossRef]
21. Niu, X.; Lei, Z. Copper doped manganese oxides to produce enhanced catalytic performance for CO oxidation. *J. Environ. Chem. Engi.* **2019**, *7*, 103055. [CrossRef]
22. Liu, H.; Li, X.; Dai, Q.; Zhao, H.; Chai, G.; Guo, Y.; Guo, Y.; Wang, L.; Zhan, W. Catalytic oxidation of chlorinated volatile organic compounds over Mn-Ti composite oxides catalysts: Elucidating the influence of surface acidity. *Appl. Catal. B Environ.* **2021**, *282*, 119577. [CrossRef]
23. Ye, Z.; Giraudon, J.-M.; Nuns, N.; Simon, P.; De Geyter, N.; Morent, R.; Lamonier, J.-F. Influence of the preparation method on the activity of copper-manganese oxides for toluene total oxidation. *Appl. Catal. B Environ.* **2018**, *223*, 154–166. [CrossRef]
24. Lin, Y.; Tian, H.; Qian, J.; Yu, M.; Hu, T.; Lassi, U.; Chen, Z.; Wu, Z. Biocarbon-directed vertical δ-MnO2 nanoflakes for boosting lithium-ion diffusion kinetics. *Mater. Today Chem.* **2022**, *26*, 101023. [CrossRef]
25. Yang, W.; Su, Z.; Xu, Z.; Yang, W.; Peng, Y.; Li, J. Comparative study of α-, β-, γ- and δ-MnO2 on toluene oxidation: Oxygen vacancies and reaction intermediates. *Appl. Catal. B Environ.* **2019**, *260*, 118150. [CrossRef]
26. Hayashi, E.; Yamaguchi, Y.; Kamata, K.; Tsunoda, N.; Kumagai, Y.; Oba, F.; Hara, M. Effect of MnO2 Crystal Structure on Aerobic Oxidation of 5-Hydroxymethylfurfural to 2,5-Furandicarboxylic Acid. *J. Am. Chem. Soc.* **2019**, *141*, 890–900. [CrossRef] [PubMed]
27. Robinson, D.M.; Go, Y.B.; Mui, M.; Gardner, G.; Zhang, Z.; Mastrogiovanni, D.; Garfunkel, E.; Li, J.; Greenblatt, M.; Dismukes, G.C. Photochemical Water Oxidation by Crystalline Polymorphs of Manganese Oxides: Structural Requirements for Catalysis. *J. Am. Chem. Soc.* **2013**, *135*, 3494–3501. [CrossRef] [PubMed]

28. Liang, S.; Teng, F.; Bulgan, G.; Zong, R.; Zhu, Y. Effect of Phase Structure of MnO2 Nanorod Catalyst on the Activity for CO Oxidation. *J. Phys. Chem. C* **2008**, *112*, 5307–5315. [CrossRef]
29. Tian, F.; Li, H.; Zhu, M.; Tu, W.; Lin, D.; Han, Y. Effect of MnO_2 Polymorphs' Structure on Low-Temperature Catalytic Oxidation: Crystalline Controlled Oxygen Vacancy Formation. *ACS Appl. Mater. Interfaces* **2022**, *14*, 18525–18538. [CrossRef] [PubMed]
30. Yang, R.; Fan, Y.; Ye, R.; Tang, Y.; Cao, X.; Yin, Z.; Zeng, Z. MnO_2-Based Materials for Environmental Applications. *Adv. Mater.* **2021**, *33*, 2004862. [CrossRef] [PubMed]
31. Jokar, R.; Alavi, S.M.; Rezaei, M.; Akbari, E. Catalytic performance of copper oxide supported α-MnO_2 nanowires for the CO preferential oxidation in H_2-rich stream. *Int. J. Hydrogen Energy* **2021**, *46*, 32503–32513. [CrossRef]
32. Qian, K.; Qian, Z.; Hua, Q.; Jiang, Z.; Huang, W. Structure–activity relationship of CuO/MnO_2 catalysts in CO oxidation. *Appl. Surf. Sci.* **2013**, *273*, 357–363. [CrossRef]
33. Sun, R.L.; Zhang, S.R.; An, K.; Song, P.F.; Liu, Y. $Cu_{1.5}Mn_{1.5}O_4$ spinel type composite oxide modified with CuO for synergistic catalysis of CO oxidation. *J. Fuel Chem. Technol.* **2021**, *49*, 799–808. [CrossRef]
34. Sophiana, I.C.; Topandi, A.; Iskandar, F.; Devianto, H.; Nishiyama, N.; Budhi, Y.W. Catalytic oxidation of benzene at low temperature over novel combination of metal oxide based catalysts: CuO, MnO_2, NiO with $Ce_{0.75}Zr_{0.25}O_2$ as support. *Mater. Today Chem.* **2020**, *17*, 100305. [CrossRef]
35. Gao, J.; Jia, C.; Zhang, L.; Wang, H.; Yang, Y.; Hung, S.-F.; Hsu, Y.Y.; Liu, B. Tuning chemical bonding of MnO_2 through transition-metal doping for enhanced CO oxidation. *J. Catal.* **2016**, *341*, 82–90. [CrossRef]
36. Ivanov, K.I.; Kolentsova, E.N.; Dimitrov, D.Y.; Petrova, P.T.; Tabakova, T.T. Alumina Supported Cu-Mn-Cr Catalysts for CO and VOCs oxidation. *Int. Sch. Sci. Res. Innov.* **2015**, *9*, 605–612.
37. Krämer, M.; Schmidt, T.; Stöwe, K.; Maier, W.F. Structural and catalytic aspects of sol–gel derived copper manganese oxides as low-temperature CO oxidation catalyst. *Appl. Catal. A Gen.* **2006**, *302*, 257–263. [CrossRef]
38. Huang, N.; Qu, Z.; Dong, C.; Qin, Y.; Duan, X. Superior performance of α@β-MnO_2 for the toluene oxidation: Active interface and oxygen vacancy. *Appl. Catal. A Gen.* **2018**, *560*, 195–205. [CrossRef]
39. Wang, Y.; Wu, J.; Wang, G.; Yang, D.; Ishihara, T.; Guo, L. Oxygen vacancy engineering in Fe doped akhtenskite-type MnO_2 for low-temperature toluene oxidation. *Appl. Catal. B Environ.* **2021**, *285*, 119873. [CrossRef]
40. Rosso, J.J.; Hochella, M.F. Natural Iron and Manganese Oxide Samples by XPS. *Surf. Sci. Spectra* **1996**, *4*, 253–265. [CrossRef]
41. Li, J.; Zhu, P.; Zuo, S.; Huang, Q.; Zhou, R. Influence of Mn doping on the performance of CuO-CeO_2 catalysts for selective oxidation of CO in hydrogen-rich streams. *Appl. Catal. A Gen.* **2010**, *381*, 261–266. [CrossRef]
42. Michael, A.S. Mn_2O_3 by XPS. *Surf. Sci. Spectra* **1999**, *6*, 39–46. [CrossRef]
43. Liu, W.; Wang, S.; Cui, R.; Song, Z.; Zhang, X. Enhancement of catalytic combustion of toluene over $CuMnO_x$ hollow spheres prepared by oxidation method. *Micropor. Mesopor. Mat.* **2021**, *326*, 111370. [CrossRef]
44. Wang, Y.; Liu, K.; Wu, J.; Hu, Z.; Huang, L.; Zhou, J.; Ishihara, T.; Guo, L. Unveiling the Effects of Alkali Metal Ions Intercalated in Layered MnO_2 for Formaldehyde Catalytic Oxidation. *ACS Catal.* **2020**, *10*, 10021–10031. [CrossRef]
45. Wang, Y.; Yang, D.; Li, S.; Zhang, L.; Zheng, G.; Guo, L. Layered copper manganese oxide for the efficient catalytic CO and VOCs oxidation. *Chem. Eng. J.* **2019**, *357*, 258–268. [CrossRef]
46. Zeng, J.; Xie, H.; Zhang, H.; Huang, M.; Liu, X.; Zhou, G.; Jiang, Y. Insight into the effects of oxygen vacancy on the toluene oxidation over α-MnO_2 catalyst. *Chemosphere* **2022**, *291*, 135890. [CrossRef] [PubMed]
47. Xu, J.; Wu, W.; Gao, E.; Zhu, J.; Yao, S.; Li, J. Revealing the role of oxygen vacancies on α-MnO_2 of different morphologies in CO oxidation using operando DRIFTS-MS. *Appl. Surf. Sci.* **2023**, *618*, 156643. [CrossRef]
48. Liu, W.; Xiang, W.; Chen, X.; Song, Z.; Gao, C.; Tsubaki, C.; Zhang, X. A novel strategy to adjust the oxygen vacancy of CuO/MnO_2 catalysts toward the catalytic oxidation of toluene. *Fuel* **2022**, *312*, 122975. [CrossRef]
49. Yang, W.; Zhu, Y.; You, F.; Yan, L.; Ma, Y.; Lu, C.; Gao, P.; Hao, Q.; Li, W. Insights into the surface-defect dependence of molecular oxygen activation over birnessite-type MnO_2. *Appl. Catal. B Environ.* **2018**, *233*, 184–193. [CrossRef]
50. Mo, S.; Zhang, Q.; Li, J.; Sun, Y.; Ren, Q.; Zou, S.; Zhang, Q.; Lu, J.; Fu, M.; Mo, D.; et al. Highly efficient mesoporous MnO_2 catalysts for the total toluene oxidation: Oxygen-Vacancy defect engineering and involved intermediates using in situ DRIFTS. *Appl. Catal. B Environ.* **2020**, *264*, 118464. [CrossRef]
51. Wang, J.; Li, J.; Jiang, C.; Zhou, P.; Zhang, P.; Yu, J. The effect of manganese vacancy in birnessite-type MnO_2 on room-temperature oxidation of formaldehyde in air. *Appl. Catal. B Environ.* **2017**, *204*, 147–155. [CrossRef]
52. Papavasiliou, J.; Avgouropoulos, G.; Ioannides, T. Combined steam reforming of methanol over Cu–Mn spinel oxide catalysts. *J. Catal.* **2007**, *251*, 7–20. [CrossRef]
53. Liu, T.; Yao, Y.; Wei, L.; Shi, Z.; Han, L.; Yuan, H.; Li, B.; Dong, L.; Wang, F.; Sun, C. Preparation and Evaluation of Copper–Manganese Oxide as a High-Efficiency Catalyst for CO Oxidation and NO Reduction by CO. *J. Phys. Chem. C* **2017**, *121*, 12757–12770. [CrossRef]
54. Yuan, H.L.; Wang, Y.Q.; Zhou, S.M.; Liu, L.S.; Chen, X.L.; Lou, S.Y.; Yuan, R.J.; Hao, Y.M.; Li, N. Low-Temperature Preparation of Superparamagnetic $CoFe_2O_4$ Microspheres with High Saturation Magnetization. *Nanoscale Res. Lett.* **2010**, *5*, 1817–1821. [CrossRef] [PubMed]
55. Wu, Y.; Guo, Q.; Liu, H.; Wei, S.; Wang, L. Effect of Fe doping on the surface properties of δ-MnO_2 nanomaterials and its decomposition of formaldehyde at room temperature. *J. Environ. Chem. Eng.* **2022**, *10*, 108277. [CrossRef]

56. Yadav, R.S.; Havlica, J.; Masilko, J.; Kalina, L.; Wasserbauer, J.; Hajdúchová, M.; Enev, V.; Kuřitka, I.; Kožáková, Z. Impact of Nd^{3+} in $CoFe_2O_4$ spinel ferrite nanoparticles on cation distribution, structural and magnetic properties. *J. Magn. Magn. Mater.* **2016**, *399*, 109–117. [CrossRef]
57. Hoang, M.; Hughes, A.E.; Turney, T.W. An XPS study of Ru-promotion for Co/CeO_2 Fischer-Tropsch catalyst. *Appl. Surf. Sci.* **1993**, *72*, 55–65. [CrossRef]
58. Oh, Y.S.; Roh, H.S.; Jun, K.W.; Baek, Y.S. A highly active catalyst, $Ni/Ce–ZrO_2/\theta$-Al_2O_3, for on-site H_2 generation by steam methane reforming: Pretreatment effect. *Int. J. Hydrogen Energy* **2003**, *28*, 1387–1392. [CrossRef]
59. Hu, X.; Chen, J.; Li, S.; Chen, Y.; Qu, W.; Ma, Z.; Tang, X. The Promotional Effect of Copper in Catalytic Oxidation by Cu-Doped α-MnO_2 Nanorods. *J. Phys. Chem. C* **2019**, *124*, 701–708. [CrossRef]
60. Dong, C.; Wang, H.; Ren, Y.; Qu, Z. Effect of alkaline earth metal promoter on catalytic activity of MnO_2 for the complete oxidation of toluene. *J. Environ. Sci.* **2021**, *104*, 102–112. [CrossRef] [PubMed]
61. Feng, C.; Jiang, F.; Xiong, G.; Chen, C.; Wang, Z.; Pan, Y.; Fei, Z.; Lu, Y.; Li, X.; Zhang, R.; et al. Revelation of Mn^{4+}-O_{sur}-Mn^{3+} active site and combined Langmuir-Hinshelwood mechanism in propane total oxidation at low temperature over MnO_2. *Chem. Eng. J.* **2023**, *451*, 138868. [CrossRef]
62. Park, J.H.; Kang, D.C.; Park, S.J.; Shin, C.H. CO oxidation over MnO_2 catalysts prepared by a simple redox method: Influence of the Mn (II) precursors. *J. Ind. Eng. Chem.* **2015**, *25*, 250–257. [CrossRef]
63. Du, C.; Zhao, S.; Tang, X.; Yu, Q.; Gao, F.; Zhou, Y.; Liu, J.; Yi, H. Synthesis of ultra-thin 2D MnO_2 nanosheets rich in oxygen vacancy defects and the catalytic oxidation of n-heptane. *Appl. Surf. Sci.* **2022**, *606*, 154846. [CrossRef]
64. Cui, Y.; Song, H.; Shi, Y.; Ge, P.X.; Chen, M.D.; Xu, L.L. Enhancing the Low-Temperature CO Oxidation over CuO-Based α-MnO_2 Nanowire Catalysts. *Nanomaterials* **2022**, *12*, 2083. [CrossRef] [PubMed]

catalysts

Article

Molten Salt-Assisted Catalytic Preparation of MoS_2/α-MoO_3/Graphene as High-Performance Anode of Li-Ion Battery

Wenhui Zhu and Ali Reza Kamali *

Energy and Environmental Materials Research Centre (E2MC), School of Metallurgy, Northeastern University, Shenyang 110819, China
* Correspondence: ali@mail.neu.edu.cn

Abstract: We report on the facile and scalable catalytic conversion of natural graphite and MoS_2 minerals into α-MoO_3 nanoribbons incorporated into hexagonal MoS_2 and graphene nanosheets, and evaluate the structural, morphological and electrochemical performances of the hybrid nanostructured material obtained. Mechanochemical treatment of raw materials, followed by catalytic molten salt treatment leads to the formation of nanostructures with promising electrochemical performances. We examined the effect of processing temperature on the electrochemical performance of the products. At 1100 °C, an excellent Li-ion storage capacity of 773.5 mAh g^{-1} is obtained after 180 cycles, considerably greater than that of MoS_2 (176.8 mAh g^{-1}). The enhanced capacity and the rate performance of this electrode are attributed to the well-integrated components, characterized by the formation of interfacial molybdenum oxycarbide layer during the synthesis process, contributing to the reduced electrical/electrochemical resistance of the sample. This unique morphology promotes the charge and ions transfer through the reduction of the Li-ion diffusion coefficient (1.2×10^{-18} cm^2 s^{-1}), enhancing the pseudocapacitive performance of the electrode; 59.3% at the scan rate of 0.5 mV s^{-1}. This article provides a green and low-cost route to convert highly available natural graphite and MoS_2 minerals into nanostructured hybrid materials with promising Li-ion storage performance.

Keywords: natural graphite; MoS_2; molten NaCl; graphene; MoO_3; molybdenum oxycarbide

Citation: Zhu, W.; Kamali, A.R. Molten Salt-Assisted Catalytic Preparation of MoS_2/α-MoO_3/Graphene as High-Performance Anode of Li-Ion Battery. *Catalysts* **2023**, *13*, 499. https://doi.org/10.3390/catal13030499

Academic Editors: Georgios Bampos, Athanasia Petala and Zacharias Frontistis

Received: 6 February 2023
Revised: 24 February 2023
Accepted: 27 February 2023
Published: 28 February 2023

1. Introduction

The development of efficient and low-cost energy storage devices is a key towards the electrification of various sectors, including the transportation and grid services [1–3], supporting the sustainable development goals. Lithium-ion battery (LIB) is the state of the art energy storage device in a wide variety applications and, therefore, its modifications in terms of performance, cost and availability of raw materials are of great importance [4–6] Graphite, including synthetic graphite (SG) and modified natural graphite (NG) is a commonly used material for the fabrication of the anode of LIBs, due to its modest theoretical capacity of 372 mAh g^{-1} and high cycle stability [7,8]. SG is made by the graphitization of carbonaceous material at extremely high temperatures (≈3000 °C) [9], which is an extremely energy-intensive approach. NG originating from carbon-rich organics formed through the prolonged geological interactions [9] is considered as a promising alternative anode material mainly due to its high availability and low cost [10–12]. In fact, the world's NG reserves are huge, exceeding 800 million tons of recoverable graphite [13], with a typical low price of around USD 750 per metric ton (≈94% purity) [11] making NG highly attractive for energy applications. Despite these advantages, however, NG minerals often undergo expensive, energy intensive and time-consuming processes, including high-temperature treatments (>2300 °C), and/or treatment with hazardous chemicals such as HF [13–15]. Therefore, the direct application of NG minerals for the fabrication of anode materials without going through expensive/environmentally problematic processes is required in order to appropriately utilize NG for future developments of LIBs.

Apart from NG, other naturally available materials such molybdenum disulfide (MoS_2) have also attracted attentions due to their high availability and low-cost. In particular, the theoretical Li-ion storage capacity of MoS_2 is approximately three times greater than that of commercial graphite anodes [16,17]. Despite its potential capability, the application of MoS_2 as the anode of LIBs is challenging due to the low reversible capacity of the material at large cycle numbers. In contrast to MoS_2, other molybdenum compounds such as molybdenum trioxide (MoO_3) have also been evaluated as the anode of LIBs. However, the cycling performance of MoO_3 is not satisfactory due to the low conductivity of the compound, and its disintegration during discharge/charge cycles [18–20]. These issues might be reduced by combining MoO_3 and MoS_2 so that the latter can enhance the conductivity of the nanocomposite and prevents the rapid disintegration [21–23]. The methods suggested for the synthesis of MoS_2/MoO_3 nanocomposites comprise of chemical vapor deposition [24], hydrothermal methods [25,26], in-situ growth [23], sulfur transfer [27], anion-exchange [28], and annealing treatment [29] using the precursor materials such as $(NH_4)_6Mo_7O_{24}$ and thiourea [23,28], organic amine, and MoO_3 [24], Mo metal powder and H_2O_2 [27] as well as MoO_3 powders, sulfur and N_2 [29]. These synthesis methods are typically complicated and require long processing periods involving expensive and/or environmentally problematic raw materials, limiting their capability at large scales.

With this background, the clean and facile preparation MoS_2/MoO_3 nanocomposites with enhanced Li-ion storage performance using low-cost and highly available minerals is an interesting goal. In a recent work, we suggested the mechanochemical–molten salt treatment of pure synthetic graphite with MoS_2 as a possible green way of producing nanostructures with promising Li-ion storage performance [30]. However, the application of pure synthetic graphite would greatly influence the economic and sustainability of the process. In this study, we show that the wet high-energy ball-milling of commercial MoS_2, non-purified natural graphite and NaCl, followed by a heat-treatment at temperatures above the melting point of NaCl leads to the formation of well-integrated nanostructured hybrid materials, in which MoO_3 nanoribbons are incorporated with hexagonal MoS_2 and graphene nanosheets. We further study on the effect of heating regime on the microstructural evolution of the nanocomposite materials, and find out that at 1100 °C, the nanocomposite contains the highest amount of well-integrated α-MO_3 nanoribbons providing a promising Li-ion storage performance of 773.5 mAh g^{-1} after 180 charge/discharge cycles, considerably greater than that of MoS_2 (176.8 mAh g^{-1}). We suggest the formation of interfacial molybdenum oxycarbide that reduces the internal impedance, and promotes the ion/electron transfer within the nanostructured material, and its interface with the electrolyte. Furthermore, this article reports on the molten salt synthesis of hexagonal MoS_2 nanosheets, which are well incorporated into the nanocomposite and contribute to the high performance of the electrode by improving the electron and ion transportation across the material. The formation of hexagonal MoS_2 nanosheets, presented here, is in contrast with alternative techniques reported in the literature for the fabrication of such MoS_2 morphologies based on prolonged sulfurization of $(NH_4)_2MoS_4$ using H_2S at 800 °C [31,32]. Other techniques reported on the preparation of hexagonal WS_2 nanoflakes include thermal conversion of W_xO_y nanorods in the presence of S at 750 °C (6 h) under high-vacuum [33], and the electrochemical exfoliation of WS_2 in Na_2SO_4 [34]. Also, the preparation of hexagonal nanosheets of CdI_2 [35] and SnS_2 [36] have been reported using chemical vapor deposition and steam vapor etching. In this paper, we report on the facile preparation of hexagonally-shaped MoS_2 nanosheets incorporated with MoO_3 nanoribbons and graphene nanosheets (MoO_3/MoS_2@Graphene) by a simple mechanochemical–molten salt approach using natural graphite and MoS_2 minerals, with promising Li-ion storage performance. The mechanism involved in the preparation and the electrochemical performance of the nanostructured materials are investigated.

2. Results

2.1. Preparation of Materials

In this work, MoO_3/MoS_2@Graphene samples were prepared using natural graphite mineral without conducting extensive modification/purification processes which are often performed to fabricate battery-grade materials. The details of the process are mentioned in "Materials and Methods" and summarized in Figure 1.

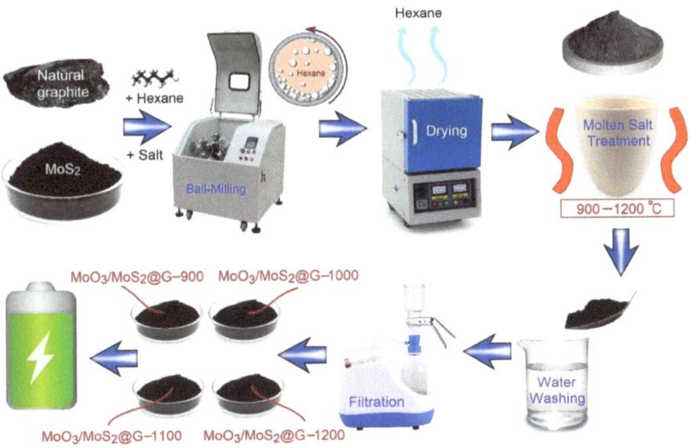

Figure 1. The process employed to convert natural graphite mineral and commercially available MoS_2 into nanostructured materials for energy-storage application.

2.2. Characterization of the Natural Graphite Mineral

It is useful to provide insights on to the chemical and microchemical composition of the NG used as the raw material in this research. Figure 2 shows the SEM/EDS element mapping analysis recorded on the sample. As can be seen from Figure 2a, the material contains graphitic flakes with lateral dimensions typically larger than 20 µm. According to the EDS analysis, in addition to carbon, other elements comprising Ca, Fe, O, S, Si, Al, and K could also be detected, indicating the impurity of the natural graphite material.

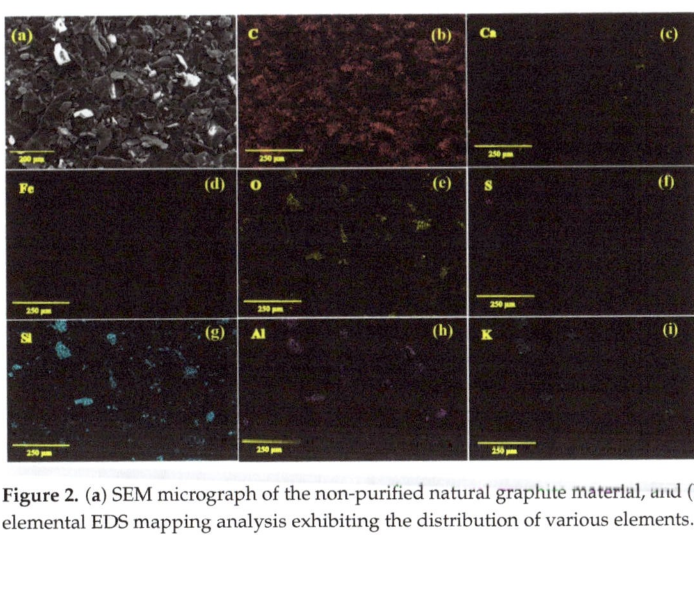

Figure 2. (a) SEM micrograph of the non-purified natural graphite material, and (b–i) corresponding elemental EDS mapping analysis exhibiting the distribution of various elements.

The presence of oxide impurities can clearly be recognized from the EDS analysis. The data obtained could be confirmed by the X-ray fluorescence (XRF) quantitative analysis recorded on the natural graphite sample, as exhibited in Table 1. The analysis performed shows that the carbon content of the material is 74.3 wt%. Other components include SiO_2 (9.27 wt%.), Al_2O_3 (6.92% wt%), CaO (2.97 wt%), Fe_2O_3 (2.29 wt%), K_2O (2.17 wt%) and S (2.08 wt%). The natural graphite mineral was employed for the synthesis of MoO_3/MoS_2@graphene samples without purification, as explained in the next section.

Table 1. Quantitative XRF analysis of the natural graphite material (wt%).

C	SiO_2	Al_2O_3	CaO	Fe_2O_3	K_2O	S
74.3	9.27	6.92	2.97	2.29	2.17	2.08

2.3. *Catalytic Phase Transitions during the Molten Salt Treatment*

MoS_2 and natural graphite precursors were subjected to a ball-milling process (2 h) in the presence of hexane, and the BMed samples were heat treated with NaCl at 900, 1000, 1100 and 1200 °C in air for 20 min. These temperatures are greater than the melting point of NaCl (~800 °C), causing the progress of the reactions in a molten salt environment. This simple molten salt process led to the formation of MoS_2/MoO_3@Graphene nanocomposites, as can be realized from Figure 1, and the XRD patterns of Figure 3. Moreover, the XRF quantitative analysis of various samples, namely MoO_3/MoS_2@G-900, MoO_3/MoS_2@G-1000, MoO_3/MoS_2@G-1100 and MoO_3/MoS_2@G-1200 are shown in Table 2. From the XRD patterns, the graphite and MoS_2 materials do not show any change of crystalline structure after ball milling, while a reduction in the thickness of their flakes can be expected under the influence of the shear forces applied during the ball milling process. However, various phases can be detected in products obtained after the heat-treatment process at different temperatures, namely MoS_2 (PDF#03-065-0160), orthorhombic-MoO_3 (PDF#01-089-5108), monoclinic-MoO_3 (PDF#01-085-2405) and graphite (PDF# 01-075-2078).

Figures S1–S3 show high resolution XRD patterns of diffraction peaks corresponding to monoclinic and orthorhombic MoO_3 which are close to each other. In addition, Table S1 shows the intensity ratios of various phases in MoO_3/MoS_2@G-900, MoO_3/MoS_2@G-1000, MoO_3/MoS_2@G-1100 and MoO_3/MoS_2@G-1200. One observation from Figure 3 and Figures S1–S3 is that the monoclinic-MoO_3 (PDF#01-085-2405) could be formed only at 1200 °C. Moreover, the combination of Table S1 and Table 2 can lead to the conclusion that the amounts of sulfur, molybdenum, carbon and other elements in the samples heated at 900 and 1000 °C are nearly similar. In the samples heated at 1100 °C, however, the amount of sulfur is less, but the contents of other elements are more than those in the samples prepared at 900 and 1000 °C. Therefore, the amount of sulfur reaches to its maximum in the sample prepared at 1200 °C, while the proportions of other elements in this sample are the lowest among all samples.

Based on the observations mentioned above, it can be concluded that the content of MoS_2 gradually decreases, while the temperature is increased from 900 to 1100 °C. On the contrary, the content of MoO_3 gradually increases upon the temperature increase. When the temperature exceeds 1100 °C, the content of MoS_2 increases, leading to the assumption that MoO_3 would further transform into MoS_2. The mechanism involved in this phase transition will be discussed in this paper. As the result, the maximum amount of MoO_3 could be achieved at the heating temperature of 1100 °C.

Further characterization of samples was performed using Raman spectroscopy, and the results are shown in Figure S4. For C-MoS_2, the Raman spectrum shows two major E^1_{2g} (380 cm^{-1}) and A_{1g} (406 cm^{-1}) activation modes. E^1_{2g} is due to the in-plane vibration of S and Mo atoms, and A_{1g} is attributed to the relative vibration of S atoms in the out of plane direction. The other two relatively weak peaks observed at 285 and 450 cm^{-1} belong to E_{1g} and longitudinal acoustic phonon mode in C-MoS_2, respectively, as shown in Figure S4a. After the ball milling process, the E_{1g} peak of C-MoS_2 disappears, and the A_{1g}-LA (M) peak is promoted. However, the structure of C-MoS_2 and natural graphite phase has no significant change, as shown in Figure S4a. The Raman spectra of the products obtained at various temperatures are shown in Figure S4b. As can be seen from Figure S4b, in addition to peaks related to graphite and MoS_2, there are also peaks related to MoO_3, which are marked with red asterisks on the spectrum of the sample prepared at 1100 °C. Therefore, Raman spectra confirm the formation of MoO_3/MoS_2@Graphene.

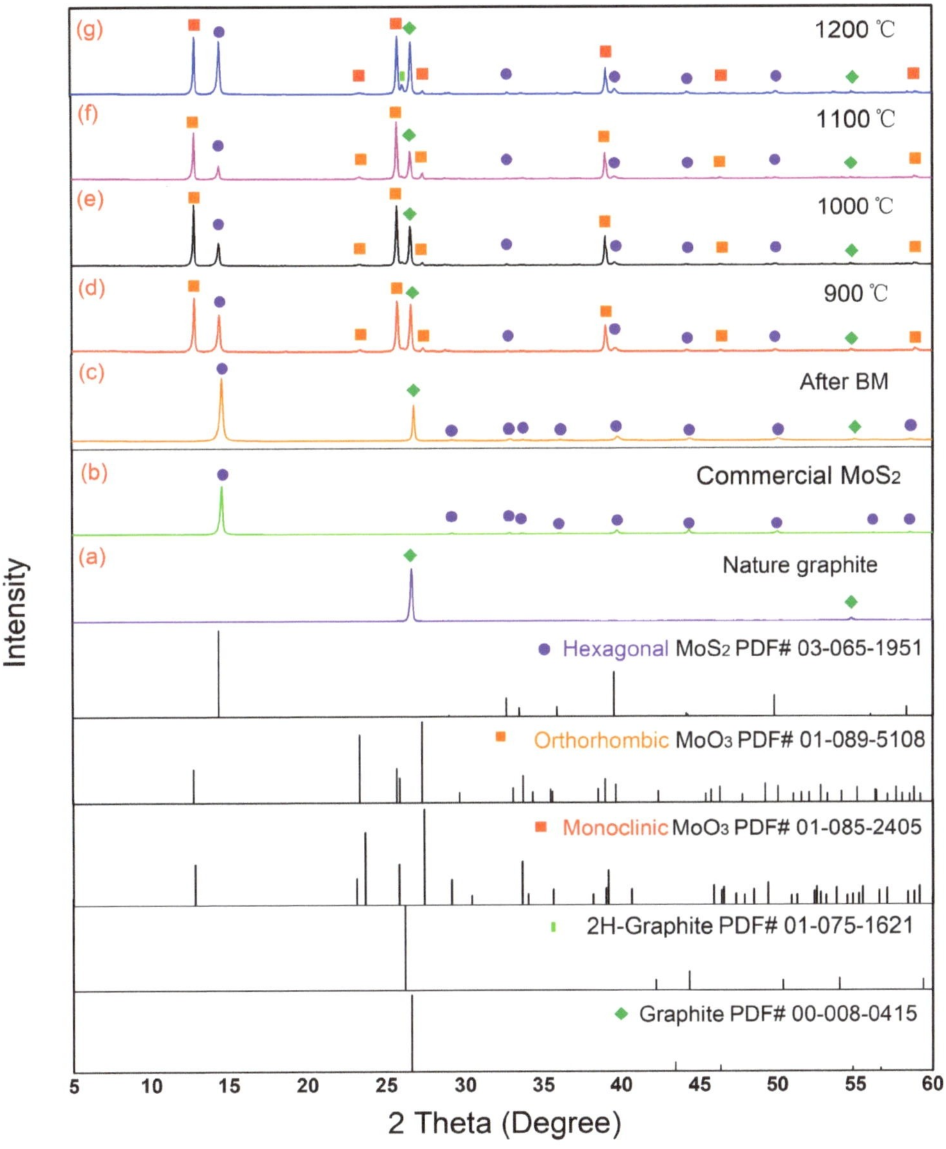

Figure 3. XRD patterns of various precursor materials and products: (**a**) the non-purified natural graphite material, (**b**) the commercial MoS_2, (**c**) the mixture after ball milling; and (**d**–**g**) nanostructured materials obtained at various temperatures.

Table 2. XRF quantitative analysis of the MoO_3/MoS_2@G samples prepared at various temperatures.

Sample	S	Mo	C	Si	Al	Fe	Ca	O
MoO_3/MoS_2@G-900	5.78	44.33	22.6	5.15	1.92	1.65	1.69	16.88
MoO_3/MoS_2@G-1000	5.59	45.18	23.3	6.36	2.42	1.15	1.74	14.26
MoO_3/MoS_2@G-1100	2.72	46.01	17.8	6.78	2.48	1.56	1.65	21.00
MoO_3/MoS_2@G-1200	10.2	48.48	15.6	5.96	2.07	1.82	1.18	14.69

On the other hand, the band gap energy indicates the energy required for the excitation of an electron to be moved from the valence band up to the conduction band. The Tauc method of evaluating the bad gap using UV Vis spectroscopy was used to measure the values of the band gap, as shown in Figures S5 and S6 and Table S2. It can be concluded that the values of band gap energy gradually increase with the increase of the content of molybdenum trioxide. The reason behind this variation is based on the fact that MoO_3 is an n-type wide band gap (≈3 eV) semiconductor [37], so the value of energy band gap in the sample produced in 1100 °C is the largest.

2.4. SEM Characterization

To investigate the effects of the processing temperature on the morphology and microstructure of the nanocomposite materials prepared in this study, SEM characterization was carried out on commercial molybdenum disulfide (C-MoS_2), natural graphite and products comprising of MoO_3/MoS_2@G-900, 1000, 1100, and 1200. The results are shown in Figure S7 and Figure 4. The SEM micrograph of C-MoS_2 is shown in Figure 4a, indicating that the material consists of irregularly-shaped flakes with the sizes of typically around 2 μm. SEM micrograph of natural graphite (Figure 4b) shows the sample is made of agglomerated flakes with typical sizes of several tens of micrometers. These two materials in combination with NaCl were used to fabricate MoO_3/MoS_2@G samples. As shown in Figure S7b–g, the morphology of the synthesized materials is different from those of starting material due to the formation of molybdenum oxides, hexagonal MoS_2 and graphene nanosheets. In MoO_3/MoS_2@G-900, the presence of orthrombic-MoO_3 was already confirmed based on XRD analysis of Figure 3c. The oxide phase can be distinguished in Figure S7b as crystals with sizes of around 1 μm, scattered within natural graphite and MoS_2 flakes. The presence of orthrombic-MoO_3 is attributed to the partial oxidation of MoS_2, leading to the formation of the molybdenum oxide and gaseous sulfur dioxide, under the influence of atmospheric air. According to Figure 2, as the temperature increases to 1000 °C, MoS_2 is almost fully oxidized into orthrombic-MoO_3. Moreover, the dimensions of resulting oxide are increased to form rod-like particles with dimensions in range of around 5–30 μm, as shown in Figure S7c. The directional growth of orthrombic-MoO_3 particles may lie on the preferential growth occurred along the-(001) direction in the orthorhombic structure of MoO_3, as further observed in TEM micrograph of Figure S8. By increasing the temperature to 1100 °C, the MoO_3 nanoribbons could grow even further, to form sheet-like particles with dimensions of around 50 μm, as shown in Figure S7d and Figure 4c,d. This observation suggests that at 1100 °C, other crystallographic directions in the orthorhombic lattice of MoO_3 nanoribbons contribute to the growth phenomenon observed, forming sheet-like particles, which is in contrast to the rod-like particles observed in Figure S7c.

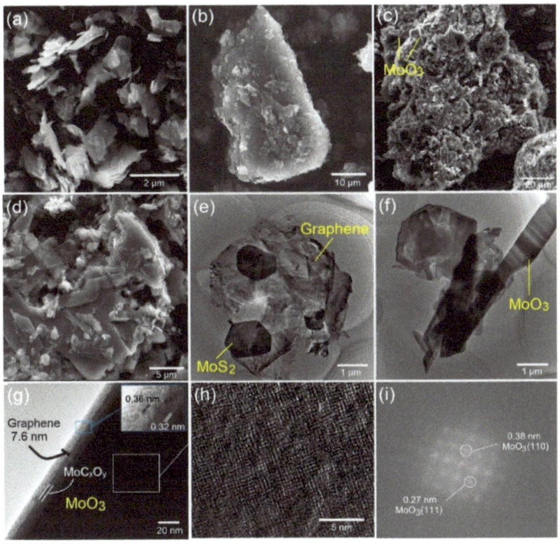

Figure 4. SEM micrographs of (**a**) commercial MoS_2, (**b**) natural graphite, and (**c**,**d**) MoO_3/MoS_2@G-1100. (**e**–**h**) TEM micrographs of MoO_3/MoS_2@G-1100. (**i**) FFT pattern recorded on the micrograph (**h**).

According to Figure S7e,f, by raising the temperature to 1200 °C, large-sized MoO_3 crystals observed in Figure S7d disappear, and instead, a large number of hexagonally-shaped MoS_2 crystals form. This morphological evolution may be described based on the possible reaction between the natural graphite flakes and sulfur dioxide released by the oxidation of the original MoS_2 flakes, leading to the formation of elemental sulfur. Then, sulfur formed reacts on the surface of MoO_3 nanoribbons particles to form hexagonal MoS_2 nanosheets. As exhibited from Figure S7f, the hexagonal MoS_2 nanosheets grow from the surface of the MoO_3 nanoribbons particles, possibly, by replacing the oxygen of MoO_3 with S [28]. The size of hexagonal MoS_2 nanosheets is typically less than around 2 μm, as shown in the high-magnification image of Figure S8g. Figure S7h shows the EDS analysis recorded on the hexagonal MoS_2 nanosheet shown in Figure 5, providing further evidence for the nature of such hexagonal crystals.

The in-situ formation of hexagonal MoS_2 nanosheets on the surfaces of MoO_3 nanoribbons leads to the disintegration of the MoO_3 into smaller particles, as can be observed in SEM micrographs of Figure S7e,f. This phenomenon also causes the phase transition of MoO_3 from the orthorhombic crystalline structure (α-MoO_3) to the monoclinic modification (β-MoO_3), as shown in Figure 2f. This hypothesized mechanism will be discussed further in next sections in this article. The combination of XRD and XRF analyses suggested that the MoO_3 content of the samples reaches its maximum value at 1100 °C. This behavior can be related to the action of molten NaCl to protect the MoO_3 particles from being excessively converted into hexagonal-MoS_2 nanosheets at 1100 °C, while this temperature allows the maximum oxidation of the original MoS_2 into MoO_3 nanoribbons. Considering the higher contentment of MoO_3 in the sample prepared at 1100 °C, the Li-ion storage capacity of this sample is expected to be greater than other samples, and this will be shortly discussed later in this article.

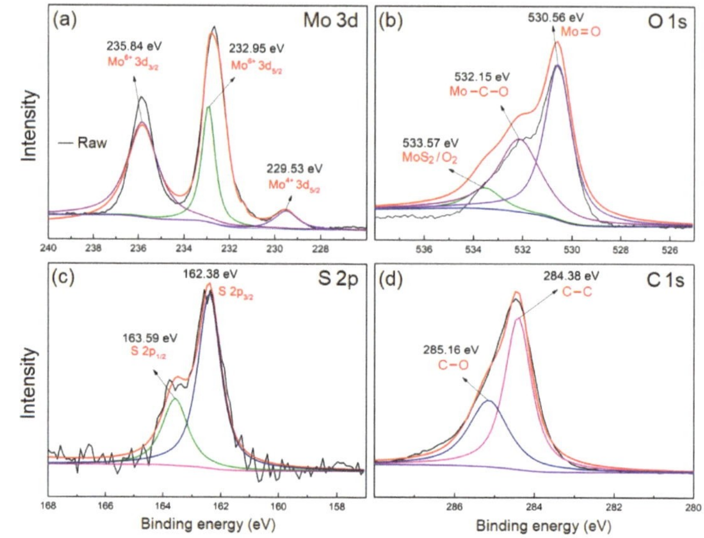

Figure 5. High resolution XPS spectra of MoO_3/MoS_2@G-1100, around (**a**) Mo 3d, (**b**) O 1s, (**c**) S 2p, and (**d**) C 1s peaks.

2.5. TEM Characterization

As observed in Figure 4c,d, the SEM morphology of MoS_2-MoO_3@G-1100 could be characterized by the presence of agglomerated nano-entities. TEM studies were performed to further investigate this morphology, and the results are shown in Figure 4e–i, where the presence of hexagonal MoS_2 nanoflakes ($\approx$1 μm), graphene nanosheets and α-MoO_3 nanoribbons (W $\approx$ 400–800 nm; L > 10 μm) can be realized. The formation of such nanoribbons can be related to the presence of oxygen vacancies on the surface of MoO_3 crystals, providing the driving force for the solid-phase growth of oxides into nanoribbons [38,39]. The MoO_3 nanoribbons appears to be formed as the result of dominant crystal growth along the [001] direction, as shown in Figure S8. In TEM images, the presence of graphene nanosheets (originating from the natural graphite material) combined with hexagonal MoS_2 nanoflakes and MoO_3 nanoribbons is evident. Figure 4g–i provide further insights into the interface between graphene and MoO_3. Figure 4g shows a MoO_3 nanoribbon, where the presence of a graphene layer with the thickness of around 7.6 nm on its surface can be seen. The high magnification TEM image shown as the inset of Figure 4g shows the presence graphene layers with the interlayer spacing of 0.36 nm, representing (002) planes of hexagonal carbon. In addition, toward the bulk of sample, the interlayer spacing of 0.32 nm can be assigned to the (021) crystalline planes of orthorhombic-MoO_3. Moreover, there is an amorphous interface between the graphene nanosheets and MoO_3 which could be assigned to molybdenum oxycarbide formed during the molten salt processing of the sample. Formation of this interfacial phase will be further discussed by XPS analysis in this article. Figure 4h shows a high resolution TEM micrograph recorded on the MoO_3 nanoribbons, where the crystalline structure of the materials can clearly be observed. Figure 4i exhibits the fast Fourier transform (FFT) recorded on the micrograph of Figure 4h, where the spots related to the crystalline planes of orthorobmhic-MoO_3 are indexed. Overall, XRD, SEM and TEM analyses suggest the formation of nanostructured MoO_3/MoS_2@G-1100, in which orthorhombic-MoO_3 nanoribbons, hexagonal MoS_2 nanoflakes and graphene nanosheets are well-integrated. The presence of interfacial molybdenum oxycarbide was also suggested.

2.6. Surface Characterization

The structural and microstructural characterizations mentioned above could demonstrate the formation of MoO_3 nanoribbons incorporated into hexagonal MoS_2 and graphene. The Li-ion storage performance of the products will be evaluated shortly in this article. However, before that, it would be interesting to shed light on to nature of the interface between the above mentioned phases. This was examined using XPS analysis, as shown in Figure 5. As can be observed, the characteristic peaks of Mo, O, S and C are debatable in the XPS spectrum of MoO_3/MoS_2@G-1100. High-resolution XPS spectra obtained on Mo 3d, O 1s and S 2p peaks are exhibited Figure 5a–c. As observed, the strong peak at 229.53 eV corresponds to Mo^{4+} $3d_{3/2}$, and peaks at about 163.1 eV and 161.5 eV to S $2p_{1/2}$ and S $2p_{3/2}$, respectively, revealing the existence of MoS_2 [40,41]. On the other hand, the peaks located at 235.79, 232.95, 530.56 and 532.15 eV correspond to the Mo^{6+} $3d_{3/2}$, Mo^{6+} $3d_{5/2}$. These peaks together with that appeared at 530.65 eV (corresponding to O1s components) indicate the presence of MoO_3 [27]. Moreover, the C 1s core level shown in Figure 5d could be deconvoluted into two peaks at 284.38 and 285.16 eV, which originate from the C–C and C-O bonds, respectively [41]. The presence of C-C bond is related to the graphene nanosheets, and C-O bond to the local interaction between graphene nanosheets and MoO_3.

Moreover, the presence of peak at 532.15 eV in Figure 5b indicates the presence of Mo-O-C [42] corresponding to the formation of interfacial molybdenum oxycarbide (MoC_xO_y) at the interface between MoO_3 and graphene nanosheets, as also suggested by the TEM micrograph of Figure 4g. Density functional theory calculations have confirmed the possibility of formation of molybdenum oxycarbide by reacting between zeolite-supported Mo and CO_2 [43]. Zhu et al. [42] suggested that MoO_xC_y can be formed from MoO_3, when vacancies available on the surface of MoO_3 are filled by carbon atoms. In our case, MoO_3 nanoribbons are wrapped with graphene nanosheets (Figure 4). Under this condition, the diffusion of carbon atoms into the surface of MoO_3 at high-temperatures such as 1100 °C is possible, leading to the formation of interfacial MoO_xC_y. It should be mentioned that molybdenum oxycarbide is generally considered to possess low electrical resistivity of 3 μΩm at room temperature [44]. The interfacial molybdenum oxycarbide is likely to promote the electrochemical performances of the nanostructured materials. This will be discussed later in this article. The well-incorporated phases in MoO_3/MoS_2@G-1100, and the possible presence of interfacial molybdenum oxycarbide, can lead to the reduced specific surface area (1.74 m^2 g^{-1}) and pore volume (0.008 cm^3 g^{-1}. In agreement with these observations, the nitrogen adsorption–desorption isotherms of MoO_3/MoS_2@G-1100, shown in Figure 6, provide features corresponding to the type III isotherms, characteristic of the non-porous materials [45].

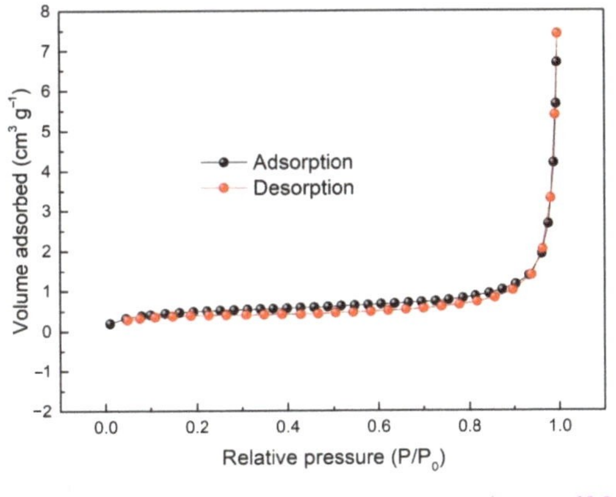

Figure 6. Nitrogen adsorption–desorption isotherms of MoO_3/MoS_2@G-1100.

2.7. Catalytic Formation of Hexagonal MoS_2 Nanocrystals and α-MoO_3

In this research, commercially-available MoS_2 flakes and natural graphite mineral were utilized as the initial material, and the product contained hexagonal MoS_2 nanocrystals. The conversion of flake MoS_2 into hexagonal nanocrystals is an interesting phenomenon discussed in this section. It is known that MoS_2 flakes start to oxidize at around 400 °C, leading to the formation of orthorhombic molybdenum trioxide (α-MoO_3) [46]. Therefore, the thermochemical reactions occurred in the process initiates with the formation of α-MoO_3 through the involvement of oxygen from the atmosphere:

$$MoS_2 + 3.5\ O_2\ (g) = \alpha\text{-}MoO_3 + 2SO_2\ (g),\ \Delta G^\circ = -922.1\ kJ\ (T = 400\ ^\circ C) \quad (1)$$

In the presence of graphite, the SO_2 gas formed based on the reaction (1) is likely to be fixed by graphite flakes to generate elemental sulfur and carbon dioxide gas, as shown in Equation (2).

$$SO_2\ (g) + C = S + CO_2\ (g),\ \Delta G^\circ = -95.6\ kJ\ (T = 400\ ^\circ C) \quad (2)$$

The combination of reactions (1) and (2) can lead to the reaction (3):

$$MoS_2 + 2.5O_2\ (g) + C = \alpha\text{-}MoO_3 + 2S + CO_2\ (g),\ \Delta G^\circ = -718.1\ kJ\ (T = 400\ ^\circ C) \quad (3)$$

The generated elemental sulfur adhered to the surface of MoO_3 is likely to in-situ reduce the MoO_3 into hexagonal MoS_2 nanocrystals:

$$\alpha\text{-}MoO_3 + 3.5S = \text{hexagonally shaped } MoS_2 + 1.5\ SO_2\ (g),\ \Delta G^\circ = -126.7\ kJ\ (T = 400\ ^\circ C) \quad (4)$$

The reactions (1)–(4) demonstrates the thermodynamic possibility of the formation of α-MoO_3 and hexagonally shaped MoS_2 at temperatures greater than 400 °C initiating with the oxidation of MoS_2. In this research, the heating process was performed at temperatures above the melting point of NaCl. Therefore, the molten salt is likely to exfoliate the graphite flakes [47,48]. The formation of graphene nanosheets is evident from the TEM micrographs of Figure 4 and Figure S7. Figure S13 illustrates the possible mechanism involved in the catalytic formation of hexagonal MoS_2 nanocrystals and MoO_3 observed in this study.

To highlight the role of NaCl in the process, the sample prepared at 1100 °C was washed to remove its NaCl content, and the product (MoO_3/MoS_2@G-1100) was heated to 1200 °C for 5 min, without the involvement of NaCl. The appearance and X-ray diffraction pattern of the resulting sample are shown in Figures S9 and S10, respectively, providing evidence that the sample heated at 1200 °C mainly contains molybdenum oxide, without the participation of molten NaCl. In this case, carbon and molybdenum disulfide content of the sample are oxidized during the heat-treatment process to form molybdenum oxide and gas species at high temperature.

It should be mentioned that oxidation in air of graphitic carbon materials typically occurs at temperatures in the range 400–800 °C, depending on their grain size, level of crystallinity and presence of impurities that can catalyze the oxidation process [49,50] leading to the formation of gashouse species and ash. It is known that the oxidation of bulk and few-layer MoS_2 flakes does not readily take place at ambient conditions due to the high energy barrier involved [51–54]. However, non-isothermal oxidation of MoS_2 flakes initiates at temperatures as low as 350 °C with a limited rate, and increases by enhancing the temperature under an apparent activation energy of $\approx$1 kcal/mol, representing the bulk oxidation event. The oxidation process leads to the increase of the molybdenum oxide content (typically MoO_3), by increasing the temperature [55,56]. This is in agreement with our observation exhibited in Figure S10. Based on the observations mentioned above, the presence of NaCl provides an essential support toward the formation of MoO_3/MoS_2@G samples at high temperatures. First, molten NaCl provides an ionic environment to enhance the chemical reactions, while preventing the oxidation of species and supporting the exfoliation of graphite into graphenen nanosheets; the latter is discussed elsewhere [57].

2.8. Electrochemical Characterization of MoO_3/MoS_2@G Samples

We further investigated the Li-ion storage performances of MoO_3/MoS_2@G samples prepared at various temperatures, namely MoO_3/MoS_2@G-900, -1000, -1100 and -1200 in comparison with those of the initial commercial MoS_2 flakes, named as C-MoS_2. In the half-cell configuration, the driving force for the Li-ion insertion/extraction into/out of the electrode during the discharge/charge processes is provided by the negative/positive polarization applied on the electrode, respectively.

The electrochemical performances of the electrodes are shown in Figure 7. The electrodes were fabricated using the water-based polystyrene acrylic-acrylate as the binder, as explained in the Experimental section. Figure 7a shows the cycle performances of MoO_3/MoS_2@G electrodes, and selected outcomes are summarized in Table 3. As can be seen, MoO_3/MoS_2@G-1100 electrode shows a greater initial coulombic efficiency (CE) of 79.58% compared with other electrodes. After 100 cycles, this electrode exhibits a reversible capacity of 616.3 mAh g^{-1} at 100 mA g^{-1}.

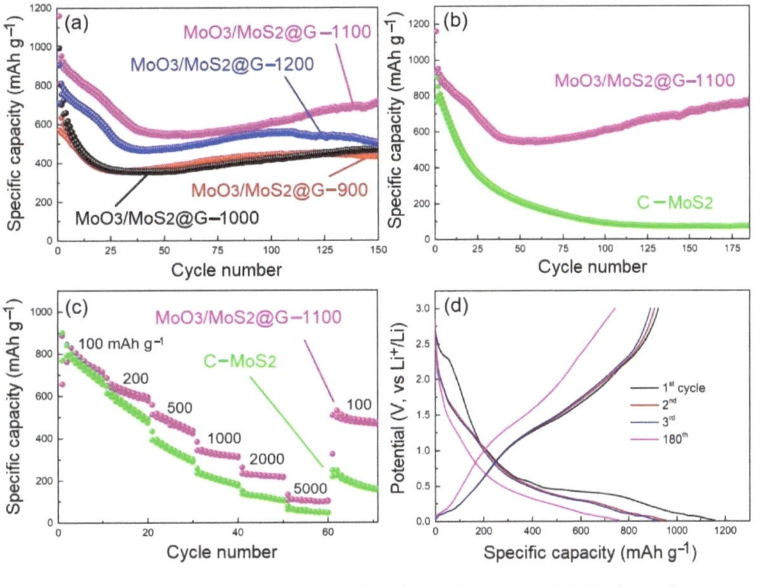

Figure 7. Li-ion storage performances of various electrodes. (**a**) Cycle performance of various electrodes. (**b**) Cycle performance, and (**c**) rate performance of C-MoS_2 in contrast with those of MoO_3/MoS_2@G-1100. (**d**) Potential-capacity curve of MoO_3/MoS_2@G-1100 electrode. All measurements were recorded at 100 mA g^{-1}, unless indicated otherwise.

Table 3. The electrochemical performance of commercial MoS_2 (C-MoS_2) in comparison with those of MoO_3/MoS_2@G nanostructured materials.

Electrode	Initial Coulombic Efficiency (%)	Specific Capacity after 100 Cycles (mAh/g)	Specific Capacity after 150 Cycles (mAh/g)
C-MoS_2	85.13	94.8	73.5
MoO_3/MoS_2@G-900	70.64	444.1	442.8
MoO_3/MoS_2@G-1000	71.13	422.2	472.7
MoO_3/MoS_2@G-1100	79.58	616.3	721.2
MoO_3/MoS_2@G-1200	78.36	560.1	505.9

The performance MoO_3/MoS_2@G-1200 electrode can be evaluated to be greater than those of -900 and -1000 electrodes, but lower than that of −1100. The greater performance of MoO_3/MoS_2@G-1100 can be attributed to the greater amount of MoO_3 in this sample as realized from Figure 2 and Table 2. The other point that deserves attention is that by increasing the temperature to 1200 °C, the amount of MoO_3 in the sample reduces to reach its minimum among all samples, as evidenced from the relative low intensity of MoO_3 reflections observed in Figure 3. Despite having the minimum amount of MoO_3, the performance of the electrode is greater than samples prepared at 900 and 1000 °C. This observation can be assigned to the formation of β-MoO_3 with monoclinic crystalline structure at 1200 °C. This is known that the metastable β-MoO_3 has a highly ionically conductive open-structure [41], enhancing the electrochemical performance of MoO_3/MoS_2@G-1200, despite the lower MoO_3 content of the sample.

As can be seen in Figure 7d, the first discharge/charge specific capacity of MoO_3/MoS_2@G-1100 is recorded at 1159.6 mAh g^{-1}/953.8 mAh g^{-1} (coulombic efficiency = 79.58%). This capacity loss can mostly be related to the generation of solid electrolyte interphase (SEI) layers on the electrode. The specific reversible capacity of the electrode could still be recorded at 773.5 mAh g^{-1} after 180 cycles, as exhibited in Figure 7d. Figure S11 shows the durability of MoS_2/MoO_3/G-1100 sample in lithium-ion battery. As seen, the specific capacity after 400 cycles is still maintained at 240 mAh g^{-1} under the current density of 500 mA g^{-1}.

Figure 8 shows the CV curve of the MoO_3/MoS_2@G-1100 electrode, providing detailed information about the Li-ion insertion and extraction processes. During the first cathodic scan, there is a cathodic peak at 0.56 V. This cathodic event cannot be observed in the subsequent cycles, providing evidence that this peak mainly corresponds to the formation of SEI. Moreover, the presence redox peaks at 1.56/1.76 V and 1.25/1.48 V in the first cycle can be related to the lithiation and delithiation of MoO_3, and to the metallization/oxidation of metallic molybdenum:

$$MoO_3 + yLi^+ + ye^- \leftrightarrow Li_yMoO_3 \quad (5)$$

$$Li_yMoO_3 + (6-y)\ Li^+ + (6-y)\ e^- \leftrightarrow 3Li_2O + Mo \quad (6)$$

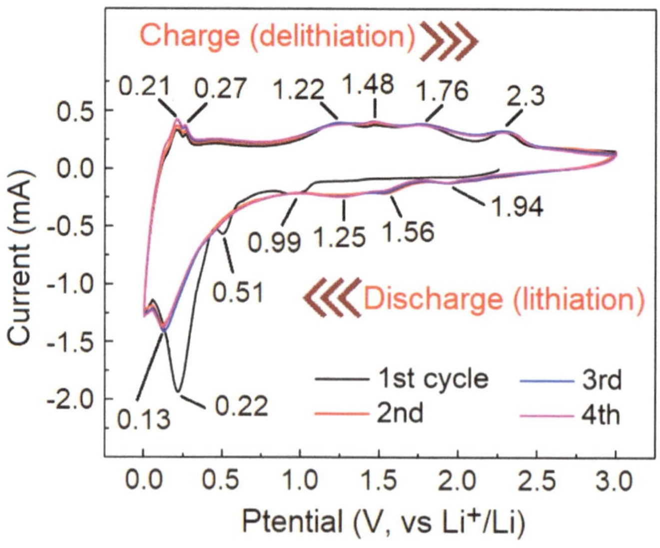

Figure 8. CV curve of MoO_3/MoS_2@G-1100 electrode recorded at 0.5 mV/s.

Moreover, the cathodic peak at 1.94 V might be attributed to the conversion from S_8 to Li_2S, and the oxidation peak at 2.3 V to the decomposition of Li_2S [58,59]. On the other hand, the redox peaks at 0.99/1.22 V corresponded to the following electrochemical reactions [60]:

$$MoS_2 + xLi^+ + xe^- \leftrightarrow LixMoS_2, \quad (7)$$

$$LixMoS_2 + (4-x)\ Li^+ + (4-x)\ e^{-1} \leftrightarrow 2Li_2S + Mo \quad (8)$$

Furthermore, the reduction/oxidation peaks at 0.21/0.22 V are related to adsorption/desorption of lithium ion on the surfaced of graphene nanosheets [61]. Overall, in agreement with Figure 7d, the CV curve of Figure 8 provides evidence for the electrochemical contribution of all the components of the electrode to the Li-ion storage performance observed.

2.9. Impedance Spectroscopy and Electrode Kinetics

Impedance spectroscopy performed on MoO_3/MoS_2@G and C-MoS_2. Figure 9a exhibits the Nyquist plots of the electrode. The plots show a semicircle in the high frequency range. The diameter of this semicircle can be related to the charge-transfer resistance (R_{ct}). In the low frequency range, the plots also show sloping straight lines. This part reflects the diffusion characteristics of lithium ions within the electrode [62]. The electrode made of MoO_3/MoS_2@G-1100 exhibits a smaller semi-circle diameter (40.71 Ω) than other samples, namely −900, −1000, −1200 and C-MoS_2; 63.65, 53.65, 62.9 and 263.8 Ω, respectively. This indicates the higher charge transfer resistance in the latter. The smaller electrochemical resistance observed in the MoO_3/MoS_2@G-1100 is consistent with the micrographs of Figure 4, indicating to the presence of integrated components with interfacial molybdenum oxycarbide that reduces the interfacial resistance across the sample.

The Li-ion diffusion (D_{Li}) involved in the electrochemical process discussed above can be calculated as follows [63]:

$$D_{Li} = \frac{R^2T^2}{2A^2n^4F^4C^2\sigma^2} \quad (9)$$

Here, R and F represent the gas and Faraday constants, respectively, T is temperature (298 K) and n is the number of participating electrons (n = 1). C represents the concentration of Li+. The electrode surface area of the electrode (A) was calculated based on the diameter of the electrode (1.1 cm^2). Moreover, σ represents the Warburg factor which is determined based on the slope of the line $Z' \sim \omega^{-1/2}$ shown in Figure 9b, considering that ω = 2Πf (f = frequency, Hz). The kinetic parameters of the electrode were determined from the equivalent circuit fitting of Nyquist plots (Figure 9a), and shown in Table 4. At low values of frequency, the MoO_3/MoS_2@G-1100 electrode exhibits a high slope (9.77 Ω $s^{-1/2}$) which demonstrates the relatively high lithium ion diffusion coefficient [64]. The lithium ion diffusion coefficient of MoO_3/MoS_2@G-900, -1000, -1100, -1200 and C-MoS_2 electrodes could be calculated to be 8.7×10^{-21}, 1.8×10^{-20}, 1.2×10^{-18}, 5.3×10^{-21} and 1.0×10^{-21}, respectively. The larger D_{Li} in the earlier can be related to the shorter diffusion pathways available in the electrode, brought about by the presence of integrated hexagonal MoS_2 nanocrystals and graphene nanosheets as shown in Figure 4.

Table 4. Electrode kinetic parameters for the MoO_3/MoS_2@G and C-MoS_2 electrodes.

Sample	R_e (Ω)	R_{ct} (Ω)	σ (Ω $s^{-1/2}$)	D_{Li} (cm^2 s^{-1})
MoO_3/MoS_2@G-900	6.90	63.65	114.61	8.7×10^{-21}
MoO_3/MoS_2@G-1000	6.56	53.56	79.39	1.8×10^{-20}
MoO_3/MoS_2@G-1100	2.21	40.71	9.77	1.2×10^{-18}
MoO_3/MoS_2@G-1200	3.92	62.9	146.53	5.3×10^{-21}
C-MoS_2	3.56	263.8	334.49	1.0×10^{-21}

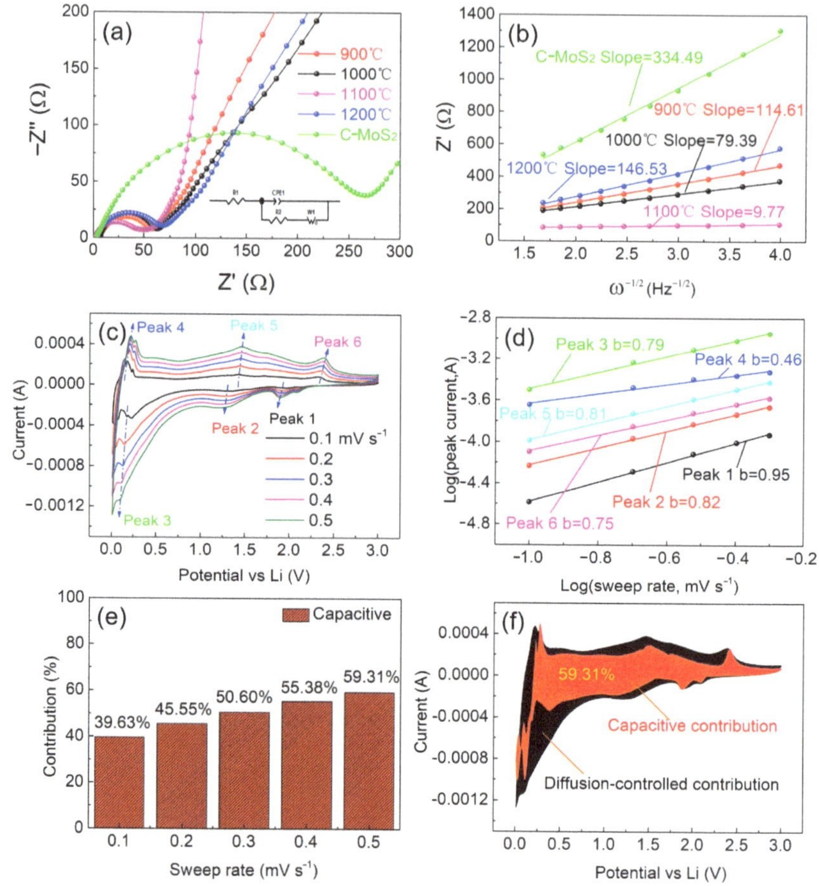

Figure 9. (**a**) AC impedance spectra of MoO_3/MoS_2@G and C-MoS_2 electrodes. (**b**) The values of Z′ vs. $\omega^{-1/2}$ ($\omega = 2\Pi f$) for MoO_3/MoS_2@G and C-MoS_2. (**c**–**f**): Contribution of pseudocapacitive Li-ion storage to the total capacity of the MoO_3/MoS_2@G-1100 electrode: (**c**) CV curves with scan rates raging 0.1 to 0.5 mV s^{-1}; (**d**) the logarithmic dependence between the peak current and sweep rate; (**e**) contribution of pseudocapacitive and diffusion-based processes to the total current at different sweep rates, and (**f**) the pseudocapacitive contribution at 0.5 mV s^{-1}.

2.10. Contribution of Diffusion- and Surface-Phenomena

We also evaluated the contribution of diffusion- and surface-phenomena to the overall Li-ion storage performance of the MoO_3/MoS_2@G-1100 electrode. Figure 9c exhibits CV curves recorded on the electrode at various scan rates from 0.1 to 0.5 mV s^{-1}. It should be mentioned that the current intensity (i) and the sweep rate (υ) can be related by Equations (10) and (11) [65]:

$$i = a \times \upsilon^{b} \tag{10}$$

$$\log i = \log a + b \times \log \upsilon \tag{11}$$

Here, a and b are dimensionless variables. Figure 9d could be established by plotting the logarithmic dependence between the current values related to various peaks shown in Figure 9c and the corresponding values of sweep rates. Based on the results obtained, the relative contributions of pseudocapacitive (vs. the diffusion-controlled events) to the total current could be found to be in the range 39–60% at different sweep rates, as exhibited in Figure 9e.

It can be seen that the contribution of pseudocapacitive processes to the total current is proportional to the scan rate: higher scan rate, the greater contribution. As such, the pseudocapacitive contribution can directly influence the high-rate performance of the electrode, as shown in Figure 7c. Figure 9f exhibits the pseudocapacitive contribution to the total current at 0.5 mV s^{-1}.

Moreover, in Equation (11), the b value depends on the Li-ion storage mechanism. The value of b = 0.5 represents a diffusion-dominated process, while b = 1 represents the pseudocapacitive processes. The be values between 0.5 and 1 represent process involving both pseudocapacitive and diffusion-based process with various contribution ratio depending of the b-value [66,67]. As can be seen from Figure 9d, the value of b for the peaks 1, 2, 3, 4, 5, and 6 are obtained to be 0.95, 0.82, 0.79, 0.46, 0.81and 0.75, respectively. It confirms that all current peaks are mainly dominated by pseudocapacitive processes, unless the anodic peak 4 with the b value of 0.46, representing diffusion based processes. The pseudocapacitive-oriented nature of the peaks mentioned above can be explained based of the integrated nanostructure of the sample (Figure 4) that promotes the Li-ion diffusion and charge transfer the electrode/electrolyte interfaces across the electrode. In contrast, the anodic peak 4 located at 0.20 V, with the b-value of 0.46, corresponds to de-intercalation of Li^+ out of the thicker graphitic flakes present in the sample.

3. Discussion

We have successfully developed a green, simple process for the conversion of the abundant and low-cost natural graphite and widely available MoS_2 into nanostructured MoO_3/MoS_2@Graphene. In this process, the oxidation of MoS_2 in the presence of graphite leads to the formation of α-MoO_3 with orthorhombic crystalline structure, as well as hexagonal MoS_2 nanocrystals. At 900 and 1000 °C, the molten salt promotes the catalytic exfoliation of the natural graphite into graphene nanosheets, incorporated with hexagonal MoS_2 crystals and α-MoO_3, denoted as MoO_3/MoS_2@G-900, and -1000, respectively. The further development of MoS_2 oxidation at 1100 °C, resulted in the formation of MoO_3/MoS_2@G-1100, containing the maximum amount of MoO_3 among all samples. The highest amount of MoO_3 in this sample combined with the presence of integrated graphene nanosheets and hexagonal MoS_2 provide the sample with the greatest value of Li-ion storage capacity about 773.5 mAh g^{-1} after 180 cycles at a current density of 100 mA g^{-1}. The formation of molybdenum oxycarbide between MoO_3 and graphene nanosheets was suggested as the cause of reduced charge-transfer resistance (53.56 Ω).

At the processing temperature of 1200 °C, the content of MoO_3 is the minimum among all samples, but the electrochemical performance of the sample (MoO_3/MoS_2@G-1200) is second only to that of the sample prepared at 1100 °C, with a Li-ion storage capacity of about 505.9 mAh g^{-1} after 180 cycles at a current density of 100 mA g^{-1} (Table 3). Interestingly, the high Li-ion storage capacity of the sample prepared at 1200 °C was corresponded to the formation of thermodynamically metastable monoclinic molybdenum trioxide (β-MoO_3), instead of conventional orthorhombic α-MoO_3 which appears at 900–1100 °C. β-MoO_3 is known to have a highly ionically conductive open structure [41], improving the electrochemical performance of the material. Selected number of methods employed in the literature for the preparation of molybdenum compounds, and their electrochemical performances are compared with those of MoO_3/MoS_2@G-1100 in Table 5. Methods shown in this table typical use expensive and/or hazardous materials such as $(NH_4)_6Mo_7O_{24}\cdot 4H_2O$ [27,28], MoO_3 [29], $(NH_4)_2MoO_4$ [32], and metallic Mo powders [33], in combination with materials such as CH_4N_2S [27], HNO_3 [28], glacial acetic acid [29], HCl [32], and H_2O_2 [33]. The excessive use of such chemicals limits the large-scale implementation of these methods. In the method reported here, ball milling is applied to incorporate natural graphite and MoS_2 in the presence of hexane and NaCl. Both hexane and NaCl can be retrieved from the mixture after ball-milling, and the final washing step, respectively. Nevertheless, the method reported here requires heating at elevated temperatures which might provide some limitations. Utilization of industrial heat waste can be an option to enhance the economic features of the method.

Table 5. A comparison between selected MoO_3/MoS_2 materials extracted from the literature, and MoO_3/MoS_2@G-1100 prepared here, in terms of the precursor materials, preparation method, and the electrochemical performance.

Mo-Oxide Based Material	Precursor Materials	Preparation Process	Performance: Capacity (mAh/g)/Cycle/Current Density (mA/g)
Core−sheath structured MoO_3@MoS_2 composite [23]	Rhododendron petals, $(NH_4)_6Mo_7O_{24}\cdot4H_2O$, CH_4N_2S	In situ growth at 550 °C (2 h) + 180 °C (24 h)	400/30th/200
Ternary $MoS_2/MoO_3/C$ nanosheets [24]	Organic amine, MoO_3	CVD at 800 °C (3 h) under N_2	961.4/120th/430
MoO_3/MoS_2 nanoribbons [25]	Molybdic acid, glacial acetic acid	Hydrothermal treatment at 800 °C (30 min)	-
MoS_2/MoO_3 nanosheets [26]	$CdCl_2$, $(NH)_2CS$, $(NH_4)_2MoO_4$, $NH_2OH.HCl$	Hydrothermal treatment at 180 °C (30 h)	-
MoS_2 coated MoO_3 nanobelts [27]	Mo metal powder 10 mL H_2O_2	Hydrothermal treatment at 180 °C (12 h)	1400/50th/100
MoS_2 nanosheets on MoO_3 nanowires [28]	$(NH_4)_6Mo_7O_{24}\cdot4H_2O$, HNO_3,thiourea	Hydrothermal treatment at 180 °C (24 h)	800/100th/100
Core-shell MoO_3/MoS_2 nanowires [29]	MoO_3 powders, sulfur, N_2	Thermal treatment at 700 °C	
MoS_2/MoO_3@graphene [This work]	MoS_2, natural graphite, NaCl	Heating at 1100 °C (20 min)	316/40th/1000 774/180th/100

In this research, the structural, morphological and optical properties of raw materials and products obtained at various temperatures were evaluated using a combination of various techniques, based on which the possible mechanism involved in the catalytic formation of hexagonal MoS_2 nanocrystals upon the thermal treatment of MoS_2 in the presence of natural graphite flakes in the molten salt environment was proposed. The formation of hexagonal MoS_2 observed in this work is interesting, in comparison with alternative methods used for the preparation of hexagonally-shaped crystals from the literature, as exhibited in Figure S12 and Table S3. These methods include high-temperature sulfidation [31,32], electrochemical exfoliation [34], and chemical vapor deposition [35,36], using chemicals such as $(NH_4)_2MoS_4$ [31], pure Mo and S [32], W_xO_y [26], and Na_2SO_4 [34]. In contrast, this work proposes a simple and facile method for the preparation of hexagonal MoS_2, and ultimately MoO_3/MoS_2@G samples using highly available and low-cost precursors comprising of MoS_2, natural graphite minerals and NaCl, as highlighted in Tables 6 and S3. The process is rapid and requires only a short-time heating at 1100 °C. Despite the simple and clean preparation method, the electrode made of MoO_3/MoS_2@G-1100 exhibits a decent Li-ion storage performance. Modification of naturally available materials for energy-storage [68–70] and other demanding applications [71–73] is a viable strategy to meet the increasing green energy demand. In this research, we explored the effect of heating at various temperatures on the ball-milled natural graphite/MoS_2/NaCl mixture for a short period of 20 min to demonstrate that the process of fabricating MoO_3/MoS_2@G samples is considerably fast. The influence of heating time on characteristics of the products needs to be explored in future studies. It should be mentioned that at 900–1200 °C, the interaction between components occurs in molten NaCl, which can effectively reduce the reaction between carbon and oxygen forming CO/CO_2 species. According to Figures S9 and S10, further heating of MoO_3/MoS_2@G-900 in the absence of NaCl to 1200 °C in air leads to the oxidation of carbon and MoS_2 to form CO/CO_2 species and MoO_3, respectively, highlighting the effect of NaCl. The influence of other salts on the process is worth investigation in future.

Table 6. Amounts of raw materials used and the resultant products obtained at various temperatures.

Raw Material	MoS_2 (10 g)-NaCl (100 g)-Graphite (5 g)			
Ball-milled sample	112.00 g			
Heat-treatment Temperature	900 °C	1000 °C	1100 °C	1200 °C
Sample	MoO_3/MoS_2@G-900	MoO_3/MoS_2@G-1000	MoO_3/MoS_2@G-1100	MoO_3/MoS_2@G-1200
Before heat-treatment	28.00 g	28.00 g	28.00 g	28.00 g
After washing	3.31	2.60	2.29	2.45

4. Materials and Methods

4.1. Synthesis of MoS_2/MoO_3@Graphene Nanocomposites

A simple and scalable ball-milling–molten salt strategy was used to synthesize MoS_2/MoO_3@Graphene nanocomposites. In a typical experiment, molybdenum disulfide (MoS_2 99.5%, Aladdin CAS-1317-33-5), natural graphite powder (collected from a graphite mine in Hunan province of China) and sodium chloride (NaCl, >99%) (10:5:100 g, respectively) were thoroughly mixed, and the mixture was placed into 500 mL polymeric jars. The ball-milling was performed using zirconia balls with ball: powder weight ratio of 10:1, in the presence of *n*-hexane. The latter was employed to prevent the occurrence of agglomeration during the ball milling process [74], and partial exfoliation of the non-purified natural graphite and MoS_2 [75,76]. The ball milling was performed for 2 h at 230 rpm using a high-energy planetary mill (MITR QM-QX-2L). After ball milling, the samples obtained were dried at 100 °C for 5 h to remove the hexane, and heat-treated at various temperatures of 900–1200 °C for 20 min in air. The heating was performed at 6 °C min^{-1}. The samples obtained were employed for various characterizations as described below. Figure 1 exhibits the process used to fabricate the samples. The amount of raw materials used and products obtained through the process are shown in Table 6.

4.2. Morphological, Structural and Surface Characterizations

A scanning electron microscope (SEM, Zeiss Ultra Plus, Oberkochen, Germany) equipped with energy dispersive X-ray spectrometer (EDS, Oxford Instruments, High Wycombe, UK) and a transmission electron microscope (TEM, JEOL JEM-F200, Tokyo, Japan) were used for morphological studies. X-Ray diffractometry (XRD, Malvern Panalytical, Malvern, UK) and fluorescence spectrometry (XRF, ZSX Primus, Rigaku, Japan) were used for phase and chemical analysis, respectively. N_2 adsorption-desorption evaluation was performed using a Micrometrics Tristar II 3020 to examine the surface characteristics of the samples, where the pore size distribution was obtained using the desorption branch of the isotherms. X-ray photoelectron spectroscopy (XPS, Thermo Scientific Instrument, East Grinstead, UK) was employed for surface evaluation using Al-K_α rays as the excitation source (1486.6 eV).

4.3. Electrochemical Characterizations

Various electrochemical evaluations were performed on the synthesized MoS_2/MoO_3@Graphene samples and MoS_2. Accordingly, cycling and rate performances as well as cyclic voltammetry were evaluated using two-electrode half-cells. For the fabrication of the working electrode, a mixture of active materials (MoS_2/MoO_3@Graphene or MoS_2), C45 conductive carbon and polystyrene acrylic-acrylate (mass ratio of 7:2:1) were thoroughly ground employing water solvent. Then, the slurry was distributed uniformly onto a copper foil (200 μm). After drying at 80 °C (10 h), the electrodes with the active material mass loading of around 1.1 mg cm^{-2} were obtained. CR-2025 coin cells were assembled using metallic Li discs as both the reference/counter electrode. For this, the solution containing 1.0 M lithium $LiPF_6$ in EC:DEC:EMC (1:1:1 wt%) was used as the electrolyte. In addition, Celgard 2400 films were employed as the separator. Coin cells were assembled in a glove-box (Mikrouna) under high purity Ar (O_2 and H_2O < 0.1 ppm). The cells were left for 10 h to equilibrate at room temperature, and subsequently evaluated using various techniques. Galvanostatic discharge/charge tests were performed within on a Land CT2001A battery testing instrument. Electrochemical impedance spectroscopy (EIS) and cyclic voltammetry at various scan rates and were conducted using a CHI 660E electrochemical interface.

5. Conclusions

In summary, we proposed a green and simple catalytic molten salt strategy for the conversion of commercial MoS_2, natural graphite and NaCl into MoO_3 nanoribbons crystals integrated with hexagonal MoS_2 nanocrystals and graphene nanosheets. The formation of interfacial molybdenum oxycarbide between graphene and MoO_3 is suggested, contributing to the reduced resistance of the hybrid material. The method is simple, involving the mechanochemical processing of the precursor materials, followed by a heat treatment process at 900–1200 °C in air for the duration of 20 min. The temperature was found to have a significant influence on the composition and morphology of the products. In this process, the thermal oxidation of MoS_2 causes the formation of molybdenum oxides, and also the formation of highly crystalline hexagonal MoS_2 nanocrystals integrated with graphene nanosheets. The Li-ion storage performances of hybrid nanostructures fabricated at various temperatures 900–1200 °C, namely MoO_3/MoS_2@G-900, -1000, -1100 and -1200 were evaluated. MoO_3/MoS_2@G-1100 was found to be capable of delivering a relatively high initial coulombic efficiency of 79.58% with a reversible capacity of 773.5 mAh g^{-1} (100 mA g^{-1}) after 180 cycles, outperforming other electrodes. This behavior was attributed to the highest content of MoO_3 in MoO_3/MoS_2@G-1100 and its unique microstructure. It was observed that MoO_3/MoS_2@G-1200 has the lowest amount of MoO_3 among all samples, but its Li-ion storage performance was second only to MoO_3/MoS_2@G-1100, recording at 505.9 mAh g^{-1} after 150 cycles at 100 mAh g^{-1}. The high performance of this sample was attributed to the presence β-MoO_3 with a highly ionically conductive open-structure monoclinic crystalline structure that only formed at 1200 °C. The formation of β-MoO_3 was concluded to contribute to the enhanced electrochemical performance of MoO_3/MoS_2@G-1200, despite its relatively low MoO_3 content. Graphene generated from the natural graphite and the hexagonal MoS_2 could be well-integrated, promoting both the electron and ion transportation across the electrode. Consequently, the contribution of pseudocapacitive Li-ion storage of the MoO_3/MoS_2@G-1100 electrode was characterized to be high at 59.31% under the scan rate of 0.5 mV s^{-1}. This resulted in the high rate performance of the electrode, providing a capacity of 216 mAh g^{-1} after 50 cycles at 2 A g^{-1}, and 106 mAh g^{-1} after 60 cycles at 5 A g^{-1}. This paper proposes a green and low-cost synthesis route for the conversion of highly available MoS_2 and non-purified natural graphite minerals into hexagonal molybdenum disulfide nanocrystals integrated with molybdenum trioxide and graphene nanosheets with enhanced Li-ion storage performance.

Supplementary Materials: The following supporting information can be downloaded at: https://www.mdpi.com/article/10.3390/catal13030499/s1, Figure S1: XRD patterns of MoO_3/MoS_2@G-900, MoO_3/MoS_2@G-1000, MoO_3/MoS_2@G-1100 and MoO_3/MoS_2@G-1200 around the (2 0 0) reflection of orthorhombic MoO_3, and the (0 0 1) reflection of monoclinic MoO_3; Figure S2: XRD pattern of MoO_3/MoS_2@G-900, MoO_3/MoS_2@G-1000, MoO_3/MoS_2@G-1100 and MoO_3/MoS_2@G-1200, around the (4 0 0) reflection of orthorhombic MoO_3, and the (0 0 2) reflection of monoclinic MoO_3; Figure S3: XRD patterns of MoO_3/MoS_2@G-900, MoO_3/MoS_2@G-1000, MoO_3/MoS_2@G-1100 and MoO_3/MoS_2@G-1200, around the (6 0 0) reflection of orthorhombic MoO_3, and the (0 0 3) reflection of monoclinic MoO_3.; Figure S4: Raman spectra of various precursor materials and products obtained at various temperatures.; Figure S5: UV-vis spectra of various precursor materials and products obtained at various temperatures; Figure S6: $(\alpha h\nu)^{1/2}$ versus hv plots according to UV–vis spectra of various precursor materials and products obtained at various temperatures; Figure S7: SEM image of (a) commercial MoS_2 and the samples prepared at (b) 900, (c) 1000, (d) 1100, and (e–f) 1200 °C. (g) High magnification image taken of the sample prepared at 1200 °C, and (h) the EDS analysis recorded on hexagonal MoS_2 shown in (g).; Figure S8: TEM micrographs of rod-like MoO_3 observed in MoO_3/MoS_2@G-900; Figure S6: Summary of methods proposed for the preparation of hexagonally-shaped nanocrystals extracted from the literature; Figure S9: Appearance of MoO_3/MoS_2@G-1100, the same sample after heating at 1200 °C for 5 min.; Figure S10: XRD patterns of MoO_3/MoS_2@G-1100 and the sample obtained by heating of MoO_3/MoS_2@G-1100 at 1200 °C for 5 min; Figure S11: Cycling performance of MoO_3/MoS_2@G-1100 at 500 mA g^{-1}; Figure S12: Summary of methods proposed for the preparation of hexagonally-shaped nanocrystals extracted from the literature; Table S1: XRD peak intensity ratio corresponding to various peaks in MoO_3/MoS_2@G-900, MoO_3/MoS_2@G-1000, MoO_3/MoS_2@G-1100 and MoO_3/MoS_2@G-1200; Table S2: Band gap values of various precursor materials and products obtained at various temperatures. and Table S3: A comparison between precursor materials and the methods used for the preparation of hexagonal crystals extracted from the literature, with those produced in the current study [31–36,77,78].

Author Contributions: Conceptualization, A.R.K.; methodology, A.R.K.; software, W.Z. and A.R.K.; validation, W.Z. and A.R.K.; formal analysis, W.Z. and A.R.K.; investigation, W.Z. and A.R.K.; resources, A.R.K.; data curation, W.Z. and A.R.K.; writing—original draft preparation, W.Z. and A.R.K.; writing—review and editing, A.R.K.; visualization, W.Z. and A.R.K.; supervision, A.R.K.; project administration, A.R.K.; funding acquisition, A.R.K. All authors have read and agreed to the published version of the manuscript.

Funding: This research was funded by the National Natural Science Foundation of China, grant number 52250610222. The APC for this invited article was waived by the journal Catalysts.

Data Availability Statement: Data is contained within the article and Supplementary Material.

Conflicts of Interest: The authors declare no conflict of interest.

References

1. Zubi, G.; Dufo-López, R.; Carvalho, M.; Pasaoglu, G. The lithium-ion battery: State of the art and future perspectives. *Renew. Sustain. Energy Rev.* **2018**, *89*, 292. [CrossRef]
2. Chen, W.D.; Liang, J.; Yang, Z.H.; Li, G. A Review of lithium-ion battery for electric vehicle applications and beyond. *Energy Procedia* **2019**, *158*, 4363. [CrossRef]
3. Tian, Y.; Zeng, G.; Rutt, A.; Shi, T.; Kim, H.; Wang, J.; Koettgen, J.; Sun, Y.; Ouyang, B.; Chen, T.; et al. Promises and challenges of next-generation "beyond Li-ion" batteries for electric vehicles and grid decarbonization. *Chem. Rev.* **2021**, *121*, 1623–1669. [CrossRef] [PubMed]
4. Li, R.; Kamali, A.R. Clean preparation of Fe_2SiO_4 coated Fe_2O_3 integrated with graphene for Li-ion storage application. *Colloids Surf. A* **2023**, *656*, 130275. [CrossRef]
5. Turcheniuk, K.; Bondarev, D.; Amatucci, G.G.; Yushin, G. Battery materials for low-cost electric transportation. *Mater. Today* **2021**, *42*, 57–72. [CrossRef]
6. Fang, S.; Bresser, D.; Passerini, S. Transition metal oxide anodes for electrochemical energy storage in lithium- and sodium-ion batteries. *Adv. Energy Mater.* **2020**, *10*, 1902485. [CrossRef]
7. Zhang, H.; Yang, Y.; Ren, D.S.; Wang, L.; He, X.M. Graphite as anode materials: Fundamental mechanism, recent progress and advances. *Energy Storage Mater.* **2021**, *36*, 147–170. [CrossRef]
8. Lee, S.; Youn, S.; Yong, C.; Chung, S.; Chui, Y.; Lee, S. High electrical and thermal conductivities of a PAN-based carbon fiber via boron-assisted catalytic graphitization. *Carbon* **2022**, *199*, 70–79. [CrossRef]
9. Bonijoly, M.; Oberlin, M.; Oberlin, A. A possible mechanism for natural graphite formation. *Int. J. Coal Geol.* **1982**, *1*, 283–312. [CrossRef]
10. Zhao, H.; Rezaei, A.; Kamali, A.R. Electrolytic conversion of natural graphite into carbon nanostructures with enhanced electrical conductivity and Na-ion storage performance. *J. Electrochem. Soc.* **2022**, *169*, 054512. [CrossRef]
11. Zhao, L.; Ding, B.C.; Qin, X.Y.; Wang, Z.J.; Lv, W.; He, Y.B.; Yang, Q.H.; Kang, F.Y. Revisiting the roles of natural graphite in ongoing lithium-ion batteries. *Adv. Mater.* **2022**, *32*, 2106704. [CrossRef] [PubMed]
12. U.S. Geological Survey. Mineral Commodity Summaries, Graphite (Natural), January 2022. Available online: https://pubs.usgs.gov/periodicals/mcs2022/mcs2022.pdf (accessed on 1 January 2022).
13. Zaghib, K.; Song, X.; Guerfi, A.; Rioux, R.; Kinoshit, K. Purification process of natural graphite as anode for Li-ion batteries: Chemical versus thermal. *J. Power Source* **2003**, *119–121*, 8–15. [CrossRef]
14. Zhao, Y.; Yang, L.; Ma, C.; Han, G. One-step fabrication of fluorine-doped graphite derived from a low-grade microcrystalline graphite ore for potassium-ion batteries. *Energy Fuels* **2020**, *34*, 8993–9001. [CrossRef]

15. Ani, T.A.; Leinonen, S.; Ahtola, T.; Salvador, D. High-grade flake graphite deposits in metamorphic schist belt, central Finland—Mineralogy and beneficiation of graphite for lithium-ion nattery applications. *Minerals* **2020**, *10*, 680. [CrossRef]
16. Asenbauer, J.; Eisenmann, T.; Kuenzel, M.; Kazzazi, A.; Chen, Z.; Bresser, D. The success story of graphite as a lithium-ion anode material—Fundamentals, remaining challenges, and recent developments including silicon (oxide) composites. *Sustain. Energy Fuels* **2020**, *4*, 5387–5416. [CrossRef]
17. Stephenson, T.; Li, Z.; Olsena, B.; Mitlin, D. Lithium ion battery applications of molybdenum disulfide (MoS_2) nanocomposites. *Energy Environ. Sci.* **2014**, *7*, 209. [CrossRef]
18. Wang, W.; Qin, J.W.; Yin, Z.G.; Cao, M.H. Achieving fully reversible conversion in MoO_3 for lithium ion batteries by rational introduction of $CoMoO_4$. *ACS Nano* **2016**, *10*, 10106–10116. [CrossRef]
19. Li, F.Y.; Cabrera, C.R.; Chen, Z.F. Theoretical design of MoO_3-based high-rate lithium ion battery electrodes: The effect of dimensionality reduction. *J. Mater. Chem. A* **2014**, *2*, 19180. [CrossRef]
20. Sahu, S.R.; Rikka, V.R.; Haridoss, P.; Chatterjee, A.; Gopalan, R.; Prakash, R. A novel α-MoO_3/single-walled carbon nanohorns composite as high-performance anode material for fast-charging lithium-ion battery. *Adv. Energy Mater.* **2020**, *10*, 2001627. [CrossRef]
21. Wang, T.X.; Li, J.; Zhao, G.L. Synthesis of MoS_2 and MoO_3 hierarchical nanostructures using a single-source molecular precursor. *Powder Technol.* **2014**, *253*, 347–351. [CrossRef]
22. Yang, J.; Qu, Y.D.; Lin, X.H.; Wang, L.Y.; Zheng, Z.H.; Zhuang, J.; Duan, L.F. MoO_3/MoS_2 flexible paper as sulfur cathode with synergistic suppress shuttle effect for lithium-sulfur batteries. *Electrochim. Acta* **2022**, *418*, 140378. [CrossRef]
23. Zhao, S.P.; Zha, Z.L.; Liu, X.; Tian, H.X.; Wu, Z.Y.; Li, W.F.; Sun, L.B.; Liu, B.; Chen, Z.G. Core−sheath structured MoO_3@MoS_2 composite for high performance lithium-ion battery anodes. *Energy Fuels* **2020**, *34*, 11498–11507. [CrossRef]
24. Du, J.L.; Wu, H.D.; Wang, X.R.; Qi, C.Y.; Mao, W.; Ren, T.Q.; Qiao, Q.D.; Yang, Z.X. Ternary MoS_2/MoO_3/C nanosheets as high-performance anode materials for lithium-ion batteries. *J. Electron. Mater.* **2018**, *47*, 6767. [CrossRef]
25. Vieira, L.; Neto, J.D.R.M.; Ferreira, O.P.; Torresi, R.M.; Torresi, S.I.C.D.; Alves, O.L. Template conversion of MoO_3 to MoS_2 nanoribbons: Synthesis and electrochemical properties. *RSC Adv.* **2018**, *8*, 30346–30353. [CrossRef]
26. Pareek, A.; Kim, H.G.; Paik, P.; Borse, P.H. Ultrathin MoS_2-MoO_3 nanosheets functionalized CdS photoanodes for effective charge transfer in photoelectrochemical (PEC) cells. *J. Mater. Chem. A* **2017**, *5*, 1541. [CrossRef]
27. Villevieille, C.; Wang, X.J.; Krumeich, F.; Nesper, R.; Novák, P. MoS_2 coating on MoO_3 nanobelts: A novel approach for a high specific charge electrode for rechargeable Li-ion batteries. *J. Power Source* **2015**, *279*, 636–644. [CrossRef]
28. Liu, H.; Chen, X.J.; Deng, L.; Ding, M.; Li, J.Q.; He, X.M. Perpendicular growth of few-layered MoS_2 nanosheets on MoO_3 nanowires fabricated by direct anion exchange reactions for high performance lithium-ion batteries. *J. Mater. Chem. A* **2016**, *4*, 17764. [CrossRef]
29. Li, B.; Yang, S.X.; Huo, N.J.; Li, Y.T.; Yang, J.H.; Li, R.X.; Fan, C.; Lu, F.Y. Growth of large area few-layer or monolayer MoS_2 from controllable MoO_3 nanowire nuclei. *RSC Adv.* **2014**, *4*, 26407. [CrossRef]
30. Zhu, W.; Kamali, A.R. Green preparation of nanostructured β-MoO_3/hexagonal-shaped MoS_2/graphene with enhanced lithium-ion storage performance. *J. Alloys Compd.* **2023**, *932*, 167724. [CrossRef]
31. Carlsson, A.; Brorson, M.; Topsoe, H. Supported metal sulphide nanoclusters studied by HAADF-STEM. *J. Microsc.* **2006**, *223*, 179. [CrossRef]
32. Cai, G.M.; Jian, J.K.; Chen, X.L.; Lei, M.; Wang, W.Y. Regular hexagonal MoS_2 microflakes grown from MoO_3 precursor. *Appl. Phys.* **2007**, *89*, 783. [CrossRef]
33. Li, P.G.; Lei, M.; Wang, X.F.; Tang, H.L.; Tang, W.H. Thermal conversion of tungsten oxide nanorods to tungsten disulfide nanoflakes. *J. Alloys Compd.* **2009**, *474*, 463. [CrossRef]
34. Leong, S.X.; Mayorga-Martinez, C.C.; Chia, X.Y.; Luxa, J.; Sofer, Z.; Pumera, M. 2H $\rightarrow$ 1T phase change in direct synthesis of WS_2 nanosheets via solution-based electrochemical exfoliation and their catalytic properties. *ACS Appl. Mater. Interfaces* **2017**, *9*, 26350. [CrossRef] [PubMed]
35. Ai, R.Q.; Guan, X.; Li, J.; Yao, K.K.; Chen, P.; Zhang, Z.W.; Duan, X.D.; Duan, X.F. Growth of single-crystalline cadmium iodide nanoplates, CdI_2/MoS_2 (WS_2, WSe_2) van der waals heterostructures, and patterned arrays. *ACS Nano* **2017**, *11*, 3413. [CrossRef] [PubMed]
36. Hu, Y.; Chen, T.; Wang, X.Q.; Ma, L.B.; Chen, R.P.; Zhu, H.F.; Yuan, X.; Yan, C.Z.; Zhu, G.Y.; Lv, H.L.; et al. Controlled growth and photoconductive properties of hexagonal SnS_2 nanoflakes with mesa-shaped atomic steps. *Nano Res.* **2017**, *10*, 1434. [CrossRef]
37. Qu, Q.; Zhang, W.B.; Huang, K.; Chen, H.M. Electronic structure, optical properties and band edges of layered MoO_3: A first-principles investigation. *Comput. Mater. Sci.* **2017**, *130*, 242–248. [CrossRef]
38. Kamali, A.R.; Fray, D.J. Solid phase growth of tin oxide nanostructures. *Mater. Sci. Eng. B* **2012**, *177*, 819–825. [CrossRef]
39. He, Z.K.; Sun, Q.; Shi, Z.; Xie, K.; Kamali, A.R. Molten salt synthesis of oxygen-deficient SnO_2 crystals with enhanced electrical conductivity. *Appl. Surf. Sci.* **2019**, *465*, 397–404. [CrossRef]
40. Deng, X.Q.; Zhu, M.H.; Ke, J.; Feng, Y.F.; Li, W.R.; Xiong, D.P.; He, M. Synthesis and electrochemical performances of ternary nanocomposite SnO_2@MoO_3@graphene as high-performance anode material for lithium-ion batteries. *Chem. Phys. Lett.* **2021**, *770*, 138408. [CrossRef]
41. Liu, X.X.; Wu, Y.; Wang, H.W.; Wang, Y.F.; Huang, C.F.; Liu, L.M.; Wang, Z.J. Two-dimensional β-MoO_3@C nanosheets as high-performance negative materials for supercapacitors with excellent cycling stability. *RSC Adv.* **2020**, *10*, 17497–17505. [CrossRef]

42. Zhu, J.; Uslamin, E.A.; Kosinov, N.; Hensen, E.J.M. Tuning the reactivity of molybdenum (oxy)carbide catalysts by the carburization degree: CO_2 reduction and anisole hydrodeoxygenation. *Catal. Sci. Technol.* **2020**, *10*, 3635–3645. [CrossRef]
43. Zheng, Y.; Tang, Y.; Gallagher, J.R.; Gao, J.; Miller, J.T.; Wachs, I.E.; Podkolzin, S.G. Molybdenum oxide, oxycarbide, and carbide: Controlling the dynamic composition, size, and catalytic activity of zeolite supported nanostructures. *J. Phys. Chem. C* **2019**, *123*, 22281–22292. [CrossRef]
44. Kado, T. Epitaxial molybdenum oxycarbide thin films synthesized by inductively coupled radio-frequency plasma assisted magnetron sputtering. *Thin Solid Film.* **2006**, *515*, 2481–2484. [CrossRef]
45. Shahsank, M.; Naik, H.S.B.; Sumedha, H.N.; Nagaraju, G. Implementing an in-situ carbon formation of MoO_3 nanoparticles for high performance lithium-ion battery. *Ceram. Int.* **2021**, *47*, 10261–10267. [CrossRef]
46. Yoon, A.; Kim, J.H.; Yoon, J.C.; Lee, Y.D.; Lee, Z. Van der Waals Epitaxial Formation of Atomic Layered α-MoO_3 on MoS_2 by Oxidation. *ACS Appl. Mater. Interfaces* **2020**, *12*, 19. [CrossRef] [PubMed]
47. Kamali, A.R.; Fray, D.J. Molten salt corrosion of graphite as a possible way to make carbon nanostructures. *Carbon* **2013**, *56*, 121–131. [CrossRef]
48. Kamali, A.R.; Yang, J. Effect of molten salts on the structure, morphology and electrical conductivity of PET-derived carbon nanostructures. *Polym. Degrad. Stab.* **2020**, *177*, 109184. [CrossRef]
49. Kamali, A.R.; Schwandt, C.; Fray, D.J. On the oxidation of electrolytic carbon nanomaterials. *Corros. Sci.* **2012**, *54*, 307–313. [CrossRef]
50. Kamali, A.R.; Divitini, G.; Schwandt, C.; Fray, D.J. Correlation between microstructure and thermokinetic characteristics of electrolytic carbon nanomaterials. *Corros. Sci.* **2012**, *64*, 90–97. [CrossRef]
51. Wu, J.; Li, H.; Yin, Z.; Li, H.; Liu, J.; Cao, X.; Zhang, Q.; Zhang, H. Layer thinning and etching of mechanically exfoliated MoS_2 nanosheets by thermal annealing in air. *Small* **2013**, *9*, 3314–3319.
52. Zhou, H.; Yu, F.; Liu, Y.; Zou, X.; Cong, C.; Qiu, C.; Yu, T.; Yan, Z.; Shen, X.; Sun, L.; et al. Thickness-dependent patterning of MoS_2 sheets with well-oriented triangular pits by heating in air. *Nano Res.* **2013**, *6*, 703–711. [CrossRef]
53. Rao, R.; Islam, A.E.; Campbell, P.M.; Vogel, E.M.; Maruyama, B. In situ thermal oxidation kinetics in few layer MoS_2. *2D Mater.* **2017**, *4*, 025058. [CrossRef]
54. Afanasiev, P.; Lorentz, C. Oxidation of nanodispersed MoS_2 in ambient air: The products and the mechanistic steps. *J. Phys. Chem. C* **2019**, *123*, 7486–7494. [CrossRef]
55. Spychalski, W.L.; Pisarek, M.; Szoszkiewicz, R. Microscale insight into oxidation of single MoS_2 crystals in air. *J. Phys. Chem. C* **2017**, *121*, 26027–26033. [CrossRef]
56. Szoszkiewicz, R.; Rogala, M.; Dąbrowski, P. Surface-bound and volatile Mo oxides produced during oxidation of single MoS_2 crystals in air and high relative humidity. *Materials* **2020**, *13*, 3067. [CrossRef]
57. Kamali, A.R. *Green Production of Carbon Nanomaterials in Molten Salts and Applications*; Springer Nature: Berlin/Heidelberg, Germany, 2020.
58. Hu, X.L.; Zhang, W.; Liu, X.X.; Mei, Y.N.; Huang, Y.H. Nanostructured Mo-based electrode materials for electrochemical energy storage. *Chem. Soc. Rev.* **2015**, *44*, 2376. [CrossRef] [PubMed]
59. Liu, Y.; Zhu, J.X.; Xu, J.S.; Liu, S.L.; Li, L.; Zhang, C.; Liu, T.X. High-temperature solvent-free sulfidation of MoO_3 confined in a polypyrrole shell: MoS_2 nanosheets encapsulated in a nitrogen, sulfur dual-doped carbon nanoprism for efficient lithium storage. *Nanoscale* **2018**, *10*, 7536–7543. [CrossRef]
60. Li, J.; Hou, S.; Liu, T.Z.; Wang, L.K.; Mei, C.; Guo, Y.Y.; Zhao, L.Z. Hierarchical hollow-nanocube Ni−Co skeleton@MoO_3/MoS_2 hybrids for improved-performance lithium-ion batteries. *Chem. Eur. J.* **2020**, *26*, 2013–2024. [CrossRef] [PubMed]
61. Zhou, H.; Xia, X.; Lv, P.; Zhang, J.; Hou, X.; Zhao, M.; Ao, K.; Wang, D.; Lu, K.; Qiao, H.; et al. C@TiO_2/MoO_3 composite nanofibers with 1T-phase MoS_2 nanograin dopant and stabilized interfaces as anodes for Li- and Na-ion batteries. *ChemSusChem* **2018**, *11*, 4060–4070. [CrossRef]
62. Liu, H.W.; Wang, J. One-pot synthesis of $ZnCo_2O_4$ nanorod anodes for high power Lithium ions batteries. *Electrochim. Acta* **2013**, *92*, 371. [CrossRef]
63. Zhang, J.J.; He, P.; Xia, Y.Y. Electrochemical kinetics study of Li-ion in Cu_6Sn_5 electrode of lithium batteries by PITT and EIS. *J. Electroanal. Chem.* **2008**, *624*, 161. [CrossRef]
64. Lei, D.; Shang, W.Z.; Zhang, X.; Li, Y.P.; Qiao, S.M.; Zhong, Y.P.; Deng, X.Y.; Shi, X.S.; Zhang, Q.; Hao, C.; et al. Facile synthesis of heterostructured MoS_2-MoO_3 nanosheets with active electrocatalytic sites for high-performance lithium−sulfur batteries. *ACS Nano* **2021**, *15*, 20478–20488. [CrossRef]
65. Zhao, B.; Song, D.Y.; Ding, Y.W.; Li, W.R.; Wang, Z.X.; Jiang, Y.; Zhang, J.J. Size-tunable SnS_2 nanoparticles assembled on graphene as anodes for high performance lithium/sodium-ion batteries. *Electrochim. Acta* **2020**, *354*, 13673. [CrossRef]
66. Zhang, P.C.; Cai, B.; Feng, Y.; Pan, H.; Yao, J.F. Constructing MoO_3@MoO_2 heterojunction on g-C_3N_4 nanosheets with advanced Li-ion storage ability. *J. Alloys Compd.* **2021**, *875*, 160077. [CrossRef]
67. Kamali, A.R.; Li, S. Molten salt-assisted valorization of waste PET plastics into nanostructured SnO_2@terephthalic acid with excellent Li-ion storage performance. *Appl. Energy* **2023**, *334*, 120692. [CrossRef]
68. Zhu, W.; Kamali, A.R. Green molten salt synthesis and Li-ion storage performance of sodium dimolybdate. *J. Alloys Compd.* **2020**, *831*, 154781. [CrossRef]

69. Kamali, A.R.; Ye, J. Reactive molten salt modification of ilmenite as a green approach for the preparation of inexpensive Li ion battery anode materials. *Miner. Eng.* **2021**, *172*, 107175. [CrossRef]
70. Zhuan, Y.; Ma, J.L.; Feng, W.J. Highlighting the implantation of metal particles into hollow cavity yeast-based carbon for improved electrochemical performance of lithium–sulfur batteries. *Catalysts* **2022**, *12*, 951. [CrossRef]
71. Zeng, G.; Wang, J.; Dai, M.; Meng, Y.; Luo, H.; Lin, L.; Zang, K.; Meng, Z.; Pan, X. Natural iron minerals in an electrocatalytic oxidation system and in situ pollutant removal in groundwater: Applications, mechanisms, and challenges. *Sci. Total Environ.* **2023**, *871*, 161826. [CrossRef]
72. Enneffatia, M.; Rasheed, M.; Louatia, B.; Guidaraa, K.; Shihab, S.; Barillé, R. Investigation of structural, morphology, optical properties and electrical transport conduction of $Li_{0.25}Na_{0.75}CdVO_4$ compound. *J. Phys. Conf. Ser.* **2021**, *1795*, 012050. [CrossRef]
73. Rasheed, M.; Shihab, S.; Sabah, O.W. An investigation of the Structural, Electrical and Optical Properties of Graphene-Oxide Thin Films Using Different Solvents. *J. Phys. Conf. Ser.* **2021**, *1795*, 012052. [CrossRef]
74. Nazarian-Samani, M.; Kamali, A.R.; Mobarra, R.; Nazarian-Samani, M. Phase transformations of Ni-15 wt. % B powders during mechanical alloying and annealing. *Mater. Lett.* **2010**, *64*, 309–312. [CrossRef]
75. Seifi, T.; Kamali, A.R. Enhanced dispersion and antibacterial activity of mechanically exfoliated graphite flakes in the presence of n-hexane and NaCl. *Mater. Lett.* **2021**, *304*, 130730. [CrossRef]
76. Seifi, T.; Kamali, A.R. The influence of mechanochemical treatment in hexane on dispersibility and floatability of graphite flakes with enhanced water evaporation performance. *Colloids Surf. A* **2022**, *638*, 128326. [CrossRef]
77. Wang, Z.G.; Lia, Q.; Xu, H.X.; Dahl-Petersen, C.; Yang, Q.; Cheng, D.J.; Cao, D.P.; Besenbacher, F.; Lauritsen, J.V.; Helveg, S.; et al. Controllable etching of MoS_2 basal planes for enhanced hydrogen evolution through the formation of active edge sites. *Nano Energy* **2018**, *49*, 634–643. [CrossRef]
78. Yorulmaz, B.; Özden, A.; Şar, H.; Ay, F.; Sevik, C.; Perkgöz, N.K. CVD growth of monolayer WS_2 through controlled seed formation and vapor density. *Mater. Sci. Semicond. Process.* **2019**, *93*, 158. [CrossRef]

catalysts

MDPI

Article

Unveiling the Role of In Situ Sulfidation and H_2O Excess on H_2S Decomposition to Carbon-Free H_2 over Cobalt/Ceria Catalysts

Tzouliana Kraia [1], Georgios Varvoutis [2,*], George E. Marnellos [1,2] and Michalis Konsolakis [3,*]

1 Centre for Research & Technology Hellas, Chemical Process & Energy Resources Institute, GR-57001 Thessaloniki, Greece
2 Department of Mechanical Engineering, University of Western Macedonia, GR-50100 Kozani, Greece
3 School of Production Engineering and Management, Technical University of Crete, GR-73100 Chania, Greece
* Correspondence: gvarvoutis@uowm.gr (G.V.); mkonsolakis@tuc.gr (M.K.)

Abstract: The emerging energy and environmental concerns nowadays are highlighting the need to turn to clean fuels, such as hydrogen. In this regard, hydrogen sulfide (H_2S), an abundant chemical compound found in several natural sources and industrial streams, can be considered a potential carbon-free H_2 source through its decomposition. In the present work, the H_2S decomposition performance of Co_3O_4/CeO_2 mixed oxide catalysts toward hydrogen production is investigated under excess H_2O conditions (1 *v/v*% H_2S, 90 *v/v*% H_2O, Ar as diluent), simulating the concentrated H_2S-H_2O inflow by the Black Sea deep waters. The effect of key operational parameters such as feed composition, temperature (550–850 °C), and cobalt loading (0–100 wt.%) on the catalytic performance of Co_3O_4/CeO_2 catalysts was systematically explored. In order to gain insight into potential structure-performance relationships, various characterization studies involving BET, XRD, SEM/EDX, and sulfur elemental analysis were performed over the fresh and spent samples. The experimental results showed that the 30 wt.% Co/CeO_2 catalyst demonstrated the optimum catalytic performance over the entire temperature range with a H_2 production rate of ca. 2.1 μmol $H_2\cdot g^{-1}\cdot s^{-1}$ at 850 °C and a stable behavior after 10 h on stream, ascribed mainly to the in-situ formation of highly active and stable cobalt sulfided phases.

Keywords: H_2S decomposition/reforming; H_2O excess conditions; carbon-free H_2 production; Co_3O_4/CeO_2 catalysts; cobalt sulfide phase

Citation: Kraia, T.; Varvoutis, G.; Marnellos, G.E.; Konsolakis, M. Unveiling the Role of In Situ Sulfidation and H_2O Excess on H_2S Decomposition to Carbon-Free H_2 over Cobalt/Ceria Catalysts. *Catalysts* **2023**, *13*, 504. https://doi.org/10.3390/catal13030504

Academic Editors: Georgios Bampos, Athanasia Petala and Zacharias Frontistis

Received: 11 January 2023
Revised: 24 February 2023
Accepted: 26 February 2023
Published: 28 February 2023

1. Introduction

The need to mitigate climate change induced by man-made activities over the past centuries is now a global imperative, implicitly highlighted by the latest report of the Intergovernmental Panel for Climate Change [1]. To this end, commitments under the EU Green Deal strategy and the REPowerEU plan, along with the Paris and Glasgow agreements, emphasize the necessity for the transition towards zero-carbon energy vectors [2,3]. In this regard, the partial or complete replacement of fossil fuels in the future energy mix can be realized by means of an economy based on the extensive use of CO_2-neutral hydrogen [4,5]. Indeed, H_2 is regarded as a promising energy carrier, which, depending on its origin and generation method, can be totally detached from CO_2 emissions and extensively utilized in a wide range of energy applications and chemical processes towards their full-scale decarbonization [6,7]. Hydrogen has been used in large quantities for well over 100 years as a non-energy feedstock in the production of fertilizers, methanol synthesis, petroleum refineries, and as an agent for the direct reduction of iron in the steel industry, while its current demand at global scale amounts to ca. 100–120 Mt H_2/yr [8,9]. However, in order to sufficiently meet the ambitious goals for a carbon-neutral economy, hydrogen would need to reach shares of around 15% of the world's energy demand by mid-century and increase its production capacity more than three-fold, to 300–350 Mt H_2/yr [4].

Generally, hydrogen is the most abundant element in the universe, albeit it is only available bounded in compounds such as fossil and bio-based hydrocarbons, gases (e.g., NH_3, H_2S), and water. It can be produced using a number of different methods with varying efficiencies, environmental footprint, costs, and technical maturity [8–10] and is typically classified into colors depending on the method and feedstock used, ranging from fossil-fuel-based hydrogen production (grey hydrogen) combined in several occasions with carbon capture, utilization, and storage (CCUS; blue hydrogen), to electrolytic hydrogen from renewable electricity (green hydrogen) or nuclear power plants (pink hydrogen) or grid electricity (yellow hydrogen), and to biohydrogen production by photolysis (orange hydrogen) and waste thermochemical conversion, among others [9]. In an attempt to improve the sustainability and competitiveness over other alternatives, research efforts on hydrogen production are mainly directed to: (a) increase in energy efficiency (less energy consumption per MJ of recovered/produced hydrogen), (b) decrease in the capital expenditures (CAPEX) and the levelized cost of hydrogen, (c) minimization or even elimination of the use of critical raw materials in key process components such as catalysts, electrodes, etc., (d) development of new pathways from CO_2-neutral or carbon-free sources and (e) leveling-up the readiness of the less matured technologies [11–13].

Among the conventional H_2 production pathways, an alternative route concerns the utilization of hydrogen sulfide (H_2S) as a potential carbon-free source of H_2 [14,15]. H_2S is a toxic and corrosive gas, and its concentration in the air ranges between 0.11–0.33 ppb, albeit it is found in much higher levels in several natural sources and H_2S-emitting industrial activities. Hydrogen sulfide is typically emitted from a variety of industrial processes, such as petroleum refineries, crude oil and natural gas production, wastewater treatment plants, coke ovens, tanneries, and kraft paper mills [16]. Moreover, H_2S is also naturally present in geothermal sources and springs [17], coal seams [18], agricultural fields [19], and marine sediments [20].

Industrially, H_2S emissions are predominantly managed via the Claus technology producing elemental sulfur and water, accounting for 90–95% of the total recovered sulfur. The Claus process achieves 95–98% recovery of the hydrogen sulfide feed stream [21]. On the downside, this process is energy-intensive due to the high operational temperature, commonly above 1000 °C. Additionally, secondary pollution by the unreacted H_2S and SO_2 species is a concern, whereas the H_2O by-product is undesirable for several downstream processes [22]. Alternatively, H_2S decomposition to its elemental constituents, namely H_2 and sulfur, could be a more alluring prospect. In this direction, the abatement of H_2S is combined with the simultaneous production of valuable products. Therefore, numerous technologies have been recently investigated for H_2S decomposition [23], mainly involving: thermal [24,25], catalytic [26,27], plasmochemical [28,29], electrochemical [30], and photochemical [31] decomposition. The majority of these approaches, however, present several drawbacks attributed to the particular high energy requirements and low efficiencies, rendering their practical deployment infeasible as of yet [32]. In this sense, the most commonly employed direct method is the thermal decomposition of H_2S, which, however, takes place at high temperatures in order to achieve conversions exceeding 50%. This is due to the fact that the cleavage of the H_2S molecule into its constituent elements requires a substantial amount of energy since the thermal decomposition of H_2S is an endothermic reaction with a highly positive Gibbs free energy [33,34]. Specifically, the gas-phase H_2S dissociation standard enthalpy and standard Gibbs free energy values are equal to 317.9 and 270.1 kJ/mol, respectively [14].

As mentioned above, appreciable amounts of H_2S exist in aquatic sediments, with particular interest being directed in the Black Sea waters, due to their high H_2S content that is replenished continuously [35–37]. Owing to the specific locational and biological characteristics of the Black Sea, a large amount of H_2S lies in the anoxic deeper layers with a highly toxic character [38]. Hence, an efficient method for H_2S exploitation is actively sought. Hydrogen production from H_2S in the Black Sea deep waters could play an important role in the clean energy transition of the surrounding countries offering at the same time significant environmental benefits toward the remediation of the Black Sea ecosystem.

Apart from thermal decomposition studies, research on the conversion of H_2S from the Black Sea based on electrochemical methods has also been conducted. H_2 production from H_2S contained in the Black Sea deep waters has been examined by employing a micro-structured electrochemical membrane reactor, which is assembled with optimized cell materials. CeO_2-based transition metal catalysts were employed as anode materials, with ceria-supported cobalt catalysts displaying the best performance in terms of H_2S decomposition activity and long-term stability [35,39]. In a more holistic approach, Petrov et al. [40] investigated a multistage process involving the electrochemical production of H_2 and polysulfides via direct alkaline electrolysis of H_2S. Raney-nickel, graphite, platinized carbon, CoS, and perovskites were tested as catalysts, with CoS and perovskites being highly stable and efficient. Elsewhere, perovskite-type catalysts were also tested for the thermal decomposition of H_2S [41,42], with $LaSr_{0.5}Mo_{0.5}O_3$, in particular, displaying the highest performance at temperatures ranging from 700–850 °C.

Recently, ceria-based transition metal catalysts have gained particular attention in the field of heterogeneous catalysis due to their unique solid-state properties, mainly linked to synergistic metal–support interactions [43–45]. In this regard, in our previous work [46], we thoroughly explored the impact of metal nature on the H_2S decomposition performance of a series of M_xO_y/CeO_2 oxides (M: Fe, Co, Ni, Cu) catalysts under atmospheric pressure and dry conditions (i.e., absence of H_2O). The results clearly revealed the superiority of Co_3O_4/CeO_2 composites, in terms of both activity and stability, offering H_2S conversions close to the thermodynamic predicted values (ca. 35% at 850 °C).

Motivated by these findings, the H_2S decomposition performance of Co_3O_4/CeO_2 catalysts is hereby investigated under extremely excess H_2O contents (1 *v/v*% H_2S, 90 *v/v*% H_2O, Ar as diluent) to better approach Black Sea inflow conditions. The impact of various operational parameters such as the feed composition, temperature, and Co loading (0–100 wt.%) on the performance and kinetics of as-prepared cobalt oxide–ceria catalysts was systematically investigated. In addition, the physicochemical properties of both fresh and spent catalysts were assessed by complementary characterization studies to gain insight into potential structure-performance relationships. To the best of our knowledge, this is the first study on the catalyst-aided H_2S decomposition reaction under harsh reaction conditions (water content of 90 *v/v*%), adhering to the dual purpose of environmental mitigation and carbon-free H_2 production.

2. Results and Discussion

2.1. Characterization Studies

The main textural and structural characteristics of the Co_3O_4/CeO_2 catalysts with different Co loadings (0–100 wt.%) are displayed in Table 1. As expected, bare CeO_2 possesses the highest BET surface area (71.5 m^2/g) and pore volume (0.27 cm^3/g). On the other hand, bare Co_3O_4 exhibits the lowest surface area (2.9 m^2/g) and pore volume (0.01 cm^3/g). Cobalt oxide incorporation into ceria carrier results in a decrease in surface area and pore volume and a small increase in pore size, which could be attributed to the inferior textural characteristics of bare Co_3O_4, as shown in relevant studies [47,48] and further discussed below. However, significant pore blockage phenomena upon the addition of Co_3O_4 to CeO_2 carrier were excluded by means of the Weisz–Prater criterion, which was calculated by the method followed in our previous work [49].

The XRD patterns of the examined materials are depicted in Figure 1, while the crystalline phases that were detected for each sample and the approximate crystallite sizes, as determined by Scherrer analysis, are provided in Table 1. In principle, all catalysts crystallized in the form of their respective oxides, while for bare cobalt oxide, both CoO and Co_3O_4 phases are detected. Specifically, Co is present in the form of Co_3O_4 in the mixed oxides, with its crystallite size rising slightly upon increasing cobalt content, suggesting the partial segregation of Co metal entities. Notably, the size of bare cobalt oxide crystallites is about 4.5 times higher than CeO_2 (ca. 50 vs. 11 nm), implying the significant segregation of the cobalt oxide phase by the formation of large Co_3O_4 crystallites. On the other hand, CeO_2 crystallite size remains practically unchanged upon increasing the loading of cobalt oxide, approximately 10–11 nm. Therefore, the progressive reduction of BET surface area of the mixed oxides upon increasing Co loading could be ascribed to the partial substitution of high-surface-area CeO_2 with the low-surface area Co_3O_4, in conjunction with the partial pore blockage by large cobalt oxide crystallites. Similar results have been reported in relevant works [50–53].

Table 1. Textural and structural properties of bare oxides and x-Co/CeO_2 catalysts.

Sample	N_2 Porosimetry			XRD Analysis	
	BET Surface Area (m^2/g)	Pore Volume (cm^3/g)	Average Pore Diameter (nm)	Crystal Phase	Crystallite Size (nm)
CeO_2 [a]	71.5	0.27	15.4	CeO_2	11.0
20-Co/CeO_2 [a]	33.4	0.13	16.0	CeO_2 Co_3O_4	10.2 37.7
30-Co/CeO_2	44.9	0.21	18.7	CeO_2 Co_3O_4	10.4 37.9
40-Co/CeO_2	28.4	0.10	14.7	CeO_2 Co_3O_4	10.5 41.7
60-Co/CeO_2	15.1	0.07	19.3	CeO_2 Co_3O_4	10.5 42.2
Co_3O_4	2.9	0.01	17.8	Co_3O_4 CoO	50.7 52.7

[a] Data taken from Ref. [46].

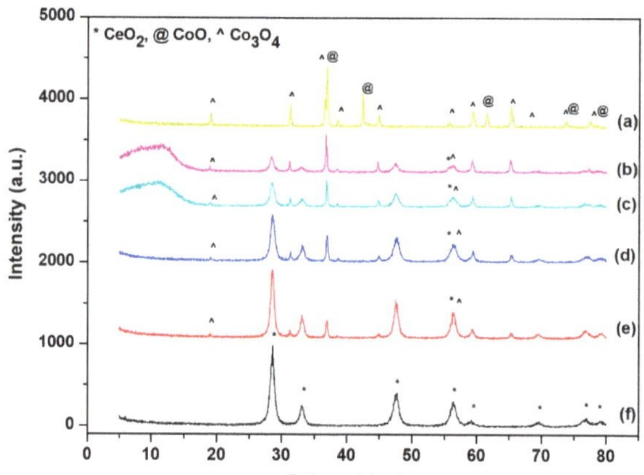

Figure 1. X-ray powder diffraction patterns of x-Co/CeO_2 catalysts with a mass Co loading of 100% (**a**), 60% (**b**), 40% (**c**), 30% (**d**), 20% (**e**) and 0% (**f**). Data for (**e**,**f**) are taken from Ref. [46].

2.2. Catalytic Evaluation

It should be stated first that as a point of reference, the H_2 production rate achieved in the absence of catalysts (blank experiments) under (i) 1 *v/v*% H_2S and 90 *v/v*% H_2O, (ii) 1.0 *v/v*% H_2S and (iii) 90 *v/v*% H_2O, balanced with Ar at the temperature range of 550–850 °C was examined. The results from these preliminary experiments, along with the corresponding experimental and equilibrium H_2S conversion values for the H_2S decomposition reaction in the absence of steam, are depicted in Figure 2. Clearly, the conversion of

H_2O and the resulting hydrogen evolution rate are negligible for the 90 *v/v*% H_2O feedstock in the entire temperature range, indicating no contribution of thermal water splitting towards hydrogen generation at these experimental conditions. This is also confirmed by the corresponding equilibrium curve, which lies at very low conversion values and is not shown for brevity. Moreover, H_2S conversion under dry conditions is lower than ca. 8% over the entire investigated temperature range, far below the corresponding thermodynamic values, implying the significant kinetic limitations of the homogeneous reaction, as further discussed below.

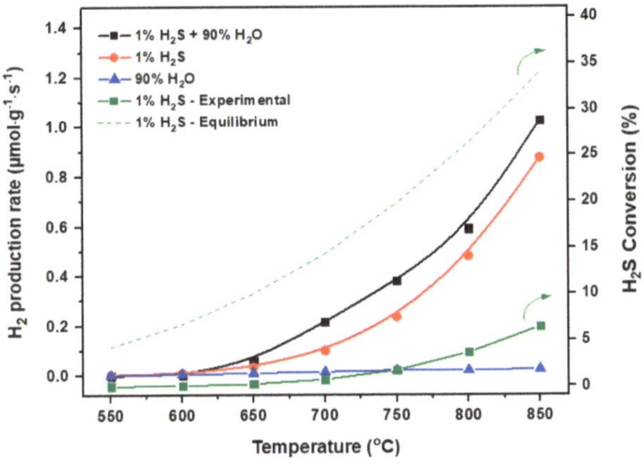

Figure 2. Left *y*-axis: Hydrogen production rate as a function of temperature and feedstock composition in the absence of catalyst (blank experiments). Right *y*-axis: Experimental and equilibrium H_2S conversion values for the dry H_2S decomposition reaction (green lines). Reaction conditions: P = 1 bar, Flowrate: 100 cm^3/min.

However, in the case of H_2S decomposition, significantly higher hydrogen production rates were obtained as compared to pure H_2O (0.87 vs. 0.01 $\mu mol \cdot g^{-1} \cdot s^{-1}$ at 850 °C, respectively). Furthermore, higher H_2 production rates were observed with an increase in reaction temperature attaining a maximum at 850 °C, as reported elsewhere for the non-catalytic thermal H_2S decomposition [14,34,54]. This behavior is thermodynamically attributable to the endothermicity and non-spontaneity of the process, evidenced by the highly positive values of reaction enthalpy and Gibbs free energy, respectively (Equation (1)). Intriguingly, the simultaneous presence of H_2S and H_2O in the reaction feed (i.e., 1 *v/v*% H_2S and 90 *v/v*% H_2O balanced in Ar), which is originally reported in this work, resulted in the maximum H_2 production (ca. 1.03 $\mu mol \cdot g^{-1} \cdot s^{-1}$ at 850 °C), suggesting the synergistic interaction of both H_2S and H_2O towards the optimal hydrogen production rate.

The above can also be corroborated by considering the thermodynamics of the gas-phase reactions of H_2S decomposition (Equation (1)) and H_2S steam reforming (Equation (2)). Indeed, regarding Equation (2), it was confirmed by a separate simulation in Aspen Plus that SO_2 and H_2 are by far the most favored products from the specific reacting mixture of hydrogen sulfide and steam in the relevant temperature regime, even when considering the production of elemental sulfur, SO_3 and/or O_2. In any case, the stoichiometry of the reaction is seen in Equation (2). Moreover, whereas both reactions are endothermic and thus H_2S conversion is favored with an increase in the reaction temperature, the beneficial role of steam lies predominantly in the significantly lower Gibbs free energy value in Equation (2) compared to dry H_2S decomposition (Equation (1)). Thus, the presence of steam essentially provides the driving force for enhancing hydrogen formation, as the stoichiometric coefficient of H_2 in Equation (2) is three times higher compared to the one in Equation (1). The above is better depicted in the thermodynamic plots in Figure 3. As can be seen from Figure 3a, the conversion of H_2S is thermodynamically favored over various H_2S/H_2O

mixtures compared to H_2S decomposition in the absence of steam, whereby H_2S conversion increases monotonically with the steam excess in the feed. Furthermore, to better assess the thermodynamic tendency of the system in the employed reaction conditions, the presence of inert gas in the reactant feed is examined over a mixture of 1% H_2S in steam (Figure 3b). Evidently, H_2S conversion increases with the increased presence of water vapor in the reactor inlet.

$$H_2S(g) \leftrightarrows S(g) + H_2(g),\ \Delta H_f^0 = +\ 317.9\ \text{kJ/mol},\ \Delta G_f^0 = +\ 270.1\ \text{kJ/mol} \quad (1)$$

$$H_2S(g) + 2H_2O(g) \leftrightarrows SO_2(g) + 3H_2(g),\ \Delta H_f^0 = +\ 157.6\ \text{kJ/mol},\ \Delta G_f^0 = +\ 190.5\ \text{kJ/mol} \quad (2)$$

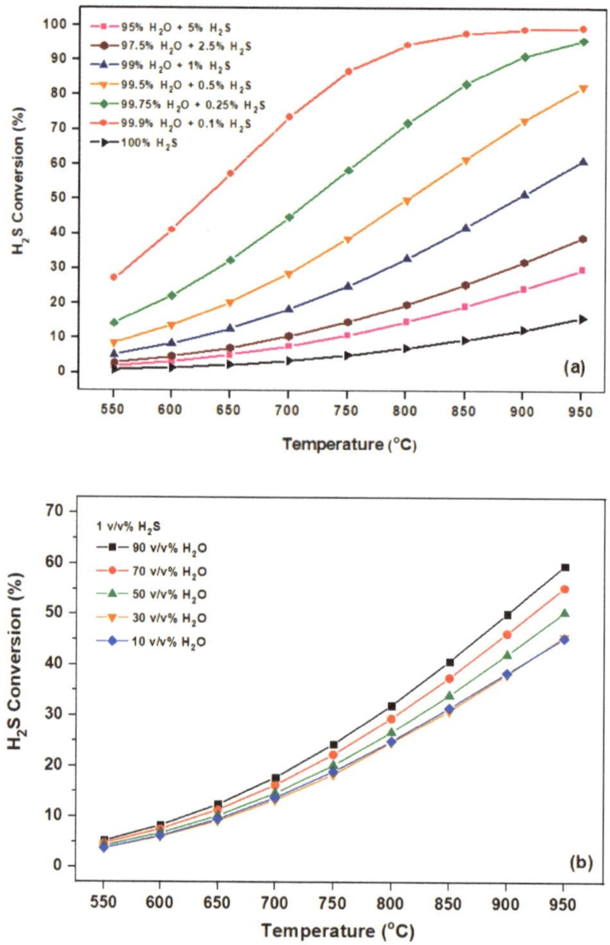

Figure 3. Effect of H_2O feed concentration on the thermodynamic H_2S conversion at ambient pressure as a function of temperature in the (**a**) absence and (**b**) presence of inert Ar.

Next, the effect of Co loading on the H_2 production rate under excess steam conditions was explored, and the corresponding results are shown in Figure 4. It is evident that within the selected temperature range, the H_2 production rate clearly increases with increasing reaction temperature, irrespective of the cobalt loading. Furthermore, in the absence of Co_3O_4 (that is, for bare CeO_2 sample), H_2 production is clearly lower than the respective values for x-Co/CeO_2 samples, while the lowest values were attained for pure cobalt oxide. These findings point towards the fact that the synergy between ceria and cobalt oxide species is a prerequisite for optimum catalytic activity. Moreover, the 30-Co/CeO_2 sample exhibited the highest performance among all x-Co/CeO_2 catalysts, showcasing the existence of an optimal cobalt oxide loading towards the achievement of maximum hydrogen production rate. Indeed, 30-Co/CeO_2 outperforms its counterparts in all the examined temperatures. For instance, at 850 °C, the 30-Co/CeO_2 sample exhibits a H_2 production rate of 2.11 μmol·g^{-1}·s^{-1} compared to 1.86 and 1.97 μmol·g^{-1}·s^{-1} over bare CeO_2 and Co_3O_4, respectively, with the superiority of mixed oxide being even more prominent at lower reaction temperatures. It should also be stated that all catalysts exhibited far superior performance compared to blank experiments (see Figure 2), clearly revealing the decisive role of the catalyst towards the progression of hydrogen generation rates.

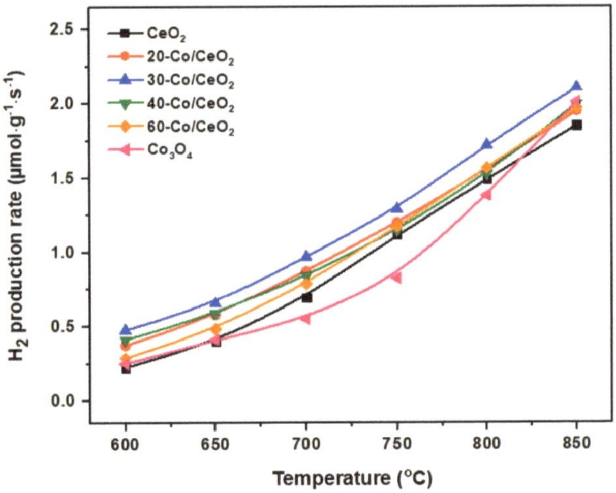

Figure 4. Hydrogen production rate as a function of temperature for x-Co/CeO_2 catalysts. Reaction conditions: 1 *v/v*% H_2S–90 *v/v*% H_2O, balanced with Ar, P = 1 bar, Flowrate: 100 cm^3/min, Catalyst mass = 250 mg, GHSV = 13,300 h^{-1}.

Subsequently, the H_2 production rate of the optimum catalyst, 30-Co/CeO_2, is compared with bare ceria, both in the absence and presence of excess H_2O (Figure 5). Interestingly, in the absence of H_2O, both bare CeO_2 and Co_3O_4/CeO_2 samples exhibited almost an identical performance in terms of H_2 production, revealing the pivotal role of CeO_2 in the deH_2S process [55]. Moreover, the significantly higher H_2S conversion under dry conditions of the catalyst-aided process compared to the homogeneous one (Figure 2) should be noted. On the other hand, the pronounced effect of H_2O excess in the feed steam is evident for both bare CeO_2 and x-Co/CeO_2 samples. The superior catalytic performance in wet conditions is in line with the thermodynamic calculations and can be attributed to the additional amount of H_2, which is produced via Equation (2).

Notably, the 30-Co/CeO_2 sample outperforms bare ceria under wet conditions, implying a synergistic effect towards higher H_2 production rates in the presence of water. This synergistic interaction, however, is not perceived under dry conditions, revealing the key role of steam in conjunction with the cobalt oxide–ceria synergy. In other words, the co-existence of cobalt oxide and ceria species is required to obtain high H_2 production rates. The latter could be possibly ascribed to the facilitation of the H_2S reforming reaction (as described by Equation (2)) at the interfacial sites, further contributing to H_2 production. This is additionally verified by the negligible rate of H_2 production during steam splitting (in the absence of H_2S) over Co_3O_4/CeO_2 samples. In view of the above, the superior reactivity of Co_3O_4/CeO_2 compared to the bare counterparts has been documented in the literature, ascribed mainly to synergistic cobalt oxide–ceria interactions [56–58].

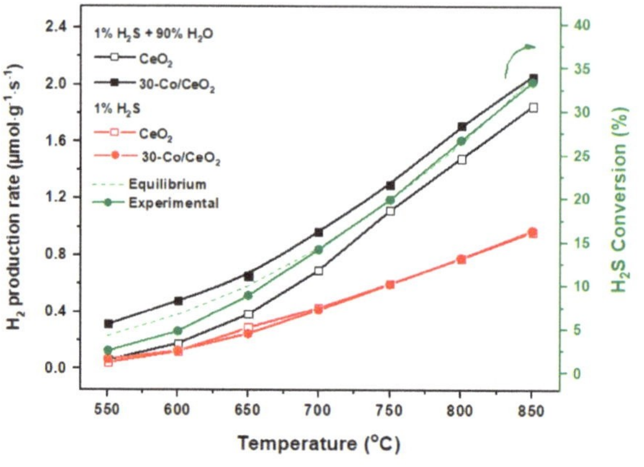

Figure 5. Left y-axis: H_2 production rate as a function of temperature for 30-Co/CeO_2 and bare CeO_2. Right y-axis: Experimental and equilibrium H_2S conversion for the dry H_2S decomposition reaction over 30-Co/CeO_2 (green lines). Reaction conditions: P = 1 bar, Flowrate: 100 cm^3/min, Catalyst mass = 250 mg, GHSV = 13,300 h^{-1}.

2.3. Apparent Activation Energies

For the approximate estimation of apparent activation energies, the differential method of analysis was followed. The possible contribution of internal mass transport phenomena was disregarded by means of the Weisz–Prater criterion (vide supra); thus, the measured apparent reaction rates were considered to be practically ascribed to the intrinsic reactivity of the catalysts. Therefore, the Arrhenius plots and the calculated activation energy values for all x-Co/CeO_2 catalysts are shown in Figure 6a and Figure 6b, respectively. The superior performance of the 30-Co/CeO_2 catalyst is again evident in terms of the apparent activation energy since it exhibits the lowest apparent activation energy value among all samples. Equally importantly, the binary cobalt oxide–ceria samples are associated with lower apparent activation energy values compared to the bare cobalt oxide and ceria, providing additional evidence of the beneficial role of the co-presence of cobalt oxide and ceria phases, which is also reflected in their catalytic activity (Figure 4).

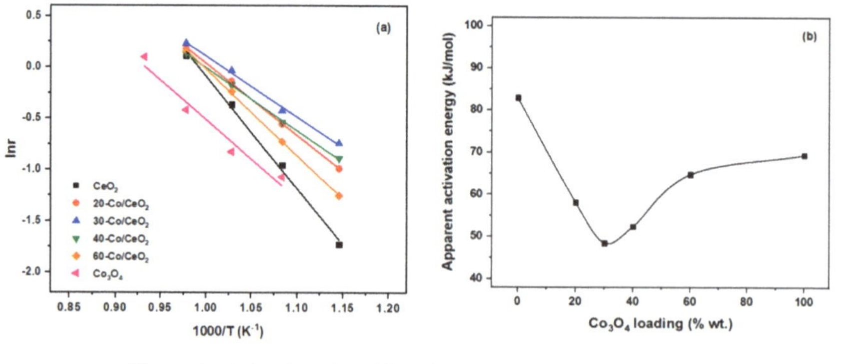

Figure 6. Arrhenius plots (a) and apparent activation energy values (b) for the as-prepared x-Co/CeO_2 catalysts. Reaction conditions: 1 v/v% H_2S–90 v/v% H_2O, balanced with Ar, P = 1 bar, Flowrate: 100 cm^3/min, Catalyst mass = 250 mg, GHSV = 13,300 h^{-1}.

A more meaningful assessment of the apparent activation energy values calculated in this study cannot be made since, to the best of our knowledge, this is the first time that such values have been reported in the literature for H_2 production through the decomposition of H_2S in extreme H_2O excess conditions. In any case, and in order to better showcase the beneficial synergistic role of the as-prepared cobalt oxide–ceria catalysts, the apparent activation energies for the H_2S decomposition reaction under both dry and wet conditions in the absence of catalysts were calculated (Arrhenius plots not shown for brevity). Specifically, apparent activation energy values equal to 124 and 137 kJ/mol were calculated for the non-catalytic thermal H_2S decomposition in the presence and absence of water, respectively. Not unexpectedly, these values are substantially higher compared to the respective values for all the employed catalysts and especially for the 30-Co/CeO_2 sample (Figure 6b). This finding showcases the significant role of catalysts, which enhance hydrogen production by offering a reaction pathway associated with a substantially decreased activation energy compared to the thermal-induced and gas-phase H_2S decomposition.

Also, with regards to the activation energy for the reaction of H_2S decomposition under dry and wet conditions for the optimum sample, i.e., 30-Co/CeO_2 (see Figure 5), the values were calculated to equal 66 and 49 kJ/mol, respectively. Therefore, in comparison with the respective values obtained in catalyst-free experiments (i.e., 137 and 124 kJ/mol in the absence and presence of H_2O, respectively), it is evident that the 30-Co/CeO_2 sample remarkably modifies the H_2S decomposition reaction pathway in the presence of water, as it decreases the apparent activation energy by a factor of 2.5. At the same time, the effect is less prominent under dry reaction conditions since the apparent activation energy in the presence of 30-Co/CeO_2 catalyst is decreased by a factor of 2.0 compared to the blank experiments. In all, and although the present results are not derived through a dedicated kinetic analysis, the obtained values can be considered as a reference point that can be used for comparison purposes. Therefore, they need to be refined and rigorously calculated in future work employing a more detailed kinetic methodology, also taking into account the reversibility of the reaction.

2.4. Stability Tests

Motivated by the superior H_2S decomposition performance of 30-Co/CeO_2 under H_2O excess conditions, short-term (10 h) stability experiments were also carried out to assess its lifetime characteristics (Figure 7). Specifically, the H_2 production rate was continuously monitored at T = 850 °C as a function of time on stream. The fresh catalyst was first heated up to 850 °C (3 °C/min) under pure Ar atmosphere, kept at this temperature for 1 hr, and then at t = 0 min, the standard feed mixture used in the activity experiments (1 *v/v*% H_2S–90 *v/v*% H_2O, balanced with Ar), was introduced into the reactor. Interestingly, H_2 production is continuously increased for the first ca. 4 h and then stabilized to a steady-state value of ca. 2.1 $\mu mol \cdot g^{-1} \cdot s^{-1}$. This value is similar to that obtained in the activity experiments and remained constant until the end of the experiment.

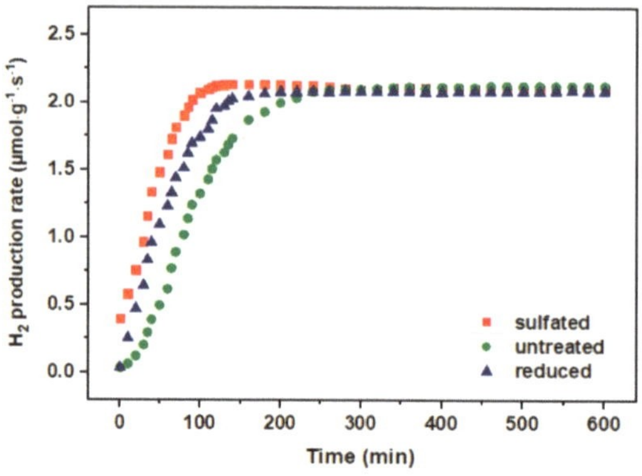

Figure 7. Hydrogen production rate as a function of time on stream for the "untreated", "sulfided" and "reduced" catalysts. Reaction conditions: 1 *v/v*% H_2S–90 *v/v*% H_2O, balanced with Ar, P = 1 bar, T = 850 °C, Flowrate: 100 cm^3/min, Catalyst mass = 250 mg, GHSV = 13,300 h^{-1}.

In an attempt to clarify whether the initial activation period for the 30-Co/CeO_2 catalyst is associated with the in situ sulfidation or reduction of the sample during the reaction, two additional short-term (10 h) stability experiments were carried out employing the same catalyst at different initial states. In particular, prior to the experiment, the "untreated" 30-Co/CeO_2 was subjected at 850 °C to a gas stream of either 1 *v/v*% H_2S–90 *v/v*% H_2O for 10 h or 10 *v/v*% H_2/Ar for 1 h in order to obtain the "sulfided" and "reduced" samples, respectively. Both the "sulfided" and "reduced" catalysts were subsequently tested under the standard reaction conditions (1 *v/v*% H_2S–90 *v/v*% H_2O, balance with Ar), and their hydrogen production rate profiles vs. time on stream were compared with the "untreated" sample.

The comparison between the different pretreated catalysts revealed the remarkable behavior of the "sulfided" catalyst. It was observed that the initial activation period of approximately 4 h of the "untreated" 30-Co/CeO_2 catalyst is reduced to less than 1.5 h for the "sulfided" sample. On the other hand, the "reduced" catalyst also demonstrated a decreased activation period (ca. 2 h) compared to the untreated material, but obviously higher compared to the "sulfided" sample.

In this point, it is worth pointing out the key role of chemisorption in the H_2S decomposition process and in the observed transient nature of hydrogen generation in this work (Figure 7). Elsewhere, on a series of well-defined rhodium crystals [59], it was found that H_2S adsorption on clean Rh(100) surfaces results in very rapid initial adsorption, followed by a very slow approach to saturation. This was followed by an increase in the surface S/Rh ratio, ascribed to the decomposition of H_2S. Even more importantly, H_2 desorption follows the same trend as the S/Rh ratio, showcasing the importance of progressive in situ sulfidation. Moreover, in a more recent study examining several transition metal sulfide catalysts [60], it was shown that at higher temperatures, the reaction could proceed under a quasi-steady state, where H_2S conversion and H_2 yield remain almost constant over elevated time on stream. Furthermore, a direct correlation between the amount of the released H_2 and sulfur was disclosed.

In light of the present findings, it can be deduced that the progressive activation is mainly due to the in situ sulfidation of the catalyst under the reaction atmosphere and, to a lesser extent, to H_2 reduction. This is better demonstrated by calculating the absolute hydrogen production (in mmol) during the activation period for each sample via the integration of the hydrogen production rate curve over the activation step. Indeed, the respective values for the "untreated", "reduced", and "sulfided" catalysts are equal to 288, 336, and 376 mmol H_2/g, respectively, demonstrating the pronounced effect of sulfidation in the total hydrogen production. In all cases ("untreated", "sulfided", and "reduced"), the H_2 production rate attained a stable value after a certain time period, reaching values similar to those obtained in the activity experiments and remaining stable until the end of the 10 h duration test. These findings clearly revealed the pronounced effect of in situ sulfidation, and to a lesser extent reduction, processes towards the progressive activation of cobalt oxide–ceria catalysts. Notably, irrespective of the catalyst's initial state, all samples reach the same steady state, exhibiting a remarkably stable performance after a certain time. The beneficial effect of H_2S during the course of the reaction between H_2S and H_2O is also worth noticing, in complete contrast to the poisoning effect of H_2S in several catalytic processes [61–63]. Similar observations on the positive effect of sulfidation have been reported elsewhere [64].

2.5. Spent Catalysts Characterization

In order to gain insight into the impact of the reaction conditions on the textural, structural, and surface features of the employed catalysts and to reveal possible structure-performance relationships, a complementary characterization study employing BET, XRD, SEM, and sulfur elemental analysis, prior ("untreated") and after ("sulfided") the stability experiments, was carried out. In terms of the textural characteristics, it is worth noticing that catalyst exposure to reaction conditions results in an almost complete collapse of the pore structure; the BET surface area and the pore volume of the "sulfided" catalyst decreased to about 2.4 m^2/g and 0.04 cm^3/g from 44.9 m^2/g and 0.21 cm^3/g of the "untreated" 30-Co/CeO_2 catalyst, respectively. Coupled with the excellent activity and stability performance of the "sulfided" catalyst, the deterioration of its porous structure clearly implies that the catalytic activity is practically independent of the textural properties.

The structural modifications induced under reaction conditions were revealed by means of XRD studies. Figure 8 depicts the XRD patterns of the "untreated" and "sulfided" catalysts. It is evident that new crystal structures were formed during the reaction, which were absent from the "untreated" samples. In particular, the XRD pattern of the "untreated" sample presents reflections that correspond to the Co_3O_4 and CeO_2 oxides, implying no chemical transformations or interactions between the parent oxides up to 600 °C (calcination temperature). However, the XRD pattern of the "sulfided" catalyst is completely different; the formation of new crystalline cobalt sulfide phases, i.e., Co_3S_4 and CoS, was clearly identified, verifying the in situ sulfidation of the catalyst under reaction conditions.

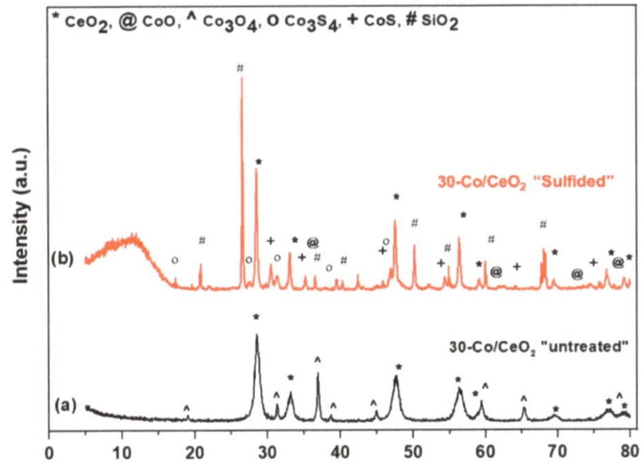

Figure 8. X-ray diffraction patterns of the (**a**) "untreated" and (**b**) "sulfided" catalysts. The "sulfided" catalyst was exposed for 10 h in a reaction mixture containing 1 *v/v*% H_2S and 90 *v/v*% H_2O, balanced in Ar, at T = 850 °C.

Surface modifications of the catalysts were further investigated by SEM/EDS analysis. Micrographs of the "untreated" and "sulfided" catalysts are presented at different magnifications in Figure 9. By comparing the SEM images of the "untreated" (Figure 9A,B) and "sulfided" (Figure 9C,D) samples, it is clear that the latter catalyst has undergone significant morphological modifications, leading to the formation of new phases, as a result of its exposure in H_2S/H_2O mixture at high temperatures. In particular, the formation of cobalt sulfide phases of polygon-like shape and cerium oxide particles of sponge-like morphology were revealed. Moreover, the average size of the cobalt-containing particles of the "sulfided" catalyst notably increased compared to the "untreated" catalyst, implying sintering under reaction conditions, in compliance with the above-discussed XRD and BET results.

In complete agreement with the identification of cobalt sulfide phases, a significant amount of sulfur (both in the form of elemental sulfur and in S-containing phases) was detected by sulfur elemental analysis of the "sulfided" sample (6.5 wt.%). However, this amount was approximately half compared to the elemental sulfur detected on the same sample when exposed to 1 *v/v*% H_2S balanced with an Ar feed mixture for 10 h (13.1 wt.%). The considerably lower amount of elemental sulfur in the case of wet conditions is most probably ascribed to the formation of sulfur-containing gaseous products, predominantly SO_2, as dictated by Equation (1). Notably, the relevant elemental mapping (not shown) demonstrated that sulfur formation and deposition was preferentially favored in the vicinity of Co-rich areas instead of CeO_2-rich regions.

Figure 9. SEM images of the "untreated" (**A**,**B**), and "sulfided" (**C**,**D**), catalysts. The "sulfided" catalyst was exposed for 10 h in a reaction mixture containing 1 *v/v*% H_2S–90 *v/v*% H_2O, balanced in Ar, at T = 850 °C.

In view of the above findings, the already established superiority of the 30-Co/CeO_2 catalyst for the H_2S decomposition reaction under extremely wet conditions (90 *v/v*% H_2O) should be mainly related to the in situ sulfidation of cobalt species (i.e., Co_3O_4 to Co_3S_4 and CoS), during its exposure to the reaction atmosphere. The previous analysis has clearly shown that during the catalytic decomposition of H_2S in the presence of H_2O, the 30-Co/CeO_2 catalyst demonstrates the optimum activity performance and, at the same time, a remarkably stable behavior. It should be therefore deduced that the produced elemental sulfur is not simply deposited on the catalyst surface but strongly interacts with the catalyst counterparts to create new active and stable phases. These promising findings could pave the way for the development of active and stable materials to be applied as heterogeneous catalysts for the thermocatalytic decomposition of H_2S in the presence of excess steam or as anode electrodes in H_2S splitting proton-conducting solid electrolyte electrochemical membrane reactors for the efficient management of H_2S emissions toward carbon-free hydrogen production.

3. Materials and Methods

3.1. Materials Preparation

A series of Co_3O_4/CeO_2 supported catalysts at different cobalt loadings (0, 20, 30, 40, 60, and 100 wt.%) was synthesized via the wet impregnation technique. The as-prepared catalysts are henceforth denoted as x-Co/CeO_2 for simplicity, with x referring to the weight percentage of Co in the composite materials. The synthetic protocol followed has been described in detail elsewhere [46]. In brief, the ceria-supported cobalt oxide catalysts were prepared by impregnation of an aqueous solution with appropriate concentrations of cobalt nitrate onto the calcined CeO_2 support. The resulting solutions were subsequently heated under continuous stirring and then dried at 100 °C overnight. Finally, the obtained dried samples were calcined in air at 600 °C for 2 h using a heating rate of 5 °C/min.

3.2. Materials Characterization

The textural, structural, and morphological characteristics of bare oxides and as-prepared cobalt oxide/ceria catalysts were determined by means of BET, XRD, and SEM/EDX methods, as previously described in detail [43]. The elemental analysis of sulfur content was conducted according to UOP 703 method on a "CHN-800" elemental analyzer (LECO Corporation, St. Joseph, MI, USA).

3.3. Catalytic Evaluation

The experimental apparatus employed for the H_2S decomposition catalytic studies under excess H_2O conditions has been described in our previous work [46] and is graphically represented in Figure 10. In particular, the experimental setup is comprised of the feed, reaction, and analysis sections. The feedstock system includes the 10 *v/v*% H_2S/Ar (Air Liquide Hellas, Athens, Greece) and pure Ar (99.998 *v/v*%, Linde Hellas, Sindos, Greece) high-pressure gas cylinders, the corresponding mass flow controllers (Brooks Instruments, Hatfield, Montgomery County, PA, USA) with a range of 0–20 cm^3/min for the H_2S/Ar mixture and 0–100 cm^3/min for Ar at an accuracy of 1% and a digital bubble flow meter (Agilent Optiflow 520, Santa Clara, CA, USA) to regulate and monitor the volumetric gas flow rates, an HPLC 305 Pump (GILSON, Middleton, WI, USA) with a flow range between 0.14–25 cm^3/min and an accuracy of 0.001 cm^3/min to control the doubly distilled liquid H_2O supply into the system, the preheater for steam generation at 130 °C and the heated stainless steel pipelines to provide the desired reactant mixture to the reactor, controlled at 130 °C by a K-type thermocouple. Appropriate flows of pure Ar and 10 *v/v*% H_2S/Ar were mixed with steam derived by a liquid water supply of ca. 0.05 cm^3/min so as to obtain the desired reactants feed mixture of 1 *v/v*% H_2S and 90 *v/v*% H_2O, balanced in Ar.

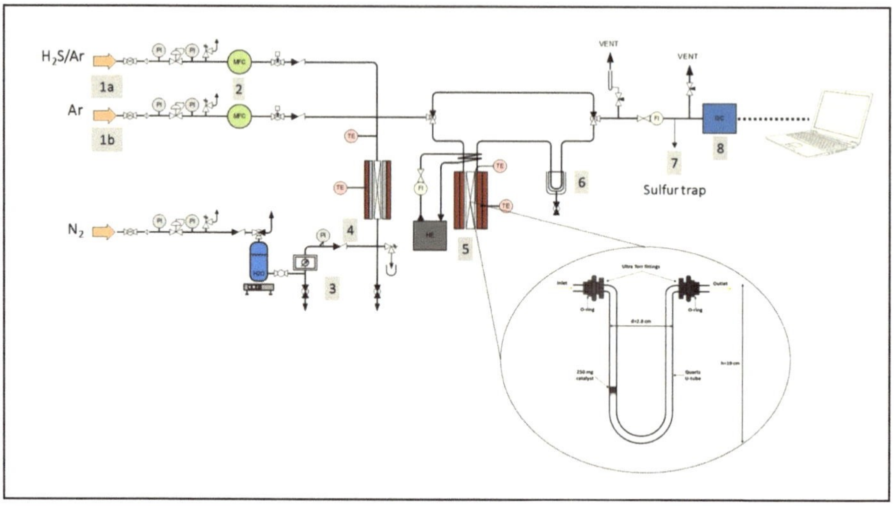

Figure 10. Schematic representation of the apparatus used in the catalytic experiments. Feed section: (1) (**1a**) 10 *v/v*% H_2S/Ar cylinder, (**1b**) Pure Ar cylinder; (2) MFC: Mass Flow Controllers for H_2S/Ar and Ar; (3) Doubly distilled liquid water pump; (4) Preheater for steam generation. Reaction section: (5) High-temperature furnace and U-tube quartz reactor; (6) Condenser; (7) Sulfur trap. Analysis section: (8) GC: Gas Chromatograph. PI: Pressure Indicator; FI: Flow Indicator; TE: Temperature Indicator; HE: Heat Exchanger. Adapted from Ref. [46].

A typical quartz U-tube fixed bed reactor with an internal diameter of 9.6 mm was employed for the activity and stability of H_2S decomposition experiments at excess H_2O conditions. In each test, the catalyst bed consisted of a mixture of 250 mg catalyst and 250 mg quartz, both in powder form, resulting in a bed volume of ~0.45 cm^3. Catalytic experiments were elaborated at atmospheric pressure and total feed rate equal to 100 cm^3/min, corresponding to a GHSV of ~13,300 h^{-1}.

The analysis of the produced hydrogen was carried out by means of a gas chromatograph (Shimadzu 14B, Kyoto, Japan) involving a thermal conductivity detector (TCD) and a molecular sieve 5A separation column. Prior to gas analysis, the reactor effluent gas mixture was flowing through a sulfur scrubber involving an aqueous KOH solution for the removal of any S-containing species.

During the catalytic activity tests, the temperature was increased at intervals of 50 °C, and the reaction was carried out in the temperature range of 550–850 °C. At the initial temperature of 550 °C, the system was left to stabilize for more than 4 h, and then for each temperature step, the system was allowed to reach a steady state (~1 h) before the effluent gas analysis was performed. Short-term stability tests were also carried out isothermally at 850 °C, where the hydrogen production rate was continuously monitored. For comparison purposes, blank experiments (without catalyst) were also carried out under specified reaction conditions. In the present work, the catalytic performance of the Co_3O_4/CeO_2 mixed oxide samples is expressed in terms of hydrogen production rate (μmol $H_2 \cdot g^{-1} \cdot s^{-1}$), calculated by the following equation (Equation (3)):

$$r_{H_2\ (\mu mol\ H_2\ \cdot\ g^{-1}\ \cdot\ s^{-1})} = \frac{10^6 \cdot F_{out} \cdot [H_2]_{out}}{60 \cdot V_M \cdot g_{cat}} \quad (3)$$

where F_{out} (cm^3/min) is the reactor outlet flowrate, $[H_2]_{out}$ is the molar fraction of hydrogen at the reactor outlet, V_M (cm^3/mol) is the molar volume of hydrogen at 1 bar and 25 °C, equal to 24,465 cm^3/mol, and g_{cat} is the catalyst mass. In addition, H_2S conversion under dry reaction conditions (i.e., absence of excess H_2O) was estimated on the basis of the following expression (Equation (4)):

$$X_{H_2S}\ (\%) = \frac{\text{moles of produced } H_2 \text{ at the outlet}}{H_2S \text{ moles at the inlet}} \cdot 100 \quad (4)$$

Moreover, thermodynamic equilibrium values were calculated using an RGibbs block in Aspen Plus software. The components involved in the simulation were elemental sulfur, H_2S, H_2O, SO_2, H_2, SO_3, O_2, and Ar, and their thermophysical properties were obtained by applying the Peng-Robinson equation of state.

4. Conclusions

In the present work, the H_2S decomposition activity of a series of Co_3O_4/CeO_2 catalysts with variable cobalt loadings (0–100 wt.%) was evaluated toward carbon-free hydrogen production under extremely harsh reaction conditions employing a feed mixture comprising of 1 *v/v*% H_2S and 90 *v/v*% H_2O at high temperatures (550–850 °C) and atmospheric pressure. Furthermore, the analysis of the experimental results is accompanied and corroborated by thermodynamic calculations in order to gain insight into the beneficial effect of steam addition on the production rate of H_2. It was revealed that the 30 wt.% Co_3O_4/CeO_2 catalyst demonstrated the optimum activity and the lower apparent activation energy, followed by remarkable stability. Notably, Co_3O_4/CeO_2 catalysts outperform bare ceria and cobalt oxide under wet conditions, implying a synergistic cobalt oxide–ceria interaction towards higher H_2 production rates via H_2S steam reforming. The enhancement in the production of hydrogen was also assessed by comparing the performance of the optimum 30-Co/CeO_2 catalyst under dry and water-excess reaction conditions. Notably, significant augmentation in the H_2 production rate was disclosed in the presence of steam.

Equally importantly, the key role of in situ sulfidation was revealed in stability tests over 30-Co/CeO_2. Intriguingly, the pretreatment of the sample with the reacting H_2S-H_2O-Ar mixture ("sulfided") led to the formation of active cobalt sulfide phases that allowed the system to faster reach steady state compared to the "untreated" and "reduced" sample. A complementary characterization study over both "untreated" and "sulfided" samples revealed the complete collapse of pore structure after catalysts exposure to reaction conditions, followed, however, by the formation of new cobalt sulfided phases, such as Co_3S_4 and CoS, which could be considered responsible for the excellent performance and stable behavior of the as-prepared cobalt oxide–ceria catalysts.

Author Contributions: Conceptualization, T.K. and G.E.M.; methodology, T.K.; software, G.V.; validation, G.V., G.E.M. and M.K.; formal analysis, G.V. and M.K.; investigation, T.K. and G.V.; resources, G.E.M.; data curation, G.E.M. and M.K.; writing—original draft preparation, T.K.; writing—review and editing, G.V., G.E.M. and M.K.; visualization, G.V.; supervision, G.E.M. and M.K.; project administration, G.E.M.; funding acquisition, G.E.M. All authors have read and agreed to the published version of the manuscript.

Funding: The work was performed within the frame of the joint program Black-Sea ERA.NET under the FP-7 initiative of the European Commission and funded by the Greek General Secretariat for Research and Technology GSRT (grant No. 11BS_10_28) and the project "Development of New Innovative Low Carbon Footprint Energy Technologies to Enhance Excellence in the Region of Western Macedonia" (MIS 5047197) which is implemented under the Action "Reinforcement of the Research and Innovation Infrastructure", funded by the Operational Programme "Competitiveness, Entrepreneurship and Innovation" (NSRF 2014–2020) and co-financed by Greece and the European Union (European Regional Development Fund).

Data Availability Statement: Not applicable.

Conflicts of Interest: The authors declare no conflict of interest.

References

1. IPCC. *Climate Change 2021—The Physical Science Basis*; Cambridge University Press: Cambridge, UK, 2021.

2. Hulme, M. 1.5 °C and Climate Research after the Paris Agreement. *Nat. Clim. Chang.* **2016**, *6*, 222–224. [CrossRef]
3. Depledge, J.; Saldivia, M.; Peñasco, C. Glass Half Full or Glass Half Empty?: The 2021 Glasgow Climate Conference. *Clim. Policy* **2022**, *22*, 147–157. [CrossRef]
4. Capurso, T.; Stefanizzi, M.; Torresi, M.; Camporeale, S.M. Perspective of the Role of Hydrogen in the 21st Century Energy Transition. *Energy Convers. Manag.* **2022**, *251*, 114898. [CrossRef]
5. Mac Dowell, N.; Sunny, N.; Brandon, N.; Herzog, H.; Ku, A.Y.; Maas, W.; Ramirez, A.; Reiner, D.M.; Sant, G.N.; Shah, N. The Hydrogen Economy: A Pragmatic Path Forward. *Joule* **2021**, *5*, 2524–2529. [CrossRef]
6. Dincer, I.; Acar, C. Smart Energy Solutions with Hydrogen Options. *Int. J. Hydrogen Energy* **2018**, *43*, 8579–8599. [CrossRef]
7. Yue, M.; Lambert, H.; Pahon, E.; Roche, R.; Jemei, S.; Hissel, D. Hydrogen Energy Systems: A Critical Review of Technologies, Applications, Trends and Challenges. *Renew. Sustain. Energy Rev.* **2021**, *146*, 111180. [CrossRef]
8. International Energy Agency. *Global Hydrogen Review 2021*; International Energy Agency: Paris, France, 2021.
9. Hermesmann, M.; Müller, T.E. Green, Turquoise, Blue, or Grey? Environmentally Friendly Hydrogen Production in Transforming Energy Systems. *Prog. Energy Combust. Sci.* **2022**, *90*, 100996. [CrossRef]
10. Younas, M.; Shafique, S.; Hafeez, A.; Javed, F.; Rehman, F. An Overview of Hydrogen Production: Current Status, Potential, and Challenges. *Fuel* **2022**, *316*, 123317. [CrossRef]
11. Nnabuife, S.G.; Ugbeh-Johnson, J.; Okeke, N.E.; Ogbonnaya, C. Present and Projected Developments in Hydrogen Production: A Technological Review. *Carbon Capture Sci. Technol.* **2022**, *3*, 100042. [CrossRef]
12. Kayfeci, M.; Keçebaş, A.; Bayat, M. Hydrogen Production. In *Solar Hydrogen Production*; Elsevier: Amsterdam, The Netherland, 2019; pp. 45–83.
13. Ishaq, H.; Dincer, I.; Crawford, C. A Review on Hydrogen Production and Utilization: Challenges and Opportunities. *Int. J. Hydrogen Energy* **2021**, *47*, 26238–26264. [CrossRef]
14. De Crisci, A.G.; Moniri, A.; Xu, Y. Hydrogen from Hydrogen Sulfide: Towards a More Sustainable Hydrogen Economy. *Int. J. Hydrogen Energy* **2019**, *44*, 1299–1327. [CrossRef]
15. Chan, Y.H.; Loy, A.C.M.; Cheah, K.W.; Chai, S.Y.W.; Ngu, L.H.; How, B.S.; Li, C.; Lock, S.S.M.; Wong, M.K.; Yiin, C.L.; et al. Hydrogen Sulfide (H_2S) Conversion to Hydrogen (H2) and Value-Added Chemicals: Progress, Challenges and Outlook. *Chem. Eng. J.* **2023**, *458*, 141398. [CrossRef]
16. Bahadori, A. *Natural Gas Processing: Technology and Engineering Design*; Elsevier: Amsterdam, The Netherlands, 2014; ISBN 9780080999715.
17. Rodríguez, E.; Harvey, W.S.; Ásbjörnsson, E.J. Review of H_2S Abatement Methods in Geothermal Plants. In Proceedings of the Thirty-Eighth Workshop on Geothermal Reservoir Engineering, Stanford, CA, USA, 11–13 February 2013.
18. Deng, Q.; Wang, Q.; Liu, M.; Zhao, F. Geological Factors Controlling H_2S in Coal Seams. In *Mine Planning and Equipment Selection*; Springer International Publishing: Cham, Switzerland, 2014; pp. 619–627.
19. Corpas, F.J.; Palma, J.M. H_2S Signaling in Plants and Applications in Agriculture. *J. Adv. Res.* **2020**, *24*, 131–137. [CrossRef] [PubMed]
20. Jørgensen, B.B.; Findlay, A.J.; Pellerin, A. The Biogeochemical Sulfur Cycle of Marine Sediments. *Front. Microbiol.* **2019**, *10*, 849. [CrossRef] [PubMed]
21. King, M.J.; Davenport, W.G.; Moats, M.S. *Sulfuric Acid Manufacture: Analysis, Control and Optimization*; Elsevier: Amsterdam, The Netherlands, 2013; ISBN 978-0-08-098220-5.
22. Kaiser, M.J.; de Klerk, A.; Gary, J.H.; Hwerk, G.E. *Petroleum Refining: Technology, Economics, and Markets*; CRC Press: Boca Raton, FL, USA, 2019; ISBN 9780429188893.
23. Wiheeb, A.D.; Shamsudin, I.K.; Ahmad, M.A.; Murat, M.N.; Kim, J.; Othman, M.R. Present Technologies for Hydrogen Sulfide Removal from Gaseous Mixtures. *Rev. Chem. Eng.* **2013**, *29*, 449–470. [CrossRef]
24. Ghahraloud, H.; Farsi, M.; Rahimpour, M.R. Hydrogen Production through Thermal Decomposition of Hydrogen Sulfide: Modification of the Sulfur Recovery Unit To Produce Ultrapure Hydrogen. *Ind. Eng. Chem. Res.* **2018**, *57*, 14114–14123. [CrossRef]
25. Palma, V.; Vaiano, V.; Barba, D.; Colozzi, M.; Palo, E.; Barbato, L.; Cortese, S. H_2 Production by Thermal Decomposition of H_2S in the Presence of Oxygen. *Int. J. Hydrogen Energy* **2015**, *40*, 106–113. [CrossRef]
26. Burra, K.R.G.; Bassioni, G.; Gupta, A.K. Catalytic Transformation of H_2S for H_2 Production. *Int. J. Hydrogen Energy* **2018**, *43*, 22852–22860. [CrossRef]
27. Jiang, G.; Zhang, X.; Zhang, F.; Liu, Z.; Wang, Z.; Hao, Z.; Lin, C. Efficient Recovery of Hydrogen and Sulfur Resources over Non-Sulfide Based LaFexAl12-XO19 Hexaaluminate Catalysts by H_2S Catalytic Decomposition. *Appl. Catal. B Environ.* **2020**, *263*, 118354. [CrossRef]
28. Linga Reddy, E.; Biju, V.M.; Subrahmanyam, C. Production of Hydrogen and Sulfur from Hydrogen Sulfide Assisted by Nonthermal Plasma. *Appl. Energy* **2012**, *95*, 87–92. [CrossRef]
29. Dang, X.; Huang, J.; Kang, L.; Wu, T.; Zhang, Q. Research on Decomposition of Hydrogen Sulfide Using Nonthermal Plasma with Metal Oxide Catalysis. *Energy Procedia* **2012**, *16*, 856–862. [CrossRef]
30. Zhang, B.; Bai, J.; Zhang, Y.; Zhou, C.; Wang, P.; Zha, L.; Li, J.; Simchi, A.; Zhou, B. High Yield of CO and Synchronous S Recovery from the Conversion of CO_2 and H_2S in Natural Gas Based on a Novel Electrochemical Reactor. *Environ. Sci. Technol.* **2021**, *55*, 14854–14862. [CrossRef] [PubMed]

31. Yu, S.; Zhou, Y. Photochemical Decomposition of Hydrogen Sulfide. In *Advanced Catalytic Materials-Photocatalysis and Other Current Trends*; InTech: Rijeka, Croatia, 2016.
32. Zaman, J.; Chakma, A. Production of Hydrogen and Sulfur from Hydrogen Sulfide. *Fuel Process. Technol.* **1995**, *41*, 159–198. [CrossRef]
33. Chivers, T.; Lau, C. The Thermal Decomposition of Hydrogen Sulfide over Vanadium and Molybdenum Sulfides and Mixed Sulfide Catalysts in Quartz and Thermal Diffusion Column Reactors. *Int. J. Hydrogen Energy* **1987**, *12*, 235–243. [CrossRef]
34. Kaloidas, V.; Papayannakos, N. Hydrogen Production from the Decomposition of Hydrogen Sulphide. Equilibrium Studies on the System $H_2S/H_2/Si$, (i = 1, ... ,8) in the Gas Phase. *Int. J. Hydrogen Energy* **1987**, *12*, 403–409. [CrossRef]
35. Ipsakis, D.; Kraia, T.; Marnellos, G.E.; Ouzounidou, M.; Voutetakis, S.; Dittmeyer, R.; Dubbe, A.; Haas-Santo, K.; Konsolakis, M.; Figen, H.E.; et al. An Electrocatalytic Membrane-Assisted Process for Hydrogen Production from H 2 S in Black Sea: Preliminary Results. *Int. J. Hydrogen Energy* **2015**, *40*, 7530–7538. [CrossRef]
36. Demirbas, A. Hydrogen Sulfide from the Black Sea for Hydrogen Production. *Energy Sources Part A Recover. Util. Environ. Eff.* **2009**, *31*, 1866–1872. [CrossRef]
37. Ozturk, M.; Midilli, A.; Dincer, I. Effective Use of Hydrogen Sulfide and Natural Gas Resources Available in the Black Sea for Hydrogen Economy. *Int. J. Hydrogen Energy* **2021**, *46*, 10697–10707. [CrossRef]
38. Ryann, A.; Perkins, N.J. *The Black Sea: Dynamics, Ecology, and Conservation*; Nova Science Publishers, Inc.: London, UK, 2011.
39. Kraia, T.; Wachowski, S.; Vøllestad, E.; Strandbakke, R.; Konsolakis, M.; Norby, T.; Marnellos, G.E. Electrochemical Performance of Co_3O_4/CeO_2 Electrodes in H_2S/H_2O Atmospheres in a Proton-Conducting Ceramic Symmetrical Cell with BaZr0.7Ce0.2Y0.1O3 Solid Electrolyte. *Solid State Ionics* **2017**, *306*, 31–37. [CrossRef]
40. Petrov, K.; Baykara, S.Z.; Ebrasu, D.; Gulin, M.; Veziroglu, A. An Assessment of Electrolytic Hydrogen Production from H 2 S in Black Sea Waters. *Int. J. Hydrogen Energy* **2011**, *36*, 8936–8942. [CrossRef]
41. Guldal, N.O.; Figen, H.E.; Baykara, S.Z. Perovskite Catalysts for Hydrogen Production from Hydrogen Sulfide. *Int. J. Hydrogen Energy* **2018**, *43*, 1038–1046. [CrossRef]
42. Guldal, N.O.; Figen, H.E.; Baykara, S.Z. Production of Hydrogen from Hydrogen Sulfide with Perovskite Type Catalysts: $LaMO_3$. *Chem. Eng. J.* **2017**, *313*, 1354–1363. [CrossRef]
43. Lykaki, M.; Pachatouridou, E.; Iliopoulou, E.; Carabineiro, S.A.C.; Konsolakis, M. Impact of the Synthesis Parameters on the Solid State Properties and the CO Oxidation Performance of Ceria Nanoparticles. *RSC Adv.* **2017**, *7*, 6160–6169. [CrossRef]
44. Konsolakis, M. The Role of Copper–Ceria Interactions in Catalysis Science: Recent Theoretical and Experimental Advances. *Appl. Catal. B Environ.* **2016**, *198*, 49–66. [CrossRef]
45. Xu, Y.; Mofarah, S.S.; Mehmood, R.; Cazorla, C.; Koshy, P.; Sorrell, C.C. Design Strategies for Ceria Nanomaterials: Untangling Key Mechanistic Concepts. *Mater. Horizons* **2021**, *8*, 102–123. [CrossRef] [PubMed]
46. Kraia, T.; Kaklidis, N.; Konsolakis, M.; Marnellos, G.E. Hydrogen Production by H_2S Decomposition over Ceria Supported Transition Metal (Co, Ni, Fe and Cu) Catalysts. *Int. J. Hydrogen Energy* **2019**, *44*, 9753–9762. [CrossRef]
47. Jha, A.; Jeong, D.W.; Lee, Y.L.; Nah, I.W.; Roh, H.S. Enhancing the Catalytic Performance of Cobalt Oxide by Doping on Ceria in the High Temperature Water-Gas Shift Reaction. *RSC Adv.* **2015**, *5*, 103023–103029. [CrossRef]
48. Liu, Z.; Li, J.; Buettner, M.; Ranganathan, R.V.; Uddi, M.; Wang, R. Metal-Support Interactions in CeO_2- and SiO_2-Supported Cobalt Catalysts: Effect of Support Morphology, Reducibility, and Interfacial Configuration. *ACS Appl. Mater. Interfaces* **2019**, *11*, 17035–17049. [CrossRef]
49. Varvoutis, G.; Lykaki, M.; Papista, E.; Carabineiro, S.A.C.; Psarras, A.C.; Marnellos, G.E.; Konsolakis, M. Effect of Alkali (Cs) Doping on the Surface Chemistry and CO2 Hydrogenation Performance of CuO/CeO2 Catalysts. *J. CO2 Util.* **2021**, *44*, 101408. [CrossRef]
50. Díez-Ramírez, J.; Sánchez, P.; Kyriakou, V.; Zafeiratos, S.; Marnellos, G.E.; Konsolakis, M.; Dorado, F. Effect of Support Nature on the Cobalt-Catalyzed CO_2 Hydrogenation. *J. CO2 Util.* **2017**, *21*, 562–571. [CrossRef]
51. Parastaev, A.; Muravev, V.; Osta, E.H.; Kimpel, T.F.; Simons, J.F.M.; van Hoof, A.J.F.; Uslamin, E.; Zhang, L.; Struijs, J.J.C.; Burueva, D.B.; et al. Breaking Structure Sensitivity in CO_2 Hydrogenation by Tuning Metal–Oxide Interfaces in Supported Cobalt Nanoparticles. *Nat. Catal.* **2022**, *5*, 11. [CrossRef]
52. Liu, H.; Xu, S.; Zhou, G.; Xiong, K.; Jiao, Z.; Wang, S. CO_2 Hydrogenation to Methane over Co/KIT-6 Catalysts: Effect of Co Content. *Fuel* **2018**, *217*, 570–576. [CrossRef]
53. Al-Musa, A.; Al-Saleh, M.; Ioakeimidis, Z.C.; Ouzounidou, M.; Yentekakis, I.V.; Konsolakis, M.; Marnellos, G.E. Hydrogen Production by Iso-Octane Steam Reforming over Cu Catalysts Supported on Rare Earth Oxides (REOs). *Int. J. Hydrogen Energy* **2014**, *39*, 1350–1363. [CrossRef]
54. Adesina, A.A.; Meeyoo, V.; Foulds, G. Thermolysis of Hydrogen Sulphide in an Open Tubular Reactor. *Int. J. Hydrogen Energy* **1995**, *20*, 777–783. [CrossRef]
55. Shi, X.; Yu, Y.; Xue, L.; He, H. Effect of Sulfur Poisoning on Co_3O_4/CeO_2 Composite Oxide Catalyst for Soot Combustion. *Chinese J. Catal.* **2014**, *35*, 1504–1510. [CrossRef]
56. Lykaki, M.; Papista, E.; Kaklidis, N.; Carabineiro, S.; Konsolakis, M. Ceria Nanoparticles' Morphological Effects on the N2O Decomposition Performance of Co_3O_4/CeO_2 Mixed Oxides. *Catalysts* **2019**, *9*, 233. [CrossRef]

57. Carabineiro, S.A.C.; Chen, X.; Konsolakis, M.; Psarras, A.C.; Tavares, P.B.; Órfão, J.J.M.; Pereira, M.F.R.; Figueiredo, J.L. Catalytic Oxidation of Toluene on Ce–Co and La–Co Mixed Oxides Synthesized by Exotemplating and Evaporation Methods. *Catal. Today* **2015**, *244*, 161–171. [CrossRef]
58. Parastaev, A.; Muravev, V.; Huertas Osta, E.; van Hoof, A.J.F.; Kimpel, T.F.; Kosinov, N.; Hensen, E.J.M. Boosting CO2 Hydrogenation via Size-Dependent Metal–Support Interactions in Cobalt/Ceria-Based Catalysts. *Nat. Catal.* **2020**, *3*, 526–533. [CrossRef]
59. Hegde, R.I.; White, J.M. Chemisorption and Decomposition of H_2S on Rh(100). *J. Phys. Chem.* **1986**, *90*, 296–300. [CrossRef]
60. Zagoruiko, A.; Mikenin, P. Decomposition of Hydrogen Sulfide into Elements in the Cyclic Chemisorption-Catalytic Regime. *Catal. Today* **2021**, *378*, 176–188. [CrossRef]
61. Blanchard, J.; Achouri, I.; Abatzoglou, N. H 2 S Poisoning of $NiAl_2O_4/Al_2O_3$-YSZ Catalyst during Methane Dry Reforming. *Can. J. Chem. Eng.* **2016**, *94*, 650–654. [CrossRef]
62. Dou, X.; Veksha, A.; Chan, W.P.; Oh, W.-D.; Liang, Y.N.; Teoh, F.; Mohamed, D.K.B.; Giannis, A.; Lisak, G.; Lim, T.-T. Poisoning Effects of H_2S and HCl on the Naphthalene Steam Reforming and Water-Gas Shift Activities of Ni and Fe Catalysts. *Fuel* **2019**, *241*, 1008–1018. [CrossRef]
63. Wachter, P.; Gaber, C.; Raic, J.; Demuth, M.; Hochenauer, C. Experimental Investigation on H_2S and SO_2 Sulphur Poisoning and Regeneration of a Commercially Available Ni-Catalyst during Methane Tri-Reforming. *Int. J. Hydrogen Energy* **2021**, *46*, 3437–3452. [CrossRef]
64. Iqbal, S.; Amjad, A.; Javed, M.; Alfakeer, M.; Mushtaq, M.; Rabea, S.; Elkaeed, E.B.; Pashameah, R.A.; Alzahrani, E.; Farouk, A.-E. Boosted Spatial Charge Carrier Separation of Binary $ZnFe_2O_4/S\text{-}g\text{-}C_3N_4$ Heterojunction for Visible-Light-Driven Photocatalytic Activity and Antimicrobial Performance. *Front. Chem.* **2022**, *10*, 894. [CrossRef]

catalysts

MDPI

Review

Magnetic Nanomaterials as Catalysts for Syngas Production and Conversion

Natarajan Chidhambaram [1,*], Samuel Jasmine Jecintha Kay [1], Saravanan Priyadharshini [1], Rajakantham Meenakshi [2], Pandurengan Sakthivel [3], Shanmugasundar Dhanbalan [4], Shajahan Shanavas [5,6], Sathish-Kumar Kamaraj [7] and Arun Thirumurugan [8,*]

1 Department of Physics, Rajah Serfoji Government College (Autonomous), Thanjavur 613005, India
2 PG and Research Department of Physics, Cauvery College for Women (Autonomous), Tiruchirappalli 620018, India
3 Centre for Materials Science, Department of Physics, Faculty of Engineering, Karpagam Academy of Higher Education, Coimbatore 641021, India
4 Functional Materials and Microsystems Research Group, RMIT University, Melbourne, VIC 3000, Australia
5 Department of Chemistry, Khalifa University of Science and Technology, Abu Dhabi P.O. Box 127788, United Arab Emirates
6 Department of Conservative Dentistry and Endodontics, Saveetha Dental College and Hospitals, SIMATS, Chennai 600077, India
7 Instituto Politécnico Nacional (IPN)—Centro de Investigación en Ciencia Aplicada y Tecnología Avanzada, Unidad Altamira (CICATA-Altamira), Carretera Tampico-Puerto Industrial Altamira Km 14.5, C. Manzano, Industrial Altamira, Altamira 89600, Tamaulipas, Mexico
8 Sede Vallenar, Universidad de Atacama, Costanera #105, Vallenar 1612178, Chile
* Correspondence: nchidambaraselvan@gmail.com (N.C.); arunthiruvbm@gmail.com (A.T.)

Abstract: The conversion of diverse non-petroleum carbon elements, such as coal, biomass, natural/shale gas, and even CO_2, into cleaner hydrocarbon fuels and useful chemicals relies heavily on syngas, which is a combination of CO and H_2. Syngas conversions, which have been around for almost a century, will probably become even more important in the production of energy and chemicals due to the rising need for liquid fuels and chemical components derived from sources of carbon other than crude oil. Although a number of syngas-based technologies, including the production of methanol, Fischer–Tropsch (FT) synthesis, and carbonylation, have been industrialized, there is still a great need for new catalysts with enhanced activity and adjustable product selectivity. New novel materials or different combinations of materials have been investigated to utilize the synergistic effect of these materials in an effective way. Magnetic materials are among the materials with magnetic properties, which provide them with extra physical characteristics compared to other carbon-based or conventional materials. Moreover, the separation of magnetic materials after the completion of a specific application could be easily performed with a magnetic separation process. In this review, we discuss the synthesis processes of various magnetic nanomaterials and their composites, which could be utilized as catalysts for syngas production and conversion. It is reported that applying an external magnetic field could influence the outcomes of any applications of magnetic nanomaterials. Here, the possible influence of the magnetic characteristics of magnetic nanomaterials with an external magnetic field is also discussed.

Keywords: magnetic materials; catalysts; magnetic separation; syngas; CO_2 conversion

Citation: Chidhambaram, N.; Kay, S.J.J.; Priyadharshini, S.; Meenakshi, R.; Sakthivel, P.; Dhanbalan, S.; Shanavas, S.; Kamaraj, S.-K.; Thirumurugan, A. Magnetic Nanomaterials as Catalysts for Syngas Production and Conversion. *Catalysts* **2023**, *13*, 440. https://doi.org/10.3390/catal13020440

Academic Editors: Georgios Bampos, Athanasia Petala and Zacharias Frontistis

Received: 30 January 2023
Revised: 16 February 2023
Accepted: 16 February 2023
Published: 18 February 2023

1. Introduction

There has been a lot of scientific interest in the production of syngas (H_2 + CO) from biomass by gasification, a thermochemical conversion process that occurs in an oxygen-lean environment [1–3]. Syngas offers an effective alternative as a fuel for transferring energy, a raw material for creating hydrogen fuel, and a source of higher-value chemicals. Through more effective syngas-based power generation facilities and fuel cell technologies, syngas production significantly reduces greenhouse gas emissions, particularly carbon dioxide (CO_2) emissions [4]. Nevertheless, it is thought that precious metal-based catalysts are the best option for producing syngas. However, significant limitations, including massive cost and unavailability, will eventually prevent their widespread use. Because of this, new non-precious metal materials have been developed as substitutes [5]. By enabling reagent adsorption and surface reactions on magnetized Fe catalysts as well as non-magnetic materials, magnetic fields can significantly improve these metals' catalytic activity. This improves selectivity to hydrocarbons and lowers apparent activation energy [6]. Although there are various metals used for syngas production using magnetic effects, Fe, Ni, and Co metals play a significant role in the production of syngas. Additionally, they have been used to generate electrically conductive materials and enhance the mechanical performance of composites. The qualities of magnetic nanoparticles include their capacity to transport other substances, their quantum characteristics, and their high surface-area-to-volume ratio. Magnetic fields can be used to alter the characteristics of magnetic nanoparticles to make them appropriate for a variety of applications, including syngas production, on a large scale. The efficiency of magnetic fields relies on the field gradient and particle magnetic moment. Based on the material, the optimum magnetic nanoparticles are between 10 and 20 nm in size as, above a certain temperature known as the blocking temperature, these particle unite into a single domain and show superparamagnetic activity [7]. However, this also leads to intrinsic instabilities over extended periods and magnetic loss because of chemically highly active bare metallic nanoparticles. The particularly attractive characteristics of spherical and cubic magnetic nanoparticles have attracted a lot of interest. It is noteworthy that these nanoparticles have been efficiently used in the pharmaceutical field for drug delivery, high-contrast MRI, DNA detection, and stem cell labeling/separation. In an initiative aimed at improving the catalytic efficiency in syngas production by using magnetic nanoparticles, it has also shown an efficient effect. This review discusses recent progress made on the effective utilization of magnetic nanomaterials for syngas production/conversion and the effective way of utilizing an external magnetic field for the synthesis of magnetic catalysts and for the adjustment of the efficiency of the synthesis/conversion process.

2. Magnetic Materials for Syngas Production/Conversion

Magnetic nanomaterials are found to be interesting in various applications due to their unique magnetic response behavior compared to other available nanomaterials. These materials are generally classified as magnetic and non-magnetic materials. Furthermore, they have sub-classifications, including dia, para, ferro, ferri, and antiferro, depending on the magnetic characteristics. Depending on the shape and size, their magnetic behavior is further classified as soft and hard magnetic materials. As a catalyst, transition metals are considered a support substance for the generation of syngas because of their mechanical, electrochemical, and metallic-like qualities. This may change a precious metal overlayer's catalytic characteristics. Transition metals had been synthesized using a nitriding technique that was carried out in the absence of hydrocarbons and only produced residual physisorbed N that easily desorbed as N_2 off the catalyst surface, in contrast to the situations of carbide and sulfide-based substrates [8]. When selecting active metals for the conversion of syngas into various sorts of products, we may be able to use the ability of CO dissociative adsorption as a guide. Due to their low adsorption heat, transition metals on the right of the periodic chart, including Cu and Pd, are not CO dissociative [9]. With the C-O bond uncleaved during CO hydrogenation, Cu and Pd are quite well-active phases for the production of methanol. However, the transition metals on the left of the periodic table, including Mo and W, exhibit a strong capacity for CO dissociative adsorption and the formation of metal carbides [10]. In contrast to Mo and W metals, Mo and W carbides are capable of activating CO differently and exhibit CO hydrogenation activity. Fe, Co, and Ru are appropriate for CO hydrogenation (Sabatier's principle) due to the moderate efficiency of CO adsorption on these metals. Rh is a promising candidate for ethanol formation from syngas because it is situated at the boundary of the CO dissociation and non-dissociation regions. Notably, the balance between CO and H_2 dissociation capacities plays a crucial role in defining product selectivity. In Fischer–Tropsch (FT) synthesis, product selectivity is determined by the balance between both hydrogenation and C-C coupling abilities on metal surfaces. Typically, a catalyst with an excessively high H_2 dissociation or hydrogenation capacity may result in the creation of CH_4 with higher selectivity. Ni can be used, for instance, in the dissociation of CO. As stated earlier, the main active metals investigated for FT synthesis are Fe, Co, and Ru since they are appropriate for the synthesis of significant cumulative hydrocarbons. Ru is the most effective catalyst among these three metals for CO hydrogenation, and CO hydrogenation on Ru can take place at low temperatures. Ru has a higher affinity for long-chain hydrocarbons and a low affinity for CH_4; however, due to its high cost and restricted supply, Ru cannot be used as the primary active ingredient in large-scale industrial applications. Fe and Co have so far only been used in industrial FT processes as active metals. Ru is appropriate for basic study, nevertheless, as an excellent FT active metal that can offer significant insight into the catalytic reaction mechanism. Additionally, Ru is frequently employed in FT catalysts as a stimulant to speed up the reduction of Co or Fe because it is difficult to reduce Co or Fe precursors under normal circumstances [11]. Although Fe is less expensive than Co, Co catalysts are typically more active and selective to linear long-chain hydrocarbons. Co catalysts are also frequently more water-resistant than other types of catalysts. As a result, Co catalysts have drawn a lot of interest for the production of long-chain linear hydrocarbons from syngas, such as wax and diesel fuel. In contrast, Co catalysts can only function effectively at restricted temperature ranges and H_2/CO ratios, whereas Fe-based catalysts can operate under a variety of temperatures and H_2/CO proportions without significantly increasing CH_4 preference. It is important to note that, relative to Co- or Ru-based catalysts, Fe-based catalysts demonstrate vastly greater activity for the water–gas system process. This is beneficial for converting syngas made from coal or biomass that has a lower H_2/CO ratio, but it is not desired for converting H_2-rich syngas made from natural gas or shale gas. For Fe catalysts to provide acceptable activity and selectivity, more substantial changes are typically needed, and quick catalytic deactivation is a major difficulty for Fe catalysts.

Solid oxygen carriers, which can be either single metal oxides or combinations of metal oxides, can be used to provide oxygen for this partial oxidation process. Zhu Xing et al. demonstrated the synthesis of syngas and hydrogen through two-step steam reforming of methane on a CeO_2-Fe_2O_3 oxygen carrier [12]. In their work, in a fixed-bed reactor, temperature-programed interactions between methane and the oxygen carrier CeO_2-Fe_2O_3 were carried out to find the ideal temperature for the gas–solid reaction that produces syngas. They noticed that the smooth onset of CH_4 conversion starts at about 500 °C using temperature as a function of reaction. However, below 650 °C, the conversion is really low. The fact that conversion increases quickly with a temperature over 650 °C suggests that reaction temperature has a significant impact on methane oxidation by solid oxides and that a higher temperature is required to produce syngas through a gas–solid reaction. This might be attributed to the fact that the rate of lattice oxygen migration from the bulk to the surface accelerates with increasing temperature. On the CH_4 conversion curve, they also noticed a slight peak at about 600 °C. CeO_2-Fe_2O has a higher CH_4 conversion when the temperature exceeds 800 °C. Since oxidation reactions at low temperatures are typically aided by the surface adsorption of oxygen on oxides, this is expected to be caused by the well-released bulk lattice oxygen in CeO_2-Fe_2O_3. High-pressure experiments were carried out for the reduction and oxidation of oxygen carrier particles in a dedicated thermogravimetric analyzer as part of a study on Fe-oxide-based oxygen carriers for syngas production from methane by Niranjani Deshpande et al. [13]. The partial oxidation of CH_4 for syngas production using a Fe-oxide-based system is shown in Figure 1.

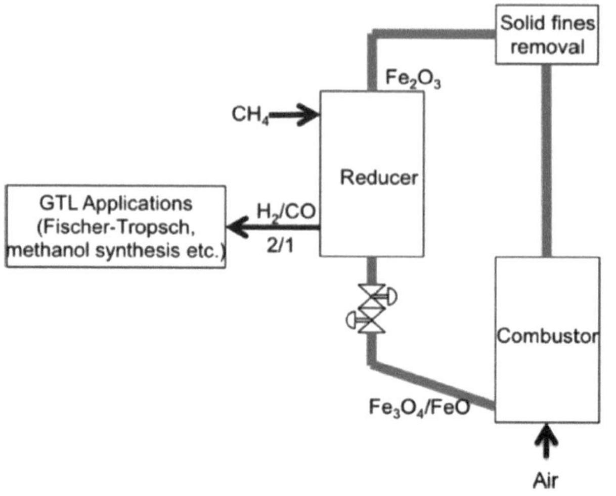

Figure 1. Illustration of the CH_4 partial oxidation-based Fe-oxide system for producing syngas. Reprinted from [13]. Copyright (2015), with permission from the American Chemical Society.

Due to their low cost and high magnetization capability, transition metals have generated interest [14]. Nevertheless, when contrasted to their respective mono-metal counterparts, bimetallic alloys have more applications and qualities, which have made them particularly attractive to researchers. Recently, there have been various attempts to use bimetallic catalysts in catalyzed reactions [15]. Bimetallic catalysts are considered successful because they are made up of two different metals with large dispersion and active sites due to the combination of their parent metals. Additionally, the creation of a solid solution improves the physical and chemical characteristics of bimetallic catalysts. A bimetallic catalyst exhibits a greater carbon yield than a monometallic catalyst, according to Pudukudy et al. [16]. Hydrothermal, pyrolytic, Sol–gel, sonochemical, radiolytic, microwave combustion, impregnation, microemulsion, and precipitation methods are a few of the physical and chemical processes that have been used to create bimetallic alloys. Researchers are still drawn to studying the synthesis of bimetallic magnetic alloys for the synthesis of syngas production [17]. The most prominent alloys used for syngas production comprise elements such as Ni, Co, and Fe. As catalysts, alloys perform better. Siddhartha Sengupta et al. synthesized (Ni/Al_2O_3), ($Ni\text{-}Co/Al_2O_3$), and (Co/Al_2O_3) catalysts with 15 % metal content. They observed that the initial turnover frequencies of reforming CH_4 for the $Ni\text{-}Co/Al_2O_3$ catalysts were higher than those for Ni/Al_2O_3, indicating that the Ni-Co alloy sites are more active than the Ni sites [18]. Through calcination and reduction of hydrotalcite-like compounds comprising Ni^{2+}, Cu^{2+}, Mg^{2+}, and Al^{3+}, Dalin Li et al. created Ni-Cu/Mg/Al bimetallic catalysts that were then tested for the steam reforming of tar developed by a low-temperature pyrolysis of biomass. Even at a low temperature of 823 K, the catalyst provided almost an complete conversion of the tar [19]. The higher metal dispersion, greater number of surface active sites, better oxygen affinity, and surface alteration brought on by the production of small Ni-Cu alloy particles all contributed to this high performance. Non-supported alloys with their distinct morphological features were synthesized by Buthainah Ali et al. and used in catalytic biogas decomposition to produce syngas and carbon bio-nanofilaments at a reaction temperature of 700 °C and 100 mL min^{-1}. The behavior of the catalysts, CH_4 and CO_2 conversion, and the carbon generated during the reaction were studied [17]. Irrespective of metals, alloys, and oxides, magnetic materials are selected as the better candidates for the catalyst process due the above discussed reasons, along with their magnetic characteristics. Various magnetic nanomaterials and their composites could be utilized as catalysts for syngas production/conversion. The next section will discuss the different synthesis methods that could be utilized for the preparation of magnetic nanocatalysts.

3. Synthesis of Magnetic Nanomaterials

To date, a variety of methods have been developed for the synthesis of magnetic nanomaterials and nanocomposites, viz. wet chemical route, microfluidic process, hydrothermal process, arc plasma discharge, chemical vapor deposition, microemulsion method, biogenic route, etc. Recently, there has been a heightened interest in synthesizing magnetic nanoparticles of uniform size for various technological applications [20]. Likewise, magnetic nanoparticle synthesis is one of the most critical challenges in tailoring particle size, shape, and crystalline structure. A wide range of magnetic nanostructures, including metals, metal alloys, oxides, and composite structures, have been synthesized using numerous techniques. Magnetic nanoparticles are prepared using three different methods, including (i) incipient wetness impregnation, (ii) precipitation, and (iii) modified sol–gel, to obtain different sizes of magnetic nanoparticles through synthesis. However, synthesis methodology plays a key role in controlling morphology, catalyst particle size, size distribution, and surface area. For example, A. Alayat et al. reported that a Fe/NS-I catalyst that was prepared using wetness impregnation and activated by CO has the highest activity and that its selectivity falls within the range of gasoline C6–C14 [21]. By contrast, the other Fe/NS catalysts show poor catalytic activity and selectivity when these Fe/NS catalysts were prepared using precipitation and modified sol–gel methods. The optimal preparation and activation methods to reach the highest catalytic activity and selectivity toward light hydrocarbons are with the Fe/NS-I catalyst activated by CO. In addition, the Fe/NS-I catalysts' selectivity favors aromatics in the C6–C14 range. In this section, we will describe some methods that provide excellent size and shape control for preparing magnetic nanomaterials, as shown in Figure 2.

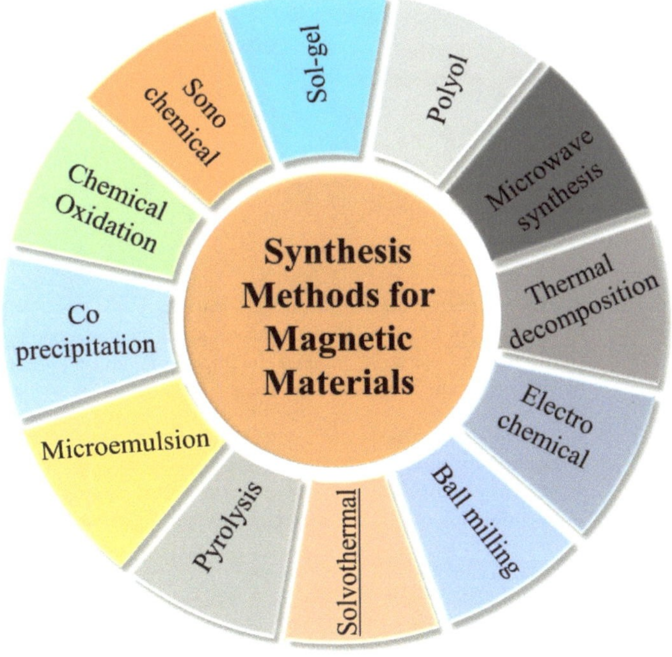

Figure 2. Different approaches for the preparation of magnetic nanoparticles.

Wet chemistry methods, which employ macro-scale equipment to carry out the reactions, enable batch production of nanoparticles. In comparison to microfluidic and biogenic synthesis, conventional MNP synthesis uses more energy [22]. Furthermore, it is more difficult to control morphology. Wet chemical routes are divided into hydraulic and non-hydraulic techniques, as shown in Figure 3, each of which has advantages and disadvantages depending on various variables, such as reaction temperature, pH value, precursor type, ratio of precursors, and nature of the base, which are the critical parameters that influence the size of magnetic nanoparticles [23]. With increasing pH, primary magnetite nanoparticles repel each other, thus becoming smaller [24]. The stoichiometry of the final magnetic nanoparticles could be altered by adjusting the ratio of precursors. The size of magnetic nanoparticles could be controlled by introducing a surfactant or by adjusting the reaction temperature. Wet chemistry method is considered one of the low-cost and simple chemical processes as it does not require any higher-cost equipment and processing. R. A. Frimpong et al. reported that the wet coprecipitation method used to prepare iron oxide nanoparticles gives a wide size distribution of magnetic nanoparticles [25] due to their chemical stability and biocompatibility compared to other metallic magnetic properties. Therefore, many investigations have focused on the development of large-scale production methods of uniform magnetic nanoparticles.

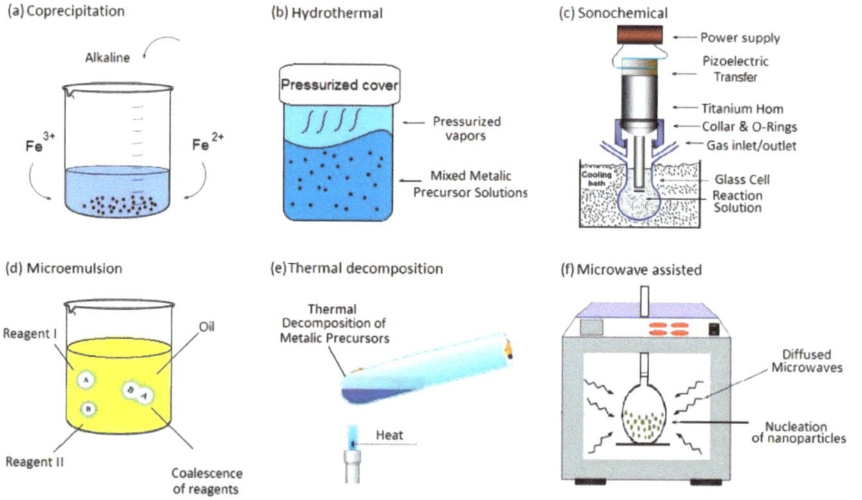

Figure 3. Schematics of distinct wet-chemistry hydrolytic and non-hydrolytic strategies for the synthesis of magnetic nanoparticles, (**a**) co-precipitation, (**b**) hydrothermal, (**c**) sonochemical, (**d**) microemulsion, (**e**) thermal decomposition, and (**f**) microwave assisted process. Reprinted with permission from [23], John Wiley & Sons, Inc.

In arc plasma technology, the plasma field provides the energy and reaction environment needed for atomic excitation, collision bonding, and more rapid formation of nanophase structures [26]. Using the arc plasma technique, high-temperature (>3000 degrees Celsius) plasma is generated between closely spaced electrodes under an inert atmosphere of helium or argon using a direct current. In this process, an electrode substance is evaporated by controlling the arc discharge, and the corresponding products are obtained upon cooling down on a cathode or a vacuum chamber wall [27]. Wang et al. reported that a minimum RL of −56.3 dB at 12.1 GHz and an absorption bandwidth of 5.2 GHz are obtained in the multi-step synthesis process [26]. Arc plasma is first used to prepare magnetic metal particles, which are then combined with carbon compounds. Methane (CH_4) is employed as a gaseous source of carbon in the reactor vessel.

Hydrothermal process is another easy and one-step way to develop hydrophilic magnetite nanomaterials using various precursors. There are many methods that have been used to develop magnetic metal oxide nanoparticles [28]. Hydrothermal synthesis of Fe_3O_4 nanoparticles in organic solvents is an advantageous process because of the homogeneity of the procedure and its comparatively low environmental impact. In a study by Misuthani et al., an aqueous solution containing ferrous and ferric ions is used as a starting solution, followed by precipitations of ferrous hydroxide ($Fe_3(OH)_2$) and goethite (a-FeOOH), which are precipitated in an alkaline solution as precursors [29]. The research by Zhang et al. demonstrated parameters that cause aqueous solutions to reach a critical or supercritical state, and these parameters are of interest because they allow the almost simultaneous reduction of two metal salts at a high temperature and high pressure [30]. They used the hydrothermal synthesis route to create magnetic CoPt alloy nanowires based on this method. Komarneni et al. first reported the synthesis of crystalline unit oxides, such as TiO_2, ZrO_2, and Fe_2O_3, as well as binary oxides, such as $KNbO_3$ and $BaTiO_3$, using the microwave hydrothermal method. Using microwave-hydrothermal conditions, the influences of several factors, such as chemical concentration, duration, and temperature, on crystallization kinetics at a microwave frequency of 2.45 GHz were investigated [31]. The reaction parameters, such as reaction temperature, reaction time, ratio of the precursors, and fraction of the surfactant, could be varied to obtain magnetic nanoparticles with different morphologies.

Solvothermal crystallization is one of the most consistent processes for crystal growth, and the resulting grains of magnetite have a significantly higher crystallinity than those made by other processes. The solvothermal fabrication process is similar to hydrothermal fabrication, which uses non-aqueous organic solvents. In solvothermal techniques, organic solvents, such as methanol, toluene, 1,4-butanediol, and amines, are frequently employed. Organic solvents are used as the solvents in these solvothermal processes. The solvothermal method's ability to control water-sensitive precursors is its most significant advantage. J. Li et al. simultaneously demonstrate that solvothermal techniques may be utilized in conjunction with microwaves and magnetic fields for the semicontinuous synthesis of materials with much increased repeatability and excellent quality [32].

The thermal decomposition process is another conventional process for the synthesis of various magnetic nanoparticles. However, the major disadvantage of this process is that it may require high-cost chemicals to produce magnetic nanoparticles, such as FePt, CoPt, and other related magnetic nanoparticles. A. G. Roca et al. evaluated the synthesis of monodispersed Fe_3O_4 nanoparticles by focusing solely on the thermal decomposition of organometallic precursors in organic solvents at high temperatures in the presence of surfactants. They yielded nanoparticles that are not hydrophilic in nature and cannot be dispersed in water, impeding their biomedical applications [33]. Other than the above discussed methods, a few more specifically designed synthesis methods for magnetic nanomaterials have also been successfully demonstrated, namely the sol–gel auto-combustion method, polyol process, chemical oxidation methods, and different deposition processes. However, each method and process has its own advantages and disadvantages. Depending on the requirements of the targeted particle's physicochemical characteristics, one can choose an appropriate synthesis process for the development of magnetic nanomaterials. Furthermore, experimental parameters, such as reaction temperature, precursors, precursors ratio, oxidation or reduction agent fraction, surfactant or stabilizing agent, reaction time, and post-thermal treatment, could be used to alter the physicochemical characteristics of magnetic nanomaterials.

4. Syngas Production and Conversion

Syngas production and conversion have been the focus of recent research. Several processes have been demonstrated for the production and conversion of syngas. Syngas can be created by gasifying waste, biomass, or coal in a high-temperature atmosphere. Syngas is a gaseous fuel created when a feedstock is partially oxidized in controlled operating conditions with a deficiency in oxygen. In a gasifier, gasification generally consists of five steps. In order to produce an incredibly clean gas at the exhaust, drying must first take place below 100 °C, with the goal of lowering the moisture content of the feedstock. The dry matter is scorched to a temperature of about 240 °C during pyrolysis as a result of the vaporization of heavy volatile chemicals (tars). In addition to hydrogen and carbon monoxide, tars contain more sophisticated compounds. A solid carbonaceous substance known as charcoal, which is composed of fixed carbon-to-carbon chains, is also produced by this process along with tar gasses. The conversion of biomass into a more useful energy form can be achieved using three main processes: (i) gasification, (ii) pyrolysis, and (iii) combustion [34]. Similar solid fuel feedstocks, such as coal, biomass, and wastes, are used in the production of syngas using traditional gasification techniques. Additionally, non-gasification technologies that employ a single reforming process or a series of catalyst-assisted reforming processes can convert liquid and gaseous feedstocks into syngas. Such procedures are anticipated to be effective in in situ syngas production techniques, as described in this section. Using chemical looping gasification in biomasses, the generation of syngas is depicted in Figure 4a. Firstly, the biomass is processed and devolatilized as it enters the fuel reactor. Additionally, there are numerous simultaneous interactions between the various pyrolysis products, the gasification agent (steam or CO_2), and the oxygen carrier. When tar is gasified with steam, it generates syngas, CO, or CO_2, as appropriate. Thirdly, during the pyrolysis and gasification of the biomass, redox reactions take place between both the active phases of the oxygen carrier (M_xO_y) and the gases that are produced. The oxygen carrier plays a vital role in the production of syngas. The oxidization of the reduced oxygen carrier and the partial combustion of the biomass yield the heat required for gasification progressions. Moreover, the oxygen carrier possesses catalytic qualities that raise syngas quality and lower tar concentration. The main material in syngas generation is an oxygen carrier containing metal oxides, which is in charge of moving heat and oxygen from the air reactor to the fuel or gasification reactor, thereby avoiding the expense of pure oxygen or significant volumes of steam [35] Higher CO_2 conversion and lower carbon formation are the results of the reaction between the surface oxygen species and the surface carbon species, which can adsorb CO_2 molecules to create bidentate carbonate species. Fuel reduction technologies based on metal oxides, such as thermal carbon reduction, metal oxide-based selective oxidation reactions, and chemical looping processes, have been demonstrated. Chemical looping reform is a method for producing syngas that partially oxidizes methane by utilizing lattice oxygen [35]. The syngas production mechanism using metal oxides is shown in Equation (1).

$$M_xO_y + \delta\ CH_4 = M_xO_{(y-x)} + \delta\ (2H_2 + CO) \quad (1)$$

Gasification is a fundamental process for converting a feedstock into synthesis gas. The process of gasification is a thermochemical conversion that uses partial oxidation of the raw material to create a gaseous mixture that is primarily composed of hydrogen and carbon monoxide [36]. There are four stages to the gasification process, depending on the type of gasifier (drying, pyrolysis, oxidation, and reduction). In Figure 4b, a graphical diagram for different Fe/Ca ratios used to synthesize Fe_2O_3/CaO with H_2-enriched syngas conversion [37].

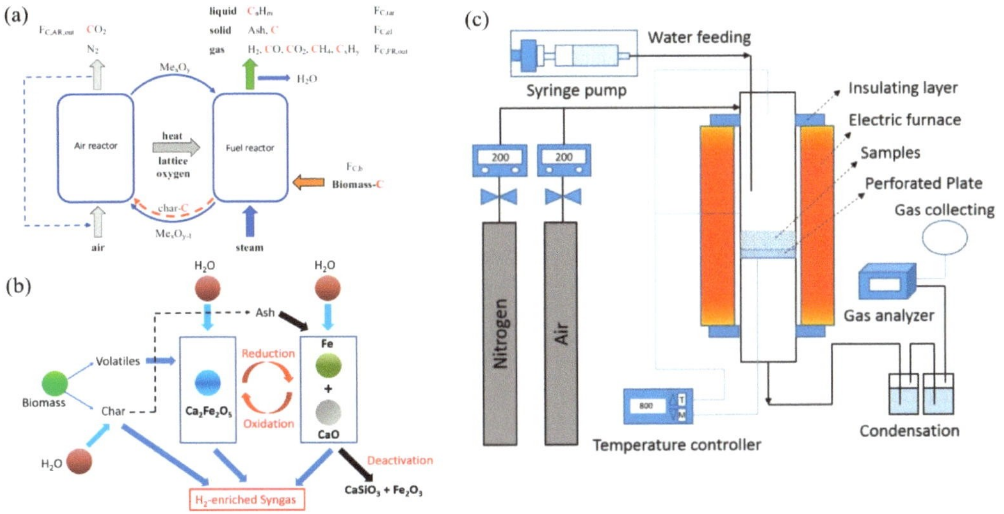

Figure 4. (**a**) The chemical looping gasification process. Reprinted from [35]. Copyright (2021), with permission from Elsevier. (**b**) Graphical diagram for different Fe/Ca ratios used to synthesize Fe_2O_3/CaO with H_2-enriched syngas. (**c**) Schematic diagram for different ratios of Fe/Ca used to synthesize Fe_2O_3/CaO with H_2-enriched syngas. Reprinted from [37]. Copyright (2020), with permission from Elsevier.

Figure 4c depicts the chemical looping gasification experimental technique, which consists of gas/water feeding, gasification reaction, condensation, syngas collection, and analysis. An electric furnace assembly with a stainless-steel tube reactor (inner diameter of 27 mm and height of 650 mm) was used as the gasification system in this experiment. Initially, the furnace was heated to a high temperature (600–900 °C). In the furnace, the stainless-steel reactor containing rice straw and OC was connected to a nitrogen flow (99.99%, 200 mL/min, and STP). A precise syringe pump was used to inject water into the reactor, and after the reactor was set up in the furnace, steam began to be produced and enter the reactor. At each run, 1 g of dried rice straw was mixed with 1 g of OC by mechanical stirring and reacted at the desired temperature for 30 min with a steam feeding rate of 0.1 g/min. The total steam was, therefore, affordable for the gasification of rice straw and the potential oxidation of reduced OCs, increasing the overall conversion efficiency. After the reaction, the solid residue was cooled down under nitrogen to room temperature and kept for further analysis. Based on this research, Quiang et al.'s study examined and compared the oxygen carrier durability, solid structure development, Fe/Ca ratio, temperature, and oxygen carrier cycling capacity to the pyrolysis and gasification processes. The trace extraction method resulted in the formation of two different kinds of calcium ferrites ($Ca_2Fe_2O_5$ and $CaFe_2O$) with various Fe/Ca ratios, with Fe and Ca being distributed equally throughout the oxygen carriers. When the Fe: Ca ratio was 1:1, the formed oxygen carrier ($Ca_2Fe_2O_5$) gave the highest hydrogen yield (23.07 mmol/g biomass) at 800 °C, benefiting from the one-step reduction and oxidation properties of $Ca_2Fe_2O_5$. A temperature of at least 800 °C was required for the complete redox reduction of $Ca_2Fe_2O_5$ during chemical vapor gasification. As a result, the high hydrogen selectivity of Fe: Ca = 1:1 ($Ca_2Fe_2O_5$) makes it a suitable oxygen carrier candidate, but the cycling stability should be improved to make biomass chemical chain gasification conversion a better application for the production of syngas [37]. Similarly, recent research by Liu et al. examined the CLG of microalgae with calcium ferrite or modified calcium ferrite in terms of syngas production and characteristics of OCs. Their findings demonstrated that $Ca_2Fe_2O_5$

was an appropriate material for microalgae because of its high selectivity for the production of synthesis gas [38].

Combustion is a chemical reaction between a fuel and an oxidant that makes an oxidized product. Generally, it is a reaction between hydrocarbons and oxygen to generate carbon dioxide, water, and heat. According to Zhang et al., the partial oxidation of fuel by chemical looping combustion can result in the formation of syngas [30]. $CaFe_2O_4$ and $Ca_2Fe_2O_5$ were discovered to be the two carriers that produced partial oxidation of solid fuels, the best among the four. The investigation on charcoal, however, may indicate a limitation with the respective carriers when using raw biomass because of the presence of various minerals, depending on its source. Besides, numerous investigations have also demonstrated the successful production of syngas using chemical looping gasification technology [39]. Likewise, Luo et al. obtained pure $Fe_2\ O_3/MgA_{12}O_4$ and iron ore with a temperature limit of less than 950 °C while using methane as a fuel [40].

Figure 5 illustrates the concept of indirect gasification technology with a combustion process [41]. In syngas production, a fuel mixture consisting mainly of hydrogen and carbon monoxide is burned, proving to be a promising fuel for combustion technology. According to Amin Paykani et al., the choice of base fuel, its physicochemical properties, and its physical conditions are all important factors that influence the synthesis of syngas produced by combustion systems [42]. M. Fiore et al. demonstrated that gasification is the only exothermic process and its heat release is necessary for water and carbon dioxide, as the products of several oxidation reactions are formed during this stage [43]. During cracking, carbonaceous particles become simpler molecules once they have been exposed to heat. To generate a gas compatible with an internal combustion engine, tars must first be broken down chemically, Otherwise, they could condense into sticky tars, which may cause damage to the valves. The reduction process, which assures the synthesis of hydrogen and carbon monoxide from water and carbon dioxide, is the last step. In order to achieve a high fraction of fuel gas at the exit, reduction is accomplished by passing the combustion products over a bed of extremely hot charcoal.

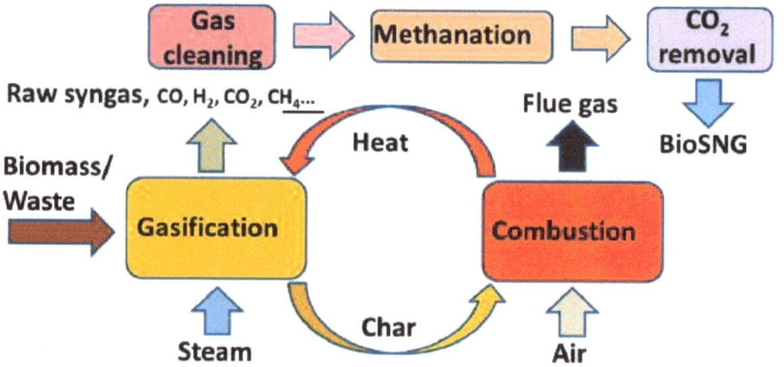

Figure 5. The concept of indirect gasification technology with the combustion process, reproduced from [41], with permission from MDPI Publisher.

Another process utilized for this purpose is pyrolysis, in which organic and plastic materials are thermally decomposed. During pyrolysis, materials undergo continuous physical and chemical changes. A wide range of applications is possible with pyrolysis technology. The chemical industry, for example, produces methanol, charcoal, and activated carbon by using this method. In addition to these, the pyrolysis process has several advantages over other energy recovery methods, such as the ability to use stone, ceramics, glass, or soil obtained during waste sorting. With this technology, a wide variety of feedstocks can be utilized. Consequently, it reduces landfill waste, landfill gas emissions, and the risk of water contamination. Moreover, building a pyrolysis plant is relatively simple [44]. Depending on the conditions associated with the process, pyrolysis can be classified into three types: slow, fast, and flash pyrolysis. A variety of thermochemical and biological processes have been used to transform biomass into products with added value. Among those processes, pyrolysis is more convenient since it has several advantages in terms of storage, transportation, and flexibility in the solicitation. This includes engines, boilers, combustion appliances, turbines, etc. Generally, it can be divided into components that generate only heat and biochar (by using slow pyrolysis), and units that produce both biochar and bio-oils (using fast pyrolysis). Additionally, flash pyrolysis is a quick pyrolysis unit that transforms biomass into bio-oil [43,44]. Typical pyrolysis systems comprise a pyrolysis reactor, an extraction unit, and equipment for pre-processing lignocellulosic residues.

5. Tuning of Experimental Parameters

When developing nanocatalysts, which are employed in numerous chemical reactions to speed up the reaction and improve product efficiency, nanomaterials are frequently used because they have special catalytic capabilities. The effects of experimental factors, such as nanoparticle size, shape, distribution, and preparational methods, on their catalytic properties have been investigated by various researchers [45]. Basically, research has been conducted to comprehend the mechanism of action of nanocatalysts or the activity of catalysts, as well as the selectivity of nanocatalysts. Due to their distinctive features, nanoparticles have attracted attention as a material for advanced biofuel processes. Nevertheless, the use of nanoparticles in the production of bioethanol is still in its infancy. The morphology, physicochemical properties, and durability of nanoparticles are significantly impacted by a number of parameters throughout the synthesis process. Therefore, operational factors have a great impact on the synthesis process. The performance of nanoparticles is influenced by a variety of factors during the production of biofuel. The synthesis method, pressure, temperature, medium pH, nanoparticle size, and other variables are crucial [46].

5.1. Temperature

Since temperature affects the geometry of nanoparticles, comprising their size, shape, and endurance, temperature is a significant factor in the synthesis of biofuel. The metabolism and proliferation of cells are greatly influenced by temperature when using biological methods to produce syngas [47]. The solubility of CO, H_2, and CO_2 in the fluid phase is influenced by temperature. For different microorganisms, different temperatures are ideal for syngas production. Physical and chemical techniques require temperatures higher than 300 °C, but biological methods utilize temperatures that are low to moderate (<100 °C), or even ambient. Besides this, based on the synthesis technique, the calcination temperatures for different metallic nanomaterials vary between 100 and 700 °C. The effects of temperature influence the geometry of nanoparticles, the rate of reaction during the catalytic reaction, the durability of nanoparticles, and the structure and size of nanoparticles. Reaction rate and temperature are tightly correlated, suggesting that reaction rate is slow at a relatively low temperature and speeds up as the temperature increases. Compared to nanoparticles produced at reduced temperatures, such as 25 °C, those produced at higher temperatures, such as 100 °C, are more stable and develop to their proper size. A higher temperature can speed up the production process [48]. Transesterification is a synthesis process that is used to prepare straight oils from renewable sources for biodiesel production [43,45]. When methanol or ethanol are present, transesterification processes begin at temperatures between 50 and 80 °C with sodium hydroxide (NaOH) functioning as a base catalyst. The impacts of catalyst calcination temperature and catalyst bed temperature on the constitution and productivity of syngas were studied by Shuangxia Yang et al. [49]. Their studies focused on the synthesis of hydrogen-enriched gas from biomass in a two-stage fixed-bed reaction system using a Fe/CaO catalyst formed from layered double hydroxides as the precursor. At a modest temperature range of 600 °C, the Fe/CaO catalysts displayed uniform morphological characteristics and crystalline size, and they had outstanding physicochemical characteristics, such as H_2 reducibility and CO_2 absorption capacity. This contributed to their high catalytic performance in terms of the production of syngas, which resulted in a syngas yield of 44.6%, an H_2 composition of 35.7%, and an H_2/CO ratio of 1.36 in the range. Although the H_2/CO molar ratio decreased to 0.99 as a result of additional CO_2 released from the decomposition of $CaCO_3$ at higher catalytic temperatures, the syngas yield and H_2 yield increased up to 63.0 weight percent and 172 mL/g of biomass, respectively, as a result of the augmented supplementary cracking and realigning reactions of volatiles in the existence of the Fe/CaO-600 catalyst. The gaseous concentration and gaseous yield of the syngas conversion of biomass pyrolysis in the presence of the Fe/CaO catalysts calcined at various temperatures are shown in Figure 6.

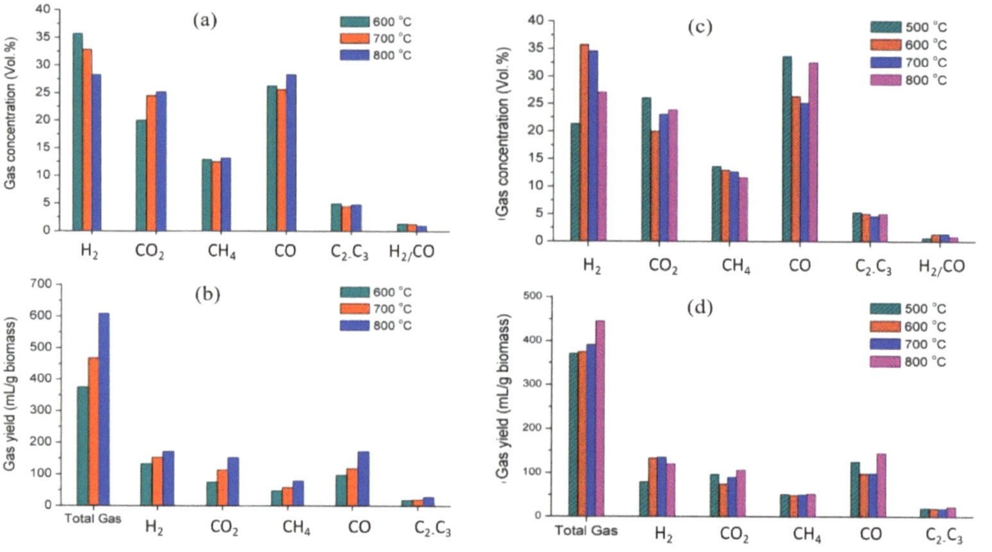

Figure 6. Bar diagram of (**a**) gaseous concentration, (**b**) gaseous yield of the syngas conversion of biomass pyrolysis, effect of catalytic bed temperature on (**c**) concentration, and (**d**) yield of the syngas. Reprinted from [49]. Copyright (2017), with permission from Elsevier.

5.2. pH

Due to the significant role that pH variation plays in the formation of nucleation centers, pH values have an effect on the shape, size, synthesizing pace, and homogeneity of particle dispersion of nanomaterials. Small nanoparticles are evident at a pH of 6, but when the pH rises from 7.0 to 11.0, the average nanoparticle size decreases [50]. The size and form of nanoparticles are influenced by the pH of the medium: a medium with a basic pH offers a quick growth rate, faster productivity, and enhanced reduction rates, and it regulates the size, synthesizing rate, homogeneity of particle distribution, and morphology of nanomaterials. The typical size of nanoparticles drops when the pH rises from 7.0 to 11.0, while microscopic nanoparticles are evident at a pH of 6. This not only impacts the regularity of particle distribution and synthesizing speed, but it also influences the growth of nucleation sites in the morphology of the particles [50–52]. Z. Fakhroueian et al. studied the impact of the pH factor on the production of nanocatalysts, nanocomposites, and nanotubes using the hydrothermal, coprecipitation, and sputtering techniques of Cu_{Ox} NiO_x for syngas processes in the petroleum sector [53]. According to their research, the pH factor can alter the original texture, surface characteristics, and material properties. For instance, 25 percent of NiCo/CeAl nanocomposites at a pH of 9 can yield nanotubes and nanocomposites with the finest adaptive modulation, 99 % CH_4 conversion, and 91 % H_2 selectivity in the syngas reactions; however, at a pH of 7, it obtains a simple nano-spherical assembly, containing 98% of CH_4 conversion and 93.40 % of H_2. The pH of the culture has a considerable impact on aerobic metabolism, cell proliferation, and product distribution in biological processes. An effective approach to promoting the formation of syngas is a pH shift.

5.3. Catalyst Preparation Process

The coprecipitation method, microemulsion, thermal breakdown, hydrothermal synthesis, synthesis employing plant materials, synthesis incorporating biological organisms, and so forth are instances of nanoparticle synthesis procedures. Every method has a unique set of benefits and drawbacks [54]. Different synthesis methods have an impact on the efficiency of nanomaterials for the production of syngas. The most efficient processes for producing biodiesel are the ultrasonic reactor, batch reaction, lipase-catalyzed technique, and supercritical process. With these techniques, transesterification is used to synthetically create pure oils from renewable resources. Enzymatic hydrolysis, preprocessing, and fermentation are several steps required for the production of bioethanol. Various pretreatment techniques, including physical, chemical, physicochemical, and biological procedures, have been employed for the conversion of biomass. Different synthesis techniques play a crucial role in the performance of nanoparticles in the conversion of biomass [55]. The generation of biofuel and the effectiveness of particles are severely affected by their synthesis methods [56].

5.4. Pressure

In the formation of nanomaterials and biofuels, pressure is a significant factor. Typically, the reaction medium is subjected to the proper pressure in order to produce the desired geometry, size, and coalescence of nanoparticles. Particle sizes rise as a result of the high pressure. In order to regulate the levels of quick deployment and decrease, pressure can also be applied [55,57]. The effects of adding ferric chloride ($FeCl_3$) to biodiesel on the performance and emission parameters of diesel engines were studied by Kannan et al. [57]. The rate of heat release and cylinder pressure were reported to have increased. While the thermal efficiency of the brakes increased by 6.3%, it was also determined that the specific fuel consumption decreased by 8.6%.

6. Utilization of Magnetic Nanomaterials

Magnetic nanomaterials of Fe, Ni, Co, and various combinations with other active nanomaterials and magnetic nanoparticles are found to be better candidates for use as the catalysts for syngas production/conversion process through various methods. This section discusses the progress made on magnetic nanomaterials containing Fe, Ni, and Co as the catalysts for syngas production/conversion process.

6.1. Fe-Based Magnetic Materials

For the synthesis of syngas, iron (Fe) is a promising active-phase material due to its lower cost than noble metal catalysts, abundance, low toxicity, and high reliability in tar breaking [58]. Due to the presence of mono-Fe and rhodium in the catalysts, conventional Rh-Fe catalysts often show greater alkane or methanol selectivity, and as a result, their specificity to higher alcohols is poor. Tong Han et al. synthesized an alternative catalyst. $YRh_{0.5}Fe_{0.5}O_3$ with a perovskite structure was layered on ZrO_2 as a solution to the conflict [59]. The homogenous and highly dispersed Rh-Fe alloys were responsible for the remarkable specificity and productivity of $YRh_{0.5}Fe_{0.5}O_3/ZrO_2$. $YRh_{0.5}Fe_{0.5}O_3/ZrO_2$ demonstrated excellent stability as well. Thomas E. L. et al. examined the hydrogenation of CO_2 into methanol using Ni-Fe-Ga alloys to generate Ni_2FeGa, and they observed that Ni_2FeGa performed the best out of all the evaluated catalysts in terms of methanol output. At low processing temperatures and pressures, the efficacy of the produced Ni_2FeGa yielded performance similar to commercial products, i.e., $Cu/ZnO/Al_2O_3/MgO$ [60].

Mixing different transition-metal elements is an efficient technique to boost catalytic activity, together with nano-structuring/crystallization to raise the bulk activity. When methane and solid oxides (oxygen carriers) combine, syngas is produced. The reduced oxygen carriers can then be replenished by a gaseous oxidant, such as water and air [61]. Chemical looping ignition is regarded as a very effective technique for capturing CO_2. Since CO_2 segregation and capture are inherent to the process, it also offers a far lower energy penalty and cheaper cost than alternative CO_2 capture devices. Combustion occurs in the chemical looping process without the fuel and air coming into contact. Metal oxide (Me_xO_y), which is converted to Me_xO_{y-1}, delivers the ignition oxygen to the fuel reactor. Air is used to replenish the metal oxide inside a network of connected air reactors. The fuel is oxidized inside the fuel reactor to CO_2 and H_2O [62]. Ral Pérez-Vega investigated the ignition of syngas using a Mn/Fe-based oxygen carrier and demonstrated complete fuel transformation to CO_2 and H_2O at fuel unit temperatures exceeding 930 °C. Moreover, the CH_4 ignition reached the maximum burning rate of 78% at 966 °C. In addition, they reported that the fuel reactor was heated to a temperature higher than 1000 °C in order to complete CH_4 combustion [63].

Jiao Zhao et al. synthesized a Fe-N-C non-precious electrocatalyst, and a flow rate of 24 mL min^{-1} of very pure CO_2 was delivered into the cathode chamber for 30 min prior to electrolysis [5]. At the same time, a magneton was used to agitate the electrolyte in the cathode chamber. Additionally, they stated that, compared to a resin made of carbonized melamine formaldehyde, the non-precious Fe-N-C electrocatalyst demonstrates well-pleasing movement of CO and H_2 generation (FE_{CO} = 74%, FE_{H2} = 25%) at a lower overpotential of 0.6 V than others. Additionally, they reported that, compared to a resin made of carbonized melamine formaldehyde, the Fe-N-C non-precious electrocatalyst exhibits well-pleasing activity of CO and H_2 generation (FE_{CO} = 74%, FE_{H2} = 25%) at a subordinate overpotential of −0.6 V, demonstrating the better selectivity of CO_2 reduction reactions [5]. Mesostructured titania and silica were synthesized by Claudio Cara and colleagues as support materials for the purification of H_2S from syngas. Fascinatingly, it was found that both ultrasmall aspects were extremely active, sensitive to H_2S, and regenerable. Under a syngas environment, the two composites' active phases differed noticeably, with the amorphous silica-based composite performing better when there were Fe infusions [64]. Aluminum magnesium spinel ($MgAl_2O_4$) was developed by G. V. Pankina et al., who employed Fe-K as the activator, and its particular surface areas were examined in relation to the physicochemical characteristics and dynamics of Fe-K [65]. They observed that syngas reduction occurs predominantly when the catalyst's magnetization is higher.

Gihoon Kwon et al. developed a Fe-impregnated bentonite, which was used as a catalyst in the CO_2-assisted pyrolysis of grass cut to increase the production of syngas, quality of bio-oil, and sorptive properties of biochar. Similar to the pyrolysis of freshly grass cut, the production of hydrogen in the pyrolysis of the biochar under the N_2 condition begins at 500 °C (Figure 7(I)a). Yet, compared to the single pyrolysis of grass cut, it creates more H_2 with a maximum intensity at 700 °C, which is >2.5 times higher. Figure 7(I)b shows the concentration of CH_4 for the biochar. From their results, it is evident that the existence of bentonite accelerates the formation of H_2, particularly between 630 and 700 °C. At the corresponding temperatures, bentonite is also found to have a similar impact on CO generation, as shown in Figure 7(I)c. Figure 7(I)d displays the outcomes of the gas investigations from the co-catalytic pyrolysis of the Fe-bentonite [66]. In the presence of N_2, the pyrolysis of the Fe-bentonite produces more H_2 than the pyrolysis of the grass cut and biochar, with two peaks at 500 and 700 °C (Figure 7(II)a). The authors concluded that the Fe-impregnated bentonite is effective in pyrolyzing biomass to increase the generation of syngas and improves the standard of pyrogenic products.

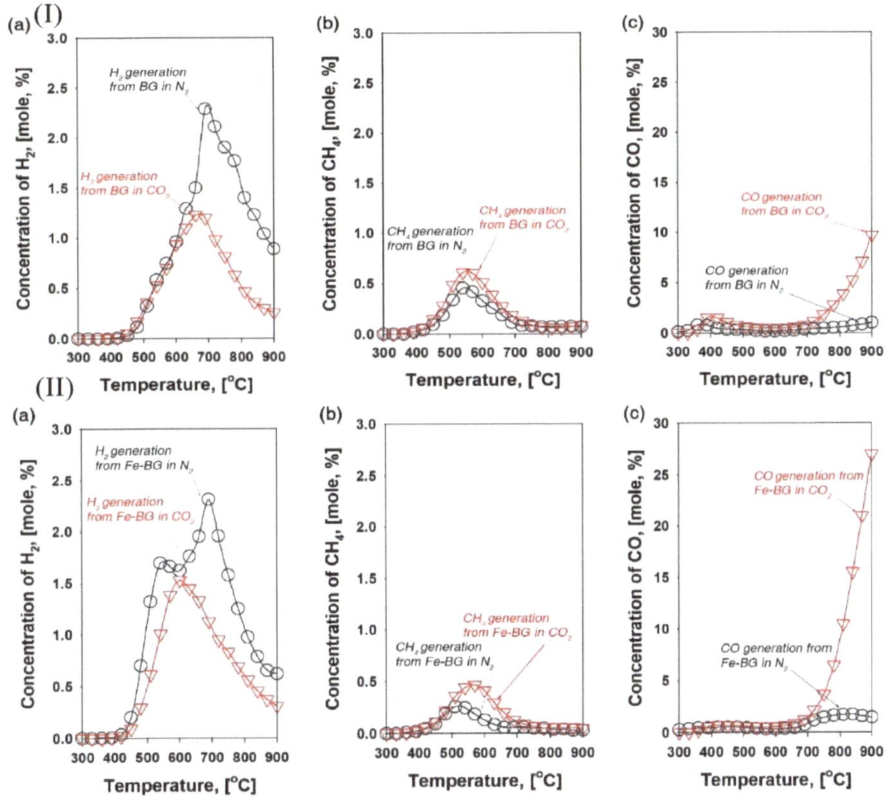

Figure 7. (**I**) Concentrations of H_2 (**a**), CH_4 (**b**), and CO (**c**) formed during the pyrolysis of grass cut under the CO_2 and N_2 conditions. (**II**) Concentrations of H_2 (**a**), CH_4 (**b**), and CO (**c**) formed during the pyrolysis of Fe-bentonite under the N_2 and CO_2 conditions. Reprinted from [66]. Copyright (2022), with permission from Elsevier.

6.2. Ni-Based Materials

The research on nickel (Ni) has focused on enhancing reforming catalysts' durability to manufacture syngas on a large scale [67]. To improve partial oxidation, hydrogenation, dry reforming, and steam reforming, Ni-based catalysts are frequently utilized. They are favored for the reforming process due to their great stability and catalytic activity, as well as the fact that they are less expensive than noble metal catalysts [68]. On account of coke formation on catalysts and Ni sintering, Ni catalysts are known to be susceptible to deactivation. Numerous research studies have proposed that improving catalyst stability might require changing the support, adding functional metals, and using the right catalyst preparation techniques [69]. Mohammad Yusuf et al. synthesized Ni- based alloy catalysts sustained on mixed oxide (Al_2O_3/MgO) with the inclusion of tungsten. They noticed that a 4% by weight of tungsten loading resulted in the best performance and that the Ni-W bimetallic alloy catalyst still produced above 90% conversion of greenhouse gasses after 12 h [70]. By studying the catalytic pyrolysis of a herbal residue using a Ni-doped Fe/Ca catalyst, Jin Deng et al. demonstrated that the strong interaction between Fe and Ca produced $Ca_2Fe_2O_5$, which prevented the formation of $CaCO_3$ and CO_2 and aided in the transition of tar and char. At 5% of the Fe-Ca catalyst, the H_2 output improved to 54.6 mL/g. Ni increased the dispersion of the catalyst and aided in the adsorption of alkaline ions while forming a stable Fe_3Ni_2 alloy with Ca. The yields of H_2 and CO improved from

54.6 mL/g to 63.4 mL/g for 0.5% of Fe/Ca and from 70.0 mL/g to 95.5 mL/g for 0.5% of Ni-Fe/Ca, respectively [71].

A multi-technique approach was used by L.B. Raberg et al. to investigate the potential relationship between support basicity, Ni/support activity, and stability of catalysts for the dry reforming of propane to synthesis gas. After being exposed to a magnetic field, ferrous metals retain a leftover magnetic moment at zero field; nonetheless, when the particle diameter falls below a particular size, the ferromagnetic particles transform into superparamagnetic particles [72]. The dry reformation of lignin with embedded Ni nanoparticles during microwave irradiation was prepared by M.V. Tsodikov et al., who adopted two distinct methods: impregnation and deposition [73]. In both instances, Ni particles with a size of 6 nm developed on the interface of the lignin. The dry reformation of lignin with microwave assistance was controlled by a variety of nanocatalysts. The first method's physical mixing of a carbon sorbent with a large dielectric loss factor and a specimen of lignin containing Ni was treated mechanically with a microwave, and the outcome was a 65-weight percent reformation of the lignin to synthesis gas with a H_2/O ratio of 1/1. The products contained 80–90 weight percent of syngas. It was observed that the microwave treatment significantly boosted syngas yield during reforming compared to convection heating. It was found that various deposition techniques produced X-ray amorphous Ni particles with noticeably varied magnetic characteristics. Superparamagnetic Ni0 particles developed by metal vapor synthesis showed the highest ability for absorption using the microwave technique, which was adequate for the plasma to be derived as well as for the reforming temperature to be attained without the need for an additional sorbent; this was in contrast to the completely paramagnetic sample made from a nickel acetate solution [73]. It is quite well known that highly distributed metals with magnetic characteristics have a large absorbency in microwave assistance.

In a two-step gasification of sewage sludge at a treatment temperature of 500, 700, and 900 °C, Chao Gai et al. showed catalyst selectivity toward the primary gas components [74]. They observed that the experiments using the hydrochar without Ni nanoparticles showed lower discernment (35–46%) of total gas composition toward H_2 and CO_2 production of 19–35%. The Ni nanoparticles synthesized with the hydrochar catalysts, on the other hand, all displayed better catalytic activity and selectivity for H_2 production. From their report, it is demonstrated that Ni nanoparticles are useful for gasifying biomass to produce H_2-enriched syngas. The authors postulate that the hydrochar-supported, evenly disseminated metallic Ni nanoparticles are responsible for the remarkable catalytic activity and selectivity toward H_2. Figure 8 shows the two-stage gasification of the sewage sludge's catalyst selectivity toward the major gas components at a treatment temperature of 500, 700, and 900 °C.

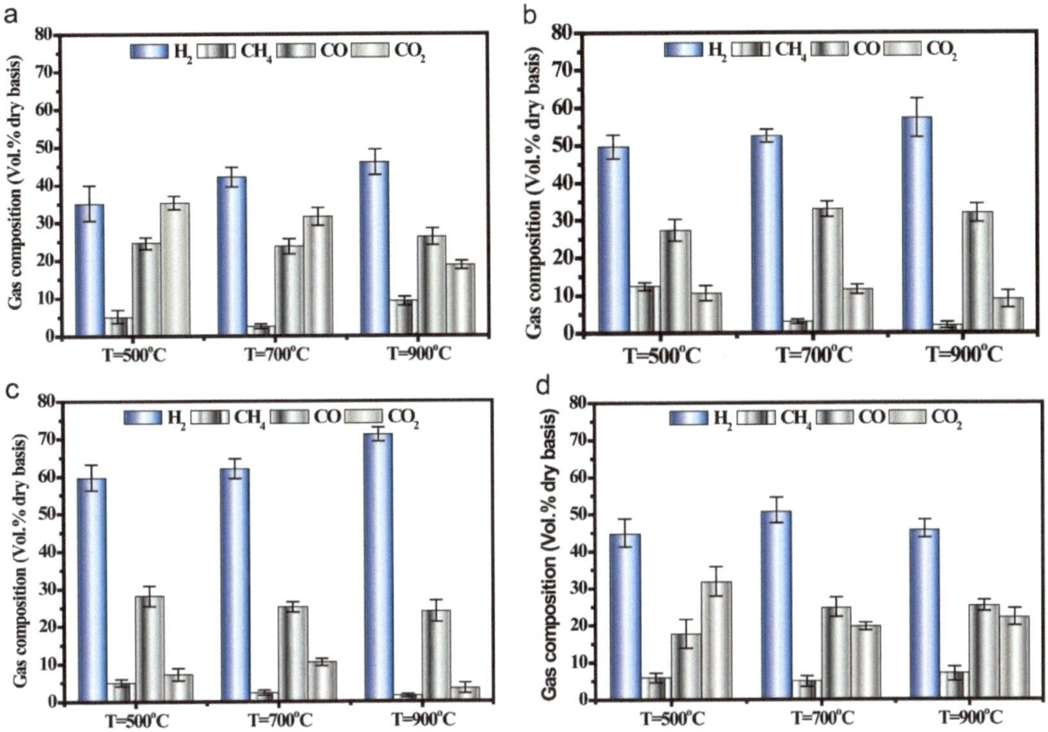

Figure 8. The gas product selectivity from the catalytic gasification of sewage sludge using catalysts from (**a**) hydrochar and (**b**–**d**) nickel@hydrochar with different concentrations of nickel (0.1, 0.5, and 1.0) at different gasification temperatures from 500 to 900 °C. Reproduced from [74]. Copyright (2019), with permission from Elsevier.

The core–shell LaMer model is largely acclaimed as being reliable, which is comprised of a hydrophobic core and a hydrophilic shell, and the model was applied by Chao Gai et al. for the synthesis of hydrochar-supported Ni nanoparticles. The interface of the hydrochar is hydrophilic with a dispersion of -OH and -C=O, such as carbonyl, hydroxyl/phenolic, and carboxylic functions. The carbonyl, hydroxyl/phenolic, and carboxylic functional groups on the surface of the hydrochar engage with Ni^{2+} ions to absorb them when they are added to the hydrothermal liquid through ion exchange interactions [74]. Figure 9 displays a schematic illustration of the hydrochar-supported nickel nanoparticles.

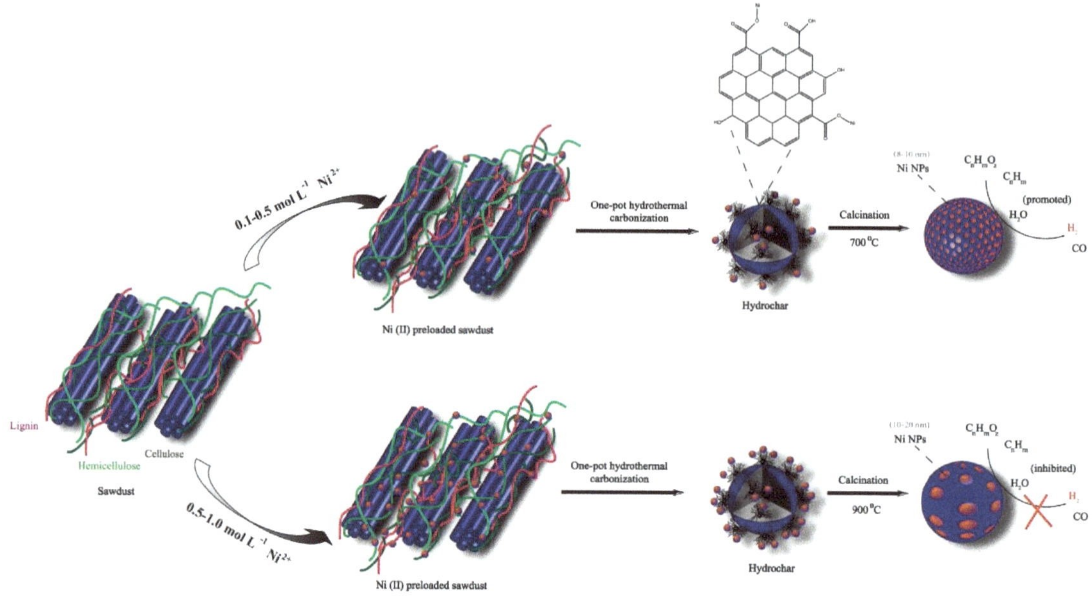

Figure 9. Schematic illustrations of hydrochar−supported nickel nanoparticles synthesized from a lignin-rich precursor biomass. Reproduced from [74]. Copyright (2019), with permission from Elsevier.

Ziwei Li et al. fabricated Ni@Ni−Mg Phyllosilicate core/shell catalysts by treating hydrothermally Ni@SiO_2 with $Mg(NO_3)_2$ [75]. The Ni@Ni-Mg Phyllosilicate core/shell catalyst treated for 10 h had the best catalytic performance out of all the Ni@Ni-Mg Phyllosilicate catalysts synthesized under different hydrothermal durations, with stable CO_2 and CH_4 conversions of around 81 and 78%, respectively. Its stability for the dry reforming of the methane reaction is shown in Figure 10(I). The most potent catalyst (treated for 10 h) was taken through testing for approximately 100 h, and it demonstrated conversions of CO_2 and CH_4 of 80 and 78%, respectively. The H_2/CO ratio was approximately 0.96. This demonstrated the stability of the catalyst, treated for 10 h for the dry reforming of the methane process, at 700 °C. The TGA analysis for the used catalysts after a 20 h process is shown in Figure 10(II). For the 20 h testing period, there is negligible weight loss for Ni@SiO_2. This may be due to the resistance influence of SiO_2 to Ni during sintering, which slows down the reaction of CH_4 breakdown. Due to the prevalence of CH_4 breakdown on sintered NiO, the catalyst used before calcination (befcal-NM-24 h) shows the most intense carbon deposition (26.1%). Despite being lower than that of the catalyst treated for 2.5 h (M-2.5 h) and 10 h (M-10 h), the weight loss for the after-calcination (aftcal-NM-24 h) catalyst is still higher at 6.3%. The M-2.5 h and M-10 h catalysts exhibit significantly less weight loss when compared to the befcal-NM-24 h catalysts, while having identical porosity and Ni exposures but higher basicity. Additionally, although having a weight loss of 10.4%, the catalyst treated for 24 h (M-24 h) loses 2.5 times less weight than the befcal-NM-24 h catalyst since it has a similar porosity and surface Ni exposures. These findings demonstrate the magnesium entity's potential for reducing the issue of carbon deposition, as illustrated in Figure 10(III). The authors performed a TEM analysis to examine the morphologies of the used catalysts. Following 20 h of reaction at 700 °C, cross-linking across silica shells emerges for both the Ni@SiO_2 and aftcal-NM-24 h catalysts, as can be seen in Figure 10(III)(a,b). This is mediated by the hydroxylation process between -(Si-O-Si)- assemblies on the silica shells, which results in the formation of three-dimensional systems when water is present. Due to the low H_2/CO ratio, this water is formed through a reverse water–gas shift reaction.

From Figure 10(III)(c), a substantial amount of carbon nanotubes can be seen. This might be a result of NiO phase sintering. For the catalysts treated for 2.5 h (M-2.5 h) and 10 h (M-10 h), there is no cross-linking within the silica shells, as shown in Figure 10(III)(d,e). In the core of these two catalysts, Ni is well diffused, with the outer layer consisting of a needle-like Ni@Ni-Mg Phyllosilicate phase. Moreover, for the catalysts treated for 2.5 h (M-2.5 h), a tough outer silica shell is still discernible, whilst for the catalysts treated for 10 h (M-10 h), the majority of the porous Ni@Ni-Mg Phyllosilicate species replace the silica shell of Ni@SiO_2. This makes it possible for the core portion to have great accessibility to active Ni while still limiting their sintering, which leads to strong CO_2 and CH_4 conversions and minimal carbon deposition. On the other hand, a significant number of carbon nanotubes and effective sintering of Ni /NiMg solution occur, as shown in Figure 10(III)(f), which lead to a lower catalytic activity and lower stability.

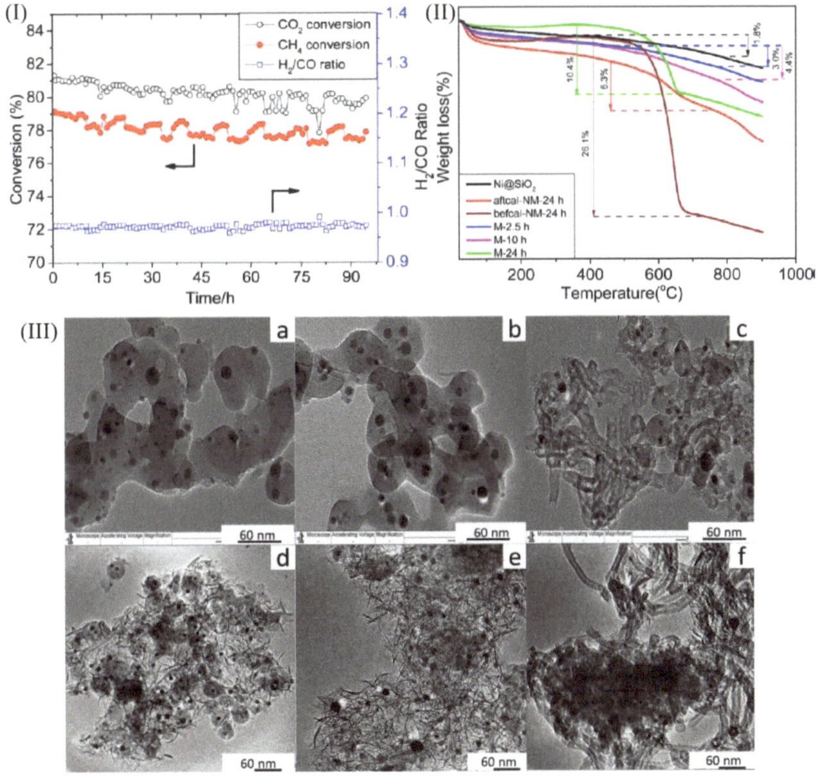

Figure 10. (**I**) Stability of the catalyst treated for 10 h and calcined at 700 °C. (**II**) Thermogravimetric analysis (TGA) for the utilized catalysts with various structures. (**III**) TEM images of the utilized catalysts after 20 h of reaction (**a**) Ni@SiO_2, (**b**) aftcal-NM-24 h, (**c**) befcal-NM-24 h, (**d**) M-2.5 h, (**e**) M-10 h, and (**f**) M-24 h. Reprinted with permission from [75]. Copyright (2014), with permission from the American Chemical Society.

6.3. *Co-Based MATERIALS*

According to a report, Co-based alloys are efficient and have a high hydrocarbon production rate, making them a suitable material for use in syngas conversion [76]. Considerable efforts have been made to develop a range of different metal alloy materials to reduce the total cost and promote the catalytic efficiency of catalysts. Co-Cu bimetallic catalysts are regarded as the most viable catalysts due to their high affinity to higher alcohol and cheap cost. Cu-decorated Co-based catalysts and Co-decorated Cu based catalysts are the two forms of Co-Cu bimetallic catalysts. Through theoretical and practical research, Prieto et al. demonstrated that a Co-Cu alloy phase having Co-rich ratios is the best catalyst interface for long-chain alcohols [77]. Using a steady-state isotopic transient kinetic analysis, the results suggest that the addition of Cu into Co significantly promotes the formation of more C^{2+} oxygenates by blocking a significant fraction of Co sites for hydrocarbon synthesis [78]. Using radio-frequency magnetic fields, Hoang M. Nguyen et al. created binary (Co-Cu)- and ternary (Co-Cu-Ni)-based alloys. All prepared catalysts were stimulated under radio-frequency ignition to transform CO_2 and co-reactants, such as H_2O and methane, into syngas (H_2 and CO) at a low temperature of 400 °C [79]. They also concluded that, in all the performance conditions, the Cu-Co sample had the best catalytic stability and performance. According to their research observations, the catalyst system's inclusion of a magnetic constituent and an electrically conductive element enables very efficient heating to occur through both hysteresis loss and Joule effect. The Cu-Co catalyst maintained its remarkable stability for about 50 h throughout the streaming test due to Joule heating. The results offer important insights into the potential of catalysts for RF-assisted chemical processes that do not require a lot of excessive input current or the usage of powerful magnetic metals to attain the optimal chemical configurations. According to Lin Chen et al., who used Co with Ni as an aerogel catalyst, the CH_4 conversion in a magnetic-assisted fluidized bed reactor increased by 12 and 7%, respectively, in comparison to a fixed bed reactor and a conventional fluidized bed reactor [80]. This was due to the catalyst's better fluidization reliability. Additionally, over the course of the 50 h reaction, the catalytic performance remained quite steady. The enhanced gas–solid association in the magnetic-assisted fluidized bed reactor and the Co-Ni aerogel catalyst's better catalytic property can both serve as examples of this higher stability.

Attapulgite-derived MFI (ADM) zeolite-encased Ni-Co alloys were developed by Defang Liang et al. utilizing a one-pot technique [81]. The 24 h of dry reforming of methane effectiveness of the 10Ni@ADM, 10Ni@ADM-0.1, and 10NilCo@ADM-0.1 catalysts were investigated at 700 °C at 20/20/60 mL/min of $CH_4/CO_2/N_2$, and the results are displayed in Figure 11a. This investigation was performed to study the influence of the structures of ADM-encased Ni-Co alloys on catalytic stability and activity. The 10Ni1Co@ADM-0.1 catalyst has the highest dry reforming of methane activity, as shown in Figure 11a–c. The stability test results are shown in Figure 11d,e. The primary conversion of CH_4 and CO_2 by the 10Ni@ADM catalyst was 52 and 95%, respectively, with an H_2/CO molar ratio of 0.88. Furthermore, the 10Ni@ADM catalyst underwent a modest deactivation after 24 h of reaction, as the rates of CH_4, CO_2 conversions, and H_2/CO molar ratio declined to 45%, 85%, and 0.88, respectively. This might be a result of the catalyst being exposed to high temperatures for a long time, which caused the encased metal grains to migrate to the interface of the zeolite, causing partial metallic Ni sintering and active area loss. It was found that the initial activity of the 10Ni@ADM-0.1 catalyst and the 10Ni@ADM catalyst was related. However, once the reaction was over, the CH_4 and CO_2 conversions only dropped to 51% and 89%, respectively, showing that the inclusion of NaOH helped increase the catalyst's reliability. In terms of catalysts, the 10Ni1Co@ADM-0.1 catalyst showed the best catalytic performance, achieving the highest initial CH_4 and CO_2 conversions of 71 and 93%, respectively, and maintaining those conversions throughout the dry reforming of methane reaction. Two mechanisms, including the presence of Co as new active sites that promote CH_4 activation/adsorption and the alloy interaction between Ni and Co that promotes CH_4 breakdown, may be responsible for the higher initial CH_4 conversion when compared to the 10Ni@ADM catalyst. In contrast to the 10Ni1Co/ADM-0.1 catalyst, as shown in Figure 11f, the 10Ni1Co@ADM-0.1 catalyst experiences little degradation over 100 h of the dry reforming of methane reaction, further demonstrating the structural resilience of the ADM zeolite-encased Ni-Co alloys.

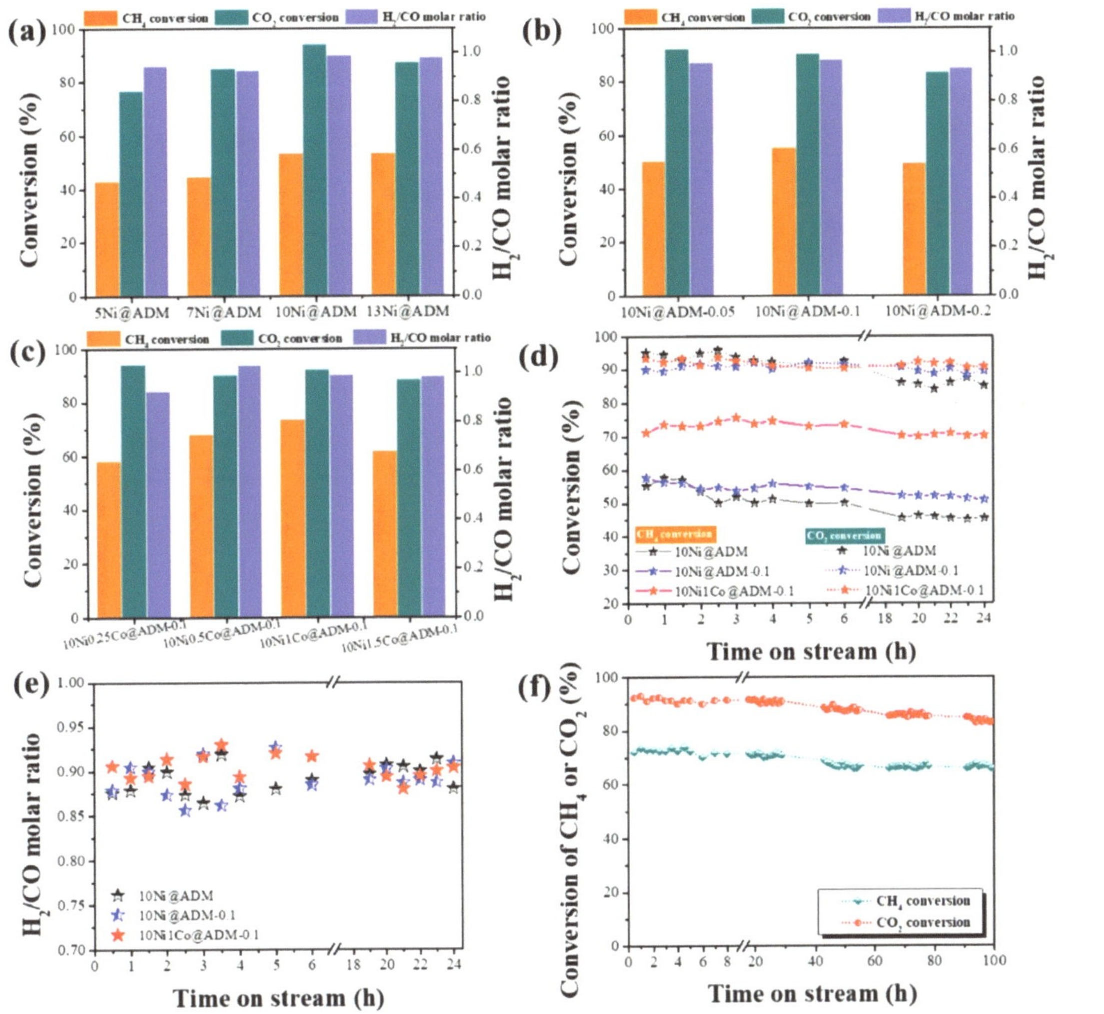

Figure 11. Dry reforming of methane reactions for 4 h over the (**a**) xNi@ADM (x = 5–13), (**b**) 10Ni@ADM-y (y = 0.05–0.2), and (**c**) 10NizCo@ADM-0.1 (z = 0.05–1.5) catalysts. (**d**) Conversions of CH_4 and CO_2 and (**e**) H_2/CO molar ratio on the attapulgite-derived MFI (ADM) zeolite-encased Ni-Co alloys. (**f**) Dry reforming of methane reactions for the 100 h test over the 10Ni1Co@ADM-0.1 catalyst. Reproduced from [81]. Copyright (2023), with permission from Elsevier.

Furthermore, the mechanism for dry reforming over the 10Ni1Co@ADM-0.1 catalyst was proposed by Defang Liang et al. [81]. By doping into the metallic Ni lattice of the 10Ni1Co@ADM-0.1 catalyst, the inclusion of Co improved the diffusion of metallic grains and aided in the emergence of Ni-Co alloys. Additionally, the generated Ni-Co alloy interfaces promoted the transfer of electrons from Co to Ni, which aided in the production of electron-rich Ni. This led to the breakdown of C-H bonds in CH_4 molecules, which generated the reactive intermediaries CHx*/C* and H* (Figure 12).

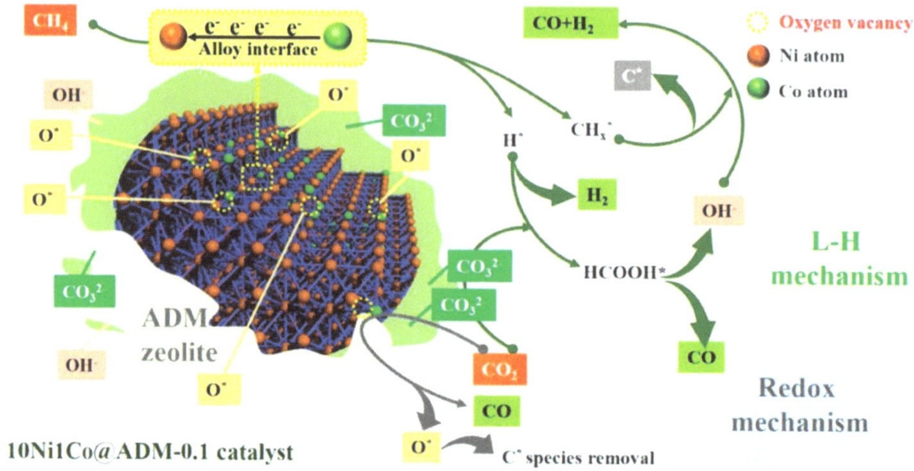

Figure 12. Synergetic effect between the attapulgite-derived MFI encapsulation layers and the Ni-Co alloys for the dry reforming of methane reaction on the 10Ni1Co@ADM − 0.1 catalyst. Reproduced from [81]. Copyright (2023), with permission from Elsevier.

Rahman Daiyan et al. show that the Co@CoNC-900 electrocatalyst, which is coated with a single Co atom and enclosed in a graphitic carbon shell, can reliably produce syngas at moderate overpotentials during CO_2 reduction processes [82]. They examined the catalyst's linear sweep voltammetry curves and the CO_2 reduction process. Figure 13a,b show that as the annealing temperature is raised, and j marginally decreases, which may be connected to the catalyst's depleting Co-N_4 centers. Additionally, these findings suggest that the reaction route is not affected by the catalysts' growing surface areas (as the surface area rises when increasing the calcination temperatures). The findings of the bulk CO_2 electrolysis support their design strategy since an increase in the annealing temperature, i.e., from 800 °C to 1000 °C, improves FE_{CO} (Figure 13b), while suppressing FE_{H2} (Figure 13c). The FE_{CO} obtained with the Co@CoNC-800, Co@CoNC-900, and Co@CoNC-1000 catalysts is 36, 45, and 61%, respectively, whereas the FE_{H2} obtained with these catalysts at the same potential is 59, 43, and 40%, respectively. The combination of a decline in Co-N_4 moieties and the potential impact of rising defects on the CO_2 reaction selectivity can be used to directly correlate this enhanced selectivity with the CO_2 reduction reaction. The syngas ratio provides a vivid demonstration of the adjustment of reaction selectivity (Figure 13d). The authors reported H_2:CO ratios of 1.5, 1, and 0.5 for the Co@CoNC-800, Co@CoNC-900, and Co@CoNC-1000 catalysts, respectively, by adjusting the annealing temperature.

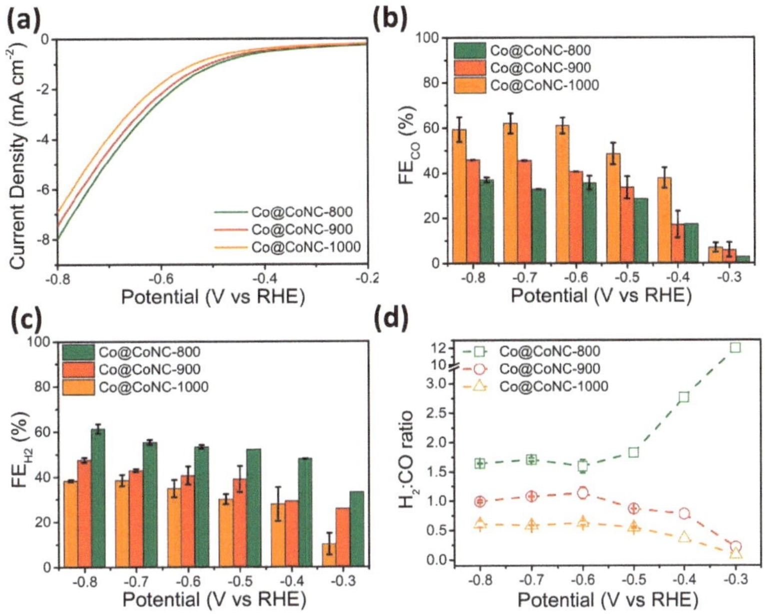

Figure 13. Modification of syngas ratio using active site manipulation. (**a**) Linear sweep voltammetry, (**b**) FE_{CO}, (**c**) FE_{H2}, and (**d**) H_2/CO ratio with applied potential for the prepared catalysts. Reproduced from [82]. Copyright (2020), with permission from the American Chemical Society.

Some of the recent progresses on syngas production/conversion by utilizing magnetic nanomaterials are compared and shown in Table 1.

Table 1. Various magnetic materials and their utilization for syngas production/conversion.

Magnetic Catalysts	Preparation Process	Process	Performance	Reference
$MgO/MgFe_2O_4$	Combustion process	Biodiesel production reaction from vegetable oil	Lowest conversion is 82.4%, and maximum conversion is 91.2%.	[83]
Ni-Fe	Power-to-gas (hydrogenation method)	CO_2 methanation	Conversion of 80% of CO_2 to methane at 150 °C. CH_4 selectivity of CO_2 methanation is 95%.	[84]
(Fe, Cu, and K) and rice husk char	Pyrolysis method	Catalytic tar conversion for improving the yield of syngas	The tar conversion efficiency is obtained at 77.1% for RHC, 82.7% for K-RHC, and for 92.6% Fe-RHC at the reforming temperature of 800 °C.	[85]
$FeSo_4$	Steam gasification and pyrolysis process	H_2-rich syngas production	The maximum overall H_2 yield and exergy efficiency for producing H_2 are estimated to be 43.63%.	[86]
Ni/A-A-CFA	Biogas dry reforming method	Syngas production	It exhibits the best catalytic activity at CH_4 and CO_2 conversion rate > 95%.	[87]
Fe_2O_3/Al_2O_3	Chemical looping dry reforming process	For CO_2 production of hydrogen and syngas	Dry reforming stage for CH_4 conversion is 3.84, and syngas yield is 98.32%.	[88]

Table 1. *Cont.*

Magnetic Catalysts	Preparation Process	Process	Performance	Reference
TiO_2/Fe	Fermentation process	For biohydrogen production	Yield increases by 24.9%.	[89]
Fe_3O_4 and Fe_5O_{12}	Hydrolysis	Biomass production	A 90% harvesting efficiency is obtained.	[90]
Fe_3O_4/CuO	Coprecipitation method	Decolorization of water	A high dye removal efficiency of 94 % is achieved with 20 wt% Fe_3O_4/CuO composition at ambient conditions and a reaction time of 90 min.	[91]
Ni, Ni-Co, Ni-Fe, and Ni-Cu	Hydrothermal synthesis	Syngas production and carbon bio nanofilament	The highest sustainability factor (0.66) and carbon yield (424%) are obtained.	[17]
Ni/Al	Coprecipitation method	Production of hydrogen-rich syngas from biomass pyrolysis	Ni/Al-700 catalyst can increase gas yield by 30–80%.	[92]
Fe_2O_3 and $MgFe_2O$	Pyrolysis	Syngas production from biomass	Tar conversion efficiency reaches 94.1% with a high gas yield of 493.5 mL/g.	[93]
Fe/Ca_xO	Simple precipitation method	Pyrolysis - gasification used for syngas production and tar removal	At an optimized composition of Ca/Fe 2/1, gasification yield efficiency (76.4%) is obtained.	[94]
Porous Ni, Ni-Co, Ni-Fe, and Ni-Cu	Precipitation method	Biomass decomposition to produce syngas	A high carbon yield efficiency of 36.43% is obtained.	[17]
Ni/Al	Ultrasonic-assisted incipient wetness impregnation method	Catalytic conversion of tar to syngas	With the addition of a catalyst, H_2 increases to 146%.	[95]
Ni-Cu/Al_2O_3	Impregnation method	Syngas production by methanol steam reforming	An increase in Ni content results in an increase in CO and a decrease in CO_2 yields.	[96]
Sc@ Ni/Fe	Pyrolysis method	Impregnation of biomass	At 600 °C, the conversion efficiency reaches 90.07%.	[97]
Fe-Ni/CNF	Pyrolysis method	Syngas production from pyrolysis gasification of biomass and plastic waste	Conversion efficiency of 87.90%.	[98]
Ni/Ru	Steam reforming process	Ru catalyst favors H_2 and CO production	90%	[99]

7. Effect of External Magnetic Field

We have discussed Fe-, Ni-, and Co-based magnetic materials for their utilization in syngas production/conversion in the previous section. This section discusses the effects of external magnetic field on magnetic nanomaterials during synthesis and applications. The effects of an external magnetic field could alter the charge transfer process during applications [100,101]. By transmitting loads of energy to the atomic and molecular dimensions of the materials, an external magnetic field can influence the microstructure and material characteristics. The production of static magnetic energy by a magnetic field affects the free energy of substances during synthesis and has an impact on their nucleation selectivity and rate, which is linked to metastable elements and innovative materials. Modern electrocatalysts can now be generated using magnetic fields, which are associated with nucleation and growth, as well as phase modulation. By enhancing the mass transfer on the working electrode, the magnetic effects of an external magnetic field, such as the magnetothermal effect, magnetohydrodynamic and micro-magnetohydrodynamic effects, Maxwell stress, and Kelvin force with spin selection effect, could indeed alter the reaction trends. This is advantageous for increasing the electrocatalytic performance of electrocatalysts with the potential utilization of external magnetic fields [102]. Various synthesis processes and applications have proven that an external magnetic field could improve the efficiency of materials, depending on their magnetic characteristics [103,104].

An external magnetic field-coupling microfluidic synthesis process consisting of a reaction liquid preheater, a reaction-nucleation stage heating and temperature controller, a thermostat, a low-temperature abrupt termination collector, and a MF controller could be utilized for the synthesis process [105]. According to Junmei Wang et al., the synthesis of a Fe_2 Pt/C catalyst under an external magnetic field has higher catalytic performance. A concurrently applied 1.4 T external magnetic field coupled with the microfluidic process results in Fe_2 Pt/C nanocrystals of larger size. This method gives better catalytic performance on both ethanol and methanol oxidation reactions compared to the samples synthesized without a magnetic field [106]. Similarly, He et al. demonstrated that distinct ferric sulfide minerals, such as greigite and marcasite, have different magnetic properties, including coercive force, remnant, saturation, and magnetization. By using an in situ MF-assisted hydrothermal process to synthesize the ferric sulfide minerals, they were able to achieve this [107].

Figure 14 shows the four steps in the magnetic field-coupling microfluidic preparation process, including the preparation of the precursor, the preheating of the precursor, the nucleation of QDs, and the collection of QDs. The magnetic field along the channel was created using the solenoid electromagnet. The products were collected under N_2 gas. Xiaoxiong Zhao et al. concluded that the magnetic fields had a significant impact on the nanoparticles' magnetic and optical properties. Regulation of the magnetic field was conveniently realized during the synthesis process. Co-doped ZnSe QD aggregates with a good size distribution were produced using a simplified microfluidic technique and a customizable magnetic field. These aggregates were created by numerous tiny nanoparticles (with diameters of about 4–6 nm) fusing together. Eventually, the authors insisted that by controlling the applied magnetic fields, particle sizes, and Co doping amounts, the Mr, Ms, Hc, net magnetization, and ferromagnetism of the QDs were adjusted. To alter the distinct magnetic characteristics of the QDs, one might alter the ratio of ferromagnetic to antiferromagnetic phases. It was realized that adding Co and altering the doping amount modified the band gaps of the QDs when estimating the band gap through using the UV-vis absorbance and R%.

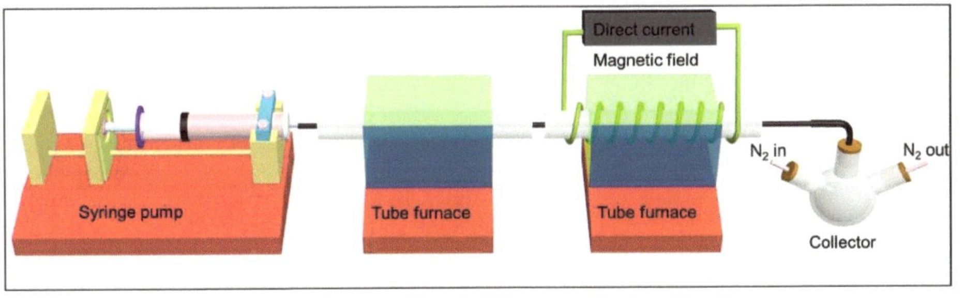

Figure 14. The four steps in the magnetic field-coupling microfluidic preparation process. Used with permission from [105]; permission conveyed through Copyright Clearance Center, Inc.

Po-Wei Lan et al. investigated various Ni/Fe catalyst ratios for increased carbon dioxide methanation by employing chemical reduction with an external magnetic field, and they showed that the Ni_8Fe_2 catalyst could convert more than 80% of CO_2 to methane at 150 °C. Furthermore, 95% of CO_2 methanation's selectivity was for CH_4 [84]. Figure 15 shows the experimental process of the magnetic field while synthesizing nanomaterials. Chemical reduction was cast off to develop a sequence of Ni/Fe catalysts with varying Fe/Ni ratios (Ni_9Fe_1, Ni_8Fe_2, and Ni_7Fe_3). The resulting solution was then transferred to a magnetic field reactor and heated to 80 °C while being subjected to a 500 G magnetic field. A neodymium magnet performed well enough to harvest the generated particles, which were subsequently purified by being washed with deionized water numerous times. Before being used, the produced Ni/Fe catalysts with various Ni/Fe ratios were dried in a desiccator.

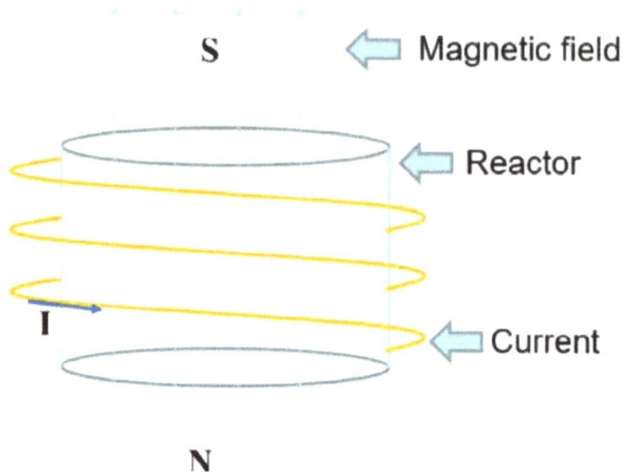

Figure 15. The magnetic field reactor for Ni-Fe catalyst preparation. Reprinted from [84], Copyright (2021), with permission from Elsevier.

Due to the simultaneous endothermic CO_2 reformation and exothermic partial oxidation of methane to syngas in the occurrence of oxygen, the methane-to-syngas transition process takes place in an energy-efficient manner, demanding little or no external source of energy. When Fe is injected into zeolite substances, Fe-containing acidic sites, iron oxides, and numerous paramagnetic O ions are all generated. The improved efficiency in a magnetic field is closely related to the magnetically triggered Fe ions and surface paramagnetic oxygen [108]. Methyl-functionalized silica and methyl-functionalized cobalt ferrite–silica nanoparticles were evaluated for their capacity to increase the generation of bioethanol in syngas fermentation via Clostridium ljungdahlii [109]. The $CoFe_2O_4/SiO_2/CH_3$ nanoparticles demonstrated higher enhancement of syngas mass transfer when these nanoparticles were utilized to promote the syngas–water mass transfer. It was verified that the nanoparticles' ability to increase mass transfer was preserved after being recovered using a magnet and employed five more times. When utilizing both varieties of nanoparticles, the syngas fermentation process produced more biomass, ethanol, and acetic acid. Due to the increased syngas mass transfer, the $CoFe_2O_4/SiO_2/CH_3$ nanoparticles were more successful for syngas fermentation. The inclusion of $CoFe_2O_4/SiO_2/CH_3$ nanoparticles boosted the generation of biomass, ethanol, and acetic acid by around 228%, 214%, and 60%, respectively, compared to a control. By using recovered nanoparticles for fermentation, the nanoparticles' reusability was demonstrated, and the results clearly show the advantages of utilizing magnetic nanomaterials in syngas production.

Using an external MF-coupling microfluidic method, homogeneous ultrasmall-sized Fe_2Pt/C nanocatalysts are synthesized [106]. The reactants could be controlled by the magnetizing force linked to the magnetic field gradient and the Lorentz force when an external magnetic field is applied. In turn, their catalytic performances in the electrochemical oxidation of methanol and ethanol can be considerably enhanced by controlling the size, composition, electronic status, and magnetic characteristics of the catalysts. This investigation can help us better understand how an external magnetic field influences nanocatalyst synthesis and offer us a good concept for enhancing the catalytic efficiency of electrochemical nanocatalysts. An attempt was made to determine whether it is possible to increase the catalytic performance of a Fe-based catalyst for the removal of AsH_3 by using a low-energy external magnetic field [110]. With a 0.1 T magnetic field present and a temperature of 120 °C, the catalytic performance for AsH_3 degradation was improved by 52%. Fe-containing acidic sites, iron oxides, and many paramagnetic O ions were all introduced when Fe was incorporated into zeolite materials. The magnetically activated Fe species and surface paramagnetic oxygen were substantially linked to the increased efficiency in a magnetic field. When a considerable magnetic field (450 mT) was introduced to the anode, electrocatalytic water oxidation in the alkaline media was revealed to be significantly strengthened by magnetism [108]. The use of an external magnetic field for water splitting investigations is suggested by the fact that magnetic enhancement works even on decorated Ni-foam electrodes with extremely high current densities, increasing their intrinsic activity by around 40% to reach over 1 A/cm^2 at low overpotentials. Thermal conversion investigations were carried out using magnetic field-assisted catalytic pyrolysis of biomass to produce H_2-rich gases from wood sawdust [111]. The outcomes of this experiment revealed that a 10 weight percent of a Ni/CaO catalyst with adequate magnetic properties had a favorable catalytic capacity to create H_2 by pyrolyzing biomass. With increased magnetic field strength, the H_2 quantity and output increased. The H_2 content and output exceeded 62 vol.% and 469 mL/g, respectively, at 650 °C, 80 m T magnetic field intensity, and 10 wt.% Ni/CaO catalyst. They improved by around 10 % and 20 %, correspondingly, in comparison to the absence of a magnetic field. As a result, a promising approach for using biomass is biomass catalytic pyrolysis supported by a magnetic field.

An external magnetic field was connected with a catalytic packed bed reactor for selective generation of methanol via CO_2 hydrogenation over Cu and Fe loaded on highly porous ZSM-5 zeolite in an effort to address both economical and environmental concerns regarding the green generation of methanol [88,112]. It was observed that the selectivity and catalytic activity of CO_2 hydrogenation significantly improved with the presence of an external magnetic field. The catalyst 10Cu-10Fe/ZSM-5 showed the maximum CO_2 conversion at a CO_2/H_2 molar ratio of 1:3 in the presence of an external magnetic field. At 220 °C, the external magnetic field enhanced the selectivity to methanol and the CO_2 conversion by factors of 1.7 and 2.24, respectively. Therefore, the use of magnetic fields promotes CO_2 adsorption and results in the generation of selective methanol, paving the way for the potential to introduce a green and sustainable technology in both petrochemical and chemical processes. An external magnetic field is used in an initiative to improve the catalytic efficiency of the CO_2 hydrogenation reaction based on green and efficient exploitation concepts [6]. Based on the activity and selectivity of Fe-based catalysts with ferro/ferrimagnetic properties, the effects of magnetic field orientation and magnetic flux density were demonstrated. In comparison to those without a magnetic field, CO_2 conversions were dramatically increased by 1.5–1.8 times with an external magnetic field, particularly in the north–south (N–S) direction, while activation energy was reduced by 1.1–1.15 times. When the magnetic flux density was changed, the rate of CO_2 conversion improved in the following order: 27.7 mT > 25.1 mT > 20.8 mT. These exceptional catalytic activities can be attributed to the magnetic field's role in facilitating reactant adsorption and surface reactions over the magnetized Fe catalysts, which reduces apparent activation energy and increases selectivity to hydrocarbons and CH_3OH.

Successful demonstrations of a magnetic field's impact on CO_2 conversion over the Cu-ZnO/ZrO_2 catalyst in the hydrogenation reaction were established [113]. The maximum CO_2 conversions were achieved at 220 °C with a magnetic field of 20.8 mT in the SN direction, which was three times greater than the results obtained without a magnetic field. Haiping Pan et al. established efficient magnetic field adjustment of the radical-pair spin states in electrocatalytic carbon dioxide reduction [114]. This work demonstrates that applying an external magnetic field for electrocatalytic CO_2 reduction to formate/formic acid considerably increases the catalytic performance of tin nanoparticle catalysts. The generation of formic acid can roughly double in comparison to zero magnetic fields when a standard Sn nanoparticle electrode is used as an example, and a constant external magnetic field of about 0.9 T is utilized. Figure 16 shows the effect of the external magnetic field on the formate yield, which depends on the applied electrode potential for different molar concentrations of $KHCO_3$ electrolytes at 1.7 V (vs. Ag/AgCl). This discovery paves the path for increased formate synthesis in the electrocatalytic reduction of CO_2 and points to the benefits of radical-pair spin states in the electron transfer process.

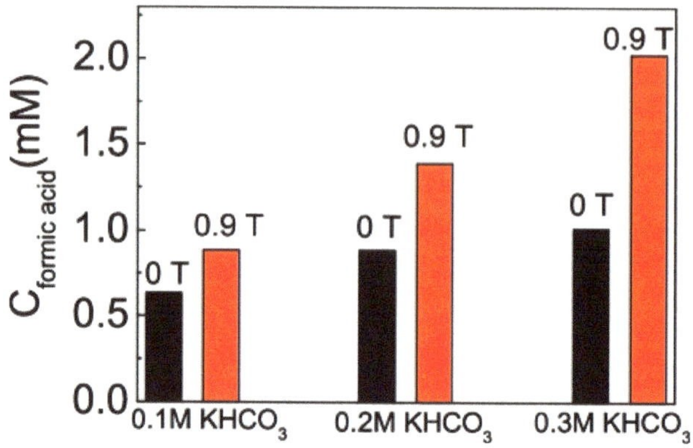

Figure 16. Magnetic field-dependent enhancement of formate yield (−1.7 V (vs/ Ag/AgCl)) with the same applied electrode potential for different concentrations of $KHCO_3$ electrolytes. Reprinted with permission from [114]. Copyright (2020), with permission from the American Chemical Society.

Magnetic field-assisted acceleration of CO_2 reforming of methane over an innovative hierarchical Co/MgO catalyst was performed [115]. The Co/MgO catalyst was developed through a hydrothermal precipitation route. The CO_2-CH_4 reforming reaction was carried out at atmospheric pressure in a bench-scale fluidized bed reactor system that included a gas feeding functionality, a fluidized bed reactor colocated with a heat furnace, and a GC analyzer. The catalyst nanoparticles were pre-reduced by utilizing H_2 (100 mL/min) at 800 °C for 1 h before the experiment. Following the reduction, the feed gas was fed into the reactor to begin the CO_2 reforming test, with a $CH_4/CO_2/N_2$ molar ratio of 1:1:1. A total of 0.2 g of the nanoparticles was added to the fluidized bed reactor for the activity. Four parallel solenoids created an axial uniform magnetic field in the case of the MFB reactor, and the current of the power source could be adjusted to modulate the magnetic field's strength (H). The magnetic field employed had a 130 Oe intensity, which is equivalent to a 344W electric energy input. The calcined CoO/MgO particle has a hierarchical architecture and gyroscopic-like structure, with an average dimension of 25 m. Many octahedral MgO nanostructures, with a crystal size of around 40 nm, self-assembled to form the particle. Figure 17 illustrates the stability test results for the Co-MgO catalyst with the MFB and CFB reactors. The first CO_2-CH_4 reformation performance of the Co/MgO catalyst in the CFB reactor was outstanding, with high CO_2 conversion (about 92%) and CH_4 conversion (nearly 90%). The conversions of CO_2 and CH_4 under the MFB and CFB reactors did not significantly decrease during the stability test of 24 h (Figure 17a,b). For the first three hours of the course of the reaction, the production of H_2 and CO for the MFB reactor displayed a modest induction period before remaining constant (Figure 17c,d). The conversions of CO_2 and CH_4 were enhanced by 22% and 30%, respectively, in the magnetic-assisted fluidized bed reactor at 800 °C, with high gas at an hourly space velocity of 15×10^4/h. The hierarchical Co/MgO catalyst's increased activity and great resistance to coke deposition are related to the development of clusters, which exhibit a high frequency of breaking and consolidating under the influence of a magnetic field.

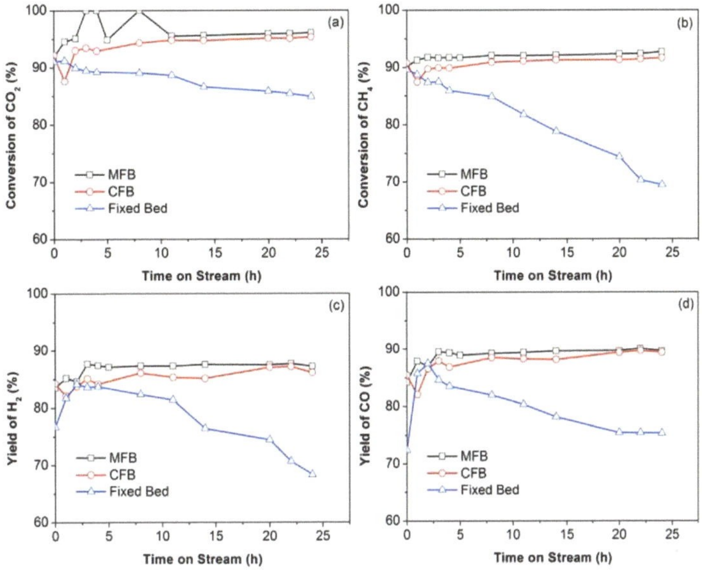

Figure 17. Catalytic stability experiments with GHSV of 60,000/ h. Conversions of (**a**) CO_2, (**b**) CH_4, the yields of (**c**) H_2 and (**d**) CO. Reprinted from [115]. Copyright (2018), with permission from Elsevier.

In general, carbon development can originate from the Boudouard reaction below 700 °C, where CO dissociates to generate surface carbon and CO_2, and CH_4 decomposition occurs beyond 557 °C, where CH_4 totally dissociates to form solid carbon on the surface of the catalyst and produce H_2. After 24 h of CO_2-CH_4 reforming at 800 °C, the used catalysts were examined by SEM and TEM to investigate the carbon deposition and changes to the catalysts' crystal size. Based on the structural examination, carbon diffraction peaks are clearly visible from the used catalysts in the CFB and MFB reactors, suggesting the presence of carbon deposition. As can be seen from the SEM image (Figure 18a–c), neither the CFB nor the MFB reactors contain any long carbon nanotubes or wires on top of the Co/MgO particles. In the conventional fluidized bed, numerous short, irregular carbon nanotubes are seen deposited on the exterior of the used catalyst in comparison to the fresh catalyst. On the catalyst that had been used up in the MFB reactor, less carbon nanotubes are seen. The fresh Co/MgO catalyst has Co NPs that are less than 5 nm in size, according to the TEM micrograph in Figure 18d. After the stability test, the average Co NP dimensions of the used Co/MgO catalysts for both the CFB and MFB reactors reveal a minor sintering compared to the fresh Co/MgO catalyst. The Co NPs after sintering show no overt response to the presence of magnetic field. This outcome contrasts with that of CO methanation in a magnetic fluidized bed reactor, where the axial magnetic field helped reduce catalyst agglomerates and prevent the sintering of NiCo particles [116]. The hierarchical Co/MgO catalyst's structure and modified particle size, which enable the hierarchical Co/MgO particle to remain an individual particle, are responsible for this variance. Following the stability test in the CFB reactor, numerous Co NPs are detached from the MgO carrier's surface and completely disseminated over the carbon nanotubes. However, due to the magnetic susceptibility of Co NPs and their interaction with one another, this transition of Co NPs is considerably diminished in the MFB reactor, as observed from the TEM micrograph in Figure 18d–f. Additional research is required to better understand the Co NPs' transition phenomena. To conclude, an axial magnetic field promotes the stability of Co particles, reduces carbon deposition, and increases the reforming rate. The suggested method successfully produces syngas via CO_2-

CH_4 reforming with improved catalytic activity through the combination of a magnetic fluidized bed and a structured nanoparticle catalyst.

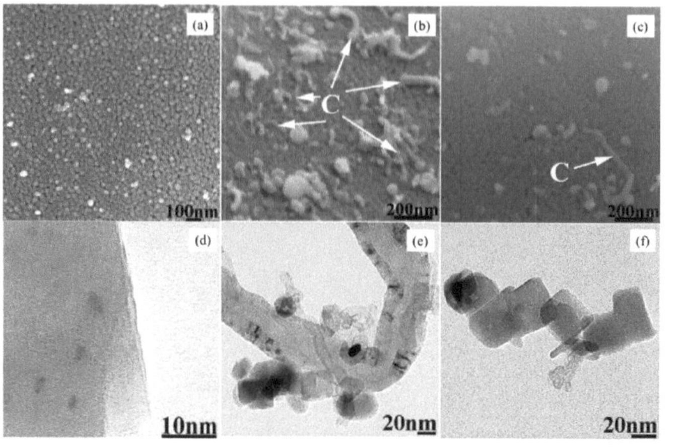

Figure 18. (**a–c**) SEM and (**d–f**) TEM micrographs of pure Co-MgO and used Co-MgO after reforming reaction in a conventional fluidized bed reactor and a magnetic-assisted fluidized bed (MFB) reactor. Reprinted from [115]. Copyright (2018), with permission from Elsevier.

The reviewed studies suggest that there are still a lot of opportunities and challenges in the effective utilization of external magnetic fields for syngas production and conversion applications. It is important to optimize magnetic nanoparticle or nanocomposite formation as the magnetic responses to external magnetic fields mainly depend on the magnetic characteristics of the materials. Furthermore, the range of the external magnetic field for selected magnetic materials also needs an optimization to obtain a better efficiency in syngas production and conversion applications. Moreover, understanding the role of external magnetic field in syngas production and conversion is also crucial, as the magnetic field itself could alter the chemical reaction kinetics even though it does not have magnetic characteristics. We could see that a non-magnetic material could also be influenced by an external magnetic field. By considering these factors, researchers could conduct a better investigation on the effect of external magnetic field on syngas production and conversion to achieve a better efficiency. The main characteristics of magnetic nanomaterials as catalysts, the types of magnetic nanomaterials, and the experimental parameters that can be tuned to improve the performance of catalysts for syngas production/conversion is schematically shown in Figure 19.

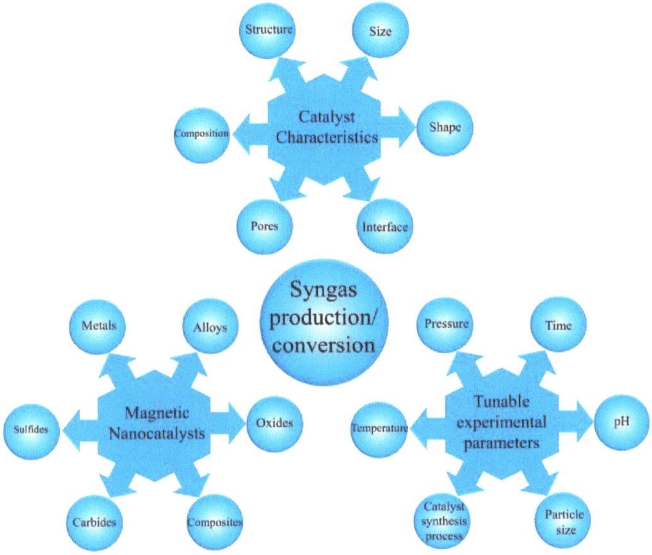

Figure 19. Schematic diagram for the typical characteristics of magnetic nanomaterials, types of magnetic nanomaterials, and experimental parameters that can be tuned to improve the performance of catalysts for syngas production/conversion.

8. Summary and Scope

We have introduced a brief discussion of magnetic materials, and the different preparation processes for the synthesis of magnetic nanomaterials are also discussed. A brief introduction to the syngas production/conversion process and the major experimental parameters, such as temperature, pH, time, and other reaction parameters that can influence the efficiency of the process, are included. The magnetic materials that this review has mainly focused on belong to Fe-, Ni-, and Co-based materials. The last section discusses the possibility of effectively utilizing an external magnetic field for the synthesis of magnetic nanocatalysts. The role of external magnetic field on the catalysis process of syngas production/conversion process is discussed with suitable examples. The advantages of magnetic nanomaterials over other available catalysts are their magnetic characteristics. One can tune the magnetic properties of magnetic materials by tuning their stoichiometry and morphology. Furthermore, the magnetic properties could be tuned by the addition of foreign elements or materials through composite formation. The required magnetic characteristics of these magnetic materials could be achieved by employing suitable experimental parameters and a suitable surface modifier. The magnetic separation process could be used to collect used magnetic nanocatalysts, which may be used for another cyclic performance. We strongly believe that the addition of an external magnetic field to the synthesis of magnetic nanoparticles and to the catalysis process could effectively alter the efficiency of the process. However, the range of the external magnetic field should be optimized, depending on the magnetic characteristics of the magnetic nanocatalysts for better performance.

Author Contributions: Conceptualization, A.T. and N.C.; writing—original draft preparation, S.J.J.K. and S.P.; writing—review and editing, P.S., S.D., S.S., S.-K.K. and R.M.; supervision and writing—review and editing, A.T. and N.C. All authors have read and agreed to the published version of the manuscript.

Funding: This study was funded by the Agencia Nacional de Investigacion y Desarollo-SA 77210070.

Data Availability Statement: The data presented in this study are available from the corresponding author upon request.

Conflicts of Interest: The authors declare no conflict of interest.

References

1. Hu, J.; Li, C.; Lee, D.-J.; Guo, Q.; Zhao, S.; Zhang, Q.; Li, D. Syngas Production from Biomass Using Fe-Based Oxygen Carrier: Optimization. *Bioresour. Technol.* **2019**, *280*, 183–187. [CrossRef]
2. Ye, Z.; Yue, W.; Tayyab, M.; Zhang, J.; Zhang, J. Simple One-Pot, High-Yield Synthesis of 2D Graphitic Carbon Nitride Nanosheets for Photocatalytic Hydrogen Production. *Dalton Trans.* **2022**, *51*, 18542–18548. [CrossRef]
3. Tayyab, M.; Liu, Y.; Liu, Z.; Xu, Z.; Yue, W.; Zhou, L.; Lei, J.; Zhang, J. A New Breakthrough in Photocatalytic Hydrogen Evolution by Amorphous and Chalcogenide Enriched Cocatalysts. *Chem. Eng. J.* **2023**, *455*, 140601. [CrossRef]
4. Mohamedali, M.; Henni, A.; Ibrahim, H. Recent Advances in Supported Metal Catalysts for Syngas Production from Methane. *ChemEngineering* **2018**, *2*, 9. [CrossRef]
5. Zhao, J.; Deng, J.; Han, J.; Imhanria, S.; Chen, K.; Wang, W. Effective Tunable Syngas Generation via CO_2 Reduction Reaction by Non-Precious Fe-N-C Electrocatalyst. *Chem. Eng. J.* **2020**, *389*, 124323. [CrossRef]
6. Kiatphuengporn, S.; Jantaratana, P.; Limtrakul, J.; Chareonpanich, M. Magnetic Field-Enhanced Catalytic CO 2 Hydrogenation and Selective Conversion to Light Hydrocarbons over Fe/MCM-41 Catalysts. *Chem. Eng. J.* **2016**, *306*, 866–875. [CrossRef]
7. Lu, A.-H.; Salabas, E.L.; Schüth, F. Magnetic Nanoparticles: Synthesis, Protection, Functionalization, and Application. *Angew. Chem. Int. Ed.* **2007**, *46*, 1222–1244. [CrossRef]
8. Liu, Y.; Tian, D.; Biswas, A.N.; Xie, Z.; Hwang, S.; Lee, J.H.; Meng, H.; Chen, J.G. Transition Metal Nitrides as Promising Catalyst Supports for Tuning CO/H_2 Syngas Production from Electrochemical CO_2 Reduction. *Angew. Chem. Int. Ed.* **2020**, *59*, 11345–11348. [CrossRef]
9. Concha, B. CO Hydrogenation on Supported Molybdenum Catalysts: Effects of Support on Specific Activities of Reduced and Sulfided Catalysts. *J. Catal.* **1984**, *89*, 536–541. [CrossRef]
10. Patterson, P. Carbon Monoxide Hydrogenation over Molybdenum and Tungsten Carbides. *Appl. Catal. A Gen.* **2003**, *251*, 449–455. [CrossRef]
11. Vo, D.-V.N.; Cooper, C.G.; Nguyen, T.-H.; Adesina, A.A.; Bukur, D.B. Evaluation of Alumina-Supported Mo Carbide Produced via Propane Carburization for the Fischer–Tropsch Synthesis. *Fuel* **2012**, *93*, 105–116. [CrossRef]
12. Zhu, X.; Wang, H.; Wei, Y.; Li, K.; Cheng, X. Hydrogen and Syngas Production from Two-Step Steam Reforming of Methane over CeO_2-Fe_2O_3 Oxygen Carrier. *J. Rare Earths* **2010**, *28*, 907–913. [CrossRef]
13. Deshpande, N.; Majumder, A.; Qin, L.; Fan, L.-S. High-Pressure Redox Behavior of Iron-Oxide-Based Oxygen Carriers for Syngas Generation from Methane. *Energy Fuels* **2015**, *29*, 1469–1478. [CrossRef]
14. Mary, J.A.; Manikandan, A.; Kennedy, L.J.; Bououdina, M.; Sundaram, R.; Vijaya, J.J. Structure and Magnetic Properties of Cu-Ni Alloy Nanoparticles Prepared by Rapid Microwave Combustion Method. *Trans. Nonferrous Met. Soc. China* **2014**, *24*, 1467–1473. [CrossRef]
15. Ray, K.; Sengupta, S.; Deo, G. Reforming and Cracking of CH_4 over Al_2O_3 Supported Ni, Ni-Fe and Ni-Co Catalysts. *Fuel Process. Technol.* **2017**, *156*, 195–203. [CrossRef]
16. Pudukudy, M.; Yaakob, Z.; Takriff, M.S. Methane Decomposition over Unsupported Mesoporous Nickel Ferrites: Effect of Reaction Temperature on the Catalytic Activity and Properties of the Produced Nanocarbon. *RSC Adv.* **2016**, *6*, 68081–68091. [CrossRef]
17. Ali, B.; Tasirin, S.; Aminayi, P.; Yaakob, Z.; Ali, N.; Noori, W. Non-Supported Nickel-Based Coral Sponge-Like Porous Magnetic Alloys for Catalytic Production of Syngas and Carbon Bio-Nanofilaments via a Biogas Decomposition Approach. *Nanomaterials* **2018**, *8*, 1053. [CrossRef]
18. Sengupta, S.; Ray, K.; Deo, G. Effects of Modifying Ni/Al2O3 Catalyst with Cobalt on the Reforming of CH_4 with CO_2 and Cracking of CH4 Reactions. *Int. J. Hydrogen Energy* **2014**, *39*, 11462–11472. [CrossRef]
19. Li, D.; Koike, M.; Chen, J.; Nakagawa, Y.; Tomishige, K. Preparation of Ni–Cu/Mg/Al Catalysts from Hydrotalcite-like Compounds for Hydrogen Production by Steam Reforming of Biomass Tar. *Int. J. Hydrogen Energy* **2014**, *39*, 10959–10970. [CrossRef]
20. Mizutani, N.; Iwasaki, T.; Watano, S.; Yanagida, T.; Kawai, T. Size Control of Magnetite Nanoparticles in Hydrothermal Synthesis by Coexistence of Lactate and Sulfate Ions. *Curr. Appl. Phys.* **2010**, *10*, 801–806. [CrossRef]
21. Alayat, A.; Mcllroy, D.N.; McDonald, A.G. Effect of Synthesis and Activation Methods on the Catalytic Properties of Silica Nanospring (NS)-Supported Iron Catalyst for Fischer-Tropsch Synthesis. *Fuel Process. Technol.* **2018**, *169*, 132–141. [CrossRef]
22. Ammar, S.; Fiévet, F. Polyol Synthesis: A Versatile Wet-Chemistry Route for the Design and Production of Functional Inorganic Nanoparticles. *Nanomaterials* **2020**, *10*, 1217. [CrossRef]
23. Mosayebi, J.; Kiyasatfar, M.; Laurent, S. Synthesis, Functionalization, and Design of Magnetic Nanoparticles for Theranostic Applications. *Adv. Healthc. Mater.* **2017**, *6*, 1700306. [CrossRef]

24. Ramalingam, V.; Harshavardhan, M.; Dinesh Kumar, S.; Malathi devi, S. Wet Chemical Mediated Hematite α-Fe_2O_3 Nanoparticles Synthesis: Preparation, Characterization and Anticancer Activity against Human Metastatic Ovarian Cancer. *J. Alloy. Compd.* **2020**, *834*, 155118. [CrossRef]
25. Frimpong, R.A.; Dou, J.; Pechan, M.; Hilt, J.Z. Enhancing Remote Controlled Heating Characteristics in Hydrophilic Magnetite Nanoparticles via Facile Co-Precipitation. *J. Magn. Magn. Mater.* **2010**, *322*, 326–331. [CrossRef]
26. Wang, H.; Dai, Y.Y.; Geng, D.Y.; Ma, S.; Li, D.; An, J.; He, J.; Liu, W.; Zhang, Z.D. Co_xNi_{100-x} Nanoparticles Encapsulated by Curved Graphite Layers: Controlled in Situ Metal-Catalytic Preparation and Broadband Microwave Absorption. *Nanoscale* **2015**, *7*, 17312–17319. [CrossRef]
27. Jia, Q.; Ghoshal, S.; Li, J.; Liang, W.; Meng, G.; Che, H.; Zhang, S.; Ma, Z.-F.; Mukerjee, S. Metal and Metal Oxide Interactions and Their Catalytic Consequences for Oxygen Reduction Reaction. *J. Am. Chem. Soc.* **2017**, *139*, 7893–7903. [CrossRef]
28. Zhou, X.; Shi, T.; Zhou, H. Hydrothermal Preparation of ZnO-Reduced Graphene Oxide Hybrid with High Performance in Photocatalytic Degradation. *Appl. Surf. Sci.* **2012**, *258*, 6204–6211. [CrossRef]
29. Mizutani, N.; Iwasaki, T.; Watano, S.; Yanagida, T.; Tanaka, H.; Kawai, T. Effect of Ferrous/Ferric Ions Molar Ratio on Reaction Mechanism for Hydrothermal Synthesis of Magnetite Nanoparticles. *Bull. Mater. Sci.* **2008**, *31*, 713–717. [CrossRef]
30. Zhang, Z.; Blom, D.A.; Gai, Z.; Thompson, J.R.; Shen, J.; Dai, S. High-Yield Solvothermal Formation of Magnetic CoPt Alloy Nanowires. *J. Am. Chem. Soc.* **2003**, *125*, 7528–7529. [CrossRef]
31. Komarneni, S.; Roy, R.; Li, Q.H. Microwave-Hydrothermal Synthesis of Ceramic Powders. *Mater. Res. Bull.* **1992**, *27*, 1393–1405. [CrossRef]
32. Li, J.; Wu, Q.; Wu, J. Synthesis of Nanoparticles via Solvothermal and Hydrothermal Methods. In *Handbook of Nanoparticles*; Aliofkhazraei, M., Ed.; Springer International Publishing: Cham, Switzerland, 2016; pp. 295–328, ISBN 978-3-319-15337-7.
33. Roca, A.G.; Morales, M.P.; O'Grady, K.; Serna, C.J. Structural and Magnetic Properties of Uniform Magnetite Nanoparticles Prepared by High Temperature Decomposition of Organic Precursors. *Nanotechnology* **2006**, *17*, 2783–2788. [CrossRef]
34. Bridgwater, A.V. Review of Fast Pyrolysis of Biomass and Product Upgrading. *Biomass Bioenergy* **2012**, *38*, 68–94. [CrossRef]
35. Condori, O.; García-Labiano, F.; de Diego, L.F.; Izquierdo, M.T.; Abad, A.; Adánez, J. Biomass Chemical Looping Gasification for Syngas Production Using LD Slag as Oxygen Carrier in a 1.5 KWth Unit. *Fuel Process. Technol.* **2021**, *222*, 106963. [CrossRef]
36. Palumbo, A.W.; Sorli, J.C.; Weimer, A.W. High Temperature Thermochemical Processing of Biomass and Methane for High Conversion and Selectivity to H_2-Enriched Syngas. *Appl. Energy* **2015**, *157*, 13–24. [CrossRef]
37. Hu, Q.; Shen, Y.; Chew, J.W.; Ge, T.; Wang, C.-H. Chemical Looping Gasification of Biomass with Fe_2O_3/CaO as the Oxygen Carrier for Hydrogen-Enriched Syngas Production. *Chem. Eng. J.* **2020**, *379*, 122346. [CrossRef]
38. Liu, G.; Liao, Y.; Wu, Y.; Ma, X. Reactivity of Co-Doped $Ca_2Fe_2O_5$ Brownmillerite Oxides as Oxygen Carriers for Microalgae Chemical Looping Gasification. *Int. J. Hydrogen Energy* **2019**, *44*, 2546–2559. [CrossRef]
39. Zhang, J.; He, T.; Wang, Z.; Zhu, M.; Zhang, K.; Li, B.; Wu, J. The Search of Proper Oxygen Carriers for Chemical Looping Partial Oxidation of Carbon. *Appl. Energy* **2017**, *190*, 1119–1125. [CrossRef]
40. Luo, M.; Wang, S.; Wang, L.; Lv, M. Reduction Kinetics of Iron-Based Oxygen Carriers Using Methane for Chemical-Looping Combustion. *J. Power Sources* **2014**, *270*, 434–440. [CrossRef]
41. Chyou, Y.-P.; Chang, D.-M.; Chen, P.-C.; Chien, H.-Y.; Wu, K.-T.; Chein, R.-Y. Development of Biomass Gasification Technology with Fluidized-Bed Reactors for Enhancing Hydrogen Generation: Part I, Hydrodynamic Characterization of Dual Fluidized-Bed Gasifiers. *Appl. Sci.* **2019**, *10*, 2. [CrossRef]
42. Paykani, A.; Khoshbakhti Saray, R.; Shervani-Tabar, M.T.; Mohammadi-Kousha, A. Effect of Exhaust Gas Recirculation and Intake Pre-Heating on Performance and Emission Characteristics of Dual Fuel Engines at Part Loads. *J. Cent. South Univ. Technol.* **2012**, *19*, 1346–1352. [CrossRef]
43. Fiore, M.; Magi, V.; Viggiano, A. Internal Combustion Engines Powered by Syngas: A Review. *Appl. Energy* **2020**, *276*, 115415. [CrossRef]
44. Lu, J.-S.; Chang, Y.; Poon, C.-S.; Lee, D.-J. Slow Pyrolysis of Municipal Solid Waste (MSW): A Review. *Bioresour. Technol.* **2020**, *312*, 123615. [CrossRef]
45. Kianfar, E. Catalytic Properties of Nanomaterials and Factors Affecting It. In *Importance & Applications of Nanotechnology*; MedDocs: Reno, NV, USA, 2020; Volume 5, pp. 22–25.
46. Kianfar, E.; Pirouzfar, V.; Sakhaeinia, H. An Experimental Study on Absorption/Stripping CO_2 Using Mono-Ethanol Amine Hollow Fiber Membrane Contactor. *J. Taiwan Inst. Chem. Eng.* **2017**, *80*, 954–962. [CrossRef]
47. Kozhushner, M.A.; Trakhtenberg, L.I.; Bodneva, V.L.; Belisheva, T.V.; Landerville, A.C.; Oleynik, I.I. Effect of Temperature and Nanoparticle Size on Sensor Properties of Nanostructured Tin Dioxide Films. *J. Phys. Chem. C* **2014**, *118*, 11440–11444. [CrossRef]
48. Vijayaraghavan, K.; Ashokkumar, T. Plant-Mediated Biosynthesis of Metallic Nanoparticles: A Review of Literature, Factors Affecting Synthesis, Characterization Techniques and Applications. *J. Environ. Chem. Eng.* **2017**, *5*, 4866–4883. [CrossRef]
49. Yang, S.; Zhang, X.; Chen, L.; Sun, L.; Xie, X.; Zhao, B. Production of Syngas from Pyrolysis of Biomass Using Fe/CaO Catalysts: Effect of Operating Conditions on the Process. *J. Anal. Appl. Pyrolysis* **2017**, *125*, 1–8. [CrossRef]
50. Marciniak, L.; Nowak, M.; Trojanowska, A.; Tylkowski, B.; Jastrzab, R. The Effect of PH on the Size of Silver Nanoparticles Obtained in the Reduction Reaction with Citric and Malic Acids. *Materials* **2020**, *13*, 5444. [CrossRef]
51. Karim, N.A.; Rubinsin, N.J.; Burukan, M.A.A.; Kamarudin, S.K. Sustainable Route of Synthesis Platinum Nanoparticles Using Orange Peel Extract. *Int. J. Green Energy* **2019**, *16*, 1518–1526. [CrossRef]

52. Jameel, M.S.; Aziz, A.A.; Dheyab, M.A. Green Synthesis: Proposed Mechanism and Factors Influencing the Synthesis of Platinum Nanoparticles. *Green Process. Synth.* **2020**, *9*, 386–398. [CrossRef]
53. Fakhroueian, Z.; Esmaeilzadeh, P.; Afroukhteh Langroudi, N.; Varmazyar, H.; Ahmadirad, M.; Zadeh, P.E.; Yousefi, M.; Karami, M.; Shafiekhani, A.; Sepehrisresht, S. Influence of Important Nanoemulsions PH on Performance of Nanostructures Catalysts for H_2 Production in Syngas Reactions. *Defect Diffus. Forum* **2011**, *312–315*, 20–26. [CrossRef]
54. Thakur, A.; Kumar, A.; Kaya, S.; Vo, D.-V.N.; Sharma, A. Suppressing Inhibitory Compounds by Nanomaterials for Highly Efficient Biofuel Production: A Review. *Fuel* **2022**, *312*, 122934. [CrossRef]
55. Rane, A.V.; Kanny, K.; Abitha, V.K.; Thomas, S. Methods for Synthesis of Nanoparticles and Fabrication of Nanocomposites. In *Synthesis of Inorganic Nanomaterials*; Elsevier: Amsterdam, The Netherlands, 2018; pp. 121–139, ISBN 978-0-08-101975-7.
56. Devi, A.; Singh, A.; Bajar, S.; Owamah, H.I. Nanomaterial in Liquid Biofuel Production: Applications and Current Status. *Environ. Sustain.* **2021**, *4*, 343–353. [CrossRef]
57. Thangavelu, S.K.; Ahmed, A.S.; Ani, F.N. Impact of Metals on Corrosive Behavior of Biodiesel–Diesel–Ethanol (BDE) Alternative Fuel. *Renew. Energy* **2016**, *94*, 1–9. [CrossRef]
58. Xu, C.; Chen, S.; Soomro, A.; Sun, Z.; Xiang, W. Hydrogen Rich Syngas Production from Biomass Gasification Using Synthesized Fe/CaO Active Catalysts. *J. Energy Inst.* **2018**, *91*, 805–816. [CrossRef]
59. Han, T.; Zhao, L.; Liu, G.; Ning, H.; Yue, Y.; Liu, Y. Rh-Fe Alloy Derived from YRh0.5Fe0.5O3/ZrO2 for Higher Alcohols Synthesis from Syngas. *Catal. Today* **2017**, *298*, 69–76. [CrossRef]
60. Smitshuysen, T.E.L.; Nielsen, M.R.; Pruessmann, T.; Zimina, A.; Sheppard, T.L.; Grunwaldt, J.; Chorkendorff, I.; Damsgaard, C.D. Optimizing Ni−Fe−Ga Alloys into Ni_2 FeGa for the Hydrogenation of CO_2 into Methanol. *ChemCatChem* **2020**, *12*, 3265–3273. [CrossRef]
61. Li, K.; Wang, H.; Wei, Y. Syngas Generation from Methane Using a Chemical-Looping Concept: A Review of Oxygen Carriers. *J. Chem.* **2013**, *2013*, 294817. [CrossRef]
62. Adánez-Rubio, I.; Bautista, H.; Izquierdo, M.T.; Gayán, P.; Abad, A.; Adánez, J. Development of a Magnetic Cu-Based Oxygen Carrier for the Chemical Looping with Oxygen Uncoupling (CLOU) Process. *Fuel Process. Technol.* **2021**, *218*, 106836. [CrossRef]
63. Pérez-Vega, R.; Abad, A.; García-Labiano, F.; Gayán, P.; de Diego, L.F.; Izquierdo, M.T.; Adánez, J. Chemical Looping Combustion of Gaseous and Solid Fuels with Manganese-Iron Mixed Oxide as Oxygen Carrier. *Energy Convers. Manag.* **2018**, *159*, 221–231. [CrossRef]
64. Cara, C.; Mameli, V.; Rombi, E.; Pinna, N.; Sanna Angotzi, M.; Nižňanský, D.; Musinu, A.; Cannas, C. Anchoring Ultrasmall FeIII-Based Nanoparticles on Silica and Titania Mesostructures for Syngas H_2S Purification. *Microporous Mesoporous Mater.* **2020**, *298*, 110062. [CrossRef]
65. Pankina, G.V.; Shumyantsev, A.V.; Chernyak, S.A.; Lunin, V.V. Influence of Structural Properties of Support on the Activation of Fe–K/$MgAl_2O_4$ Catalysts in Syngas. *Kinet. Catal.* **2019**, *60*, 672–680. [CrossRef]
66. Kwon, G.; Cho, D.-W.; Kwon, E.E.; Rinklebe, J.; Wang, H.; Song, H. Beneficial Use of Fe-Impregnated Bentonite as a Catalyst for Pyrolysis of Grass Cut into Syngas, Bio-Oil and Biochar. *Chem. Eng. J.* **2022**, *448*, 137502. [CrossRef]
67. San, X.; Cui, J.; Chu, Y.; Xu, J.; Meng, D.; Pan, Z.; Wang, G.; Qi, J.; Jin, Q. New Design and Construction of Hierarchical Porous Ni/SiO_2 Catalyst with Anti-sintering and Carbon Deposition Ability for Dry Reforming of Methane. *ChemistrySelect* **2022**, *7*, e202202258. [CrossRef]
68. Saad, J.M.; Nahil, M.A.; Wu, C.; Williams, P.T. Influence of Nickel-Based Catalysts on Syngas Production from Carbon Dioxide Reforming of Waste High Density Polyethylene. *Fuel Process. Technol.* **2015**, *138*, 156–163. [CrossRef]
69. Du, X.; Zhang, D.; Gao, R.; Huang, L.; Shi, L.; Zhang, J. Design of Modular Catalysts Derived from NiMgAl-LDH@m-SiO_2 with Dual Confinement Effects for Dry Reforming of Methane. *Chem. Commun.* **2013**, *49*, 6770. [CrossRef]
70. Yusuf, M.; Farooqi, A.S.; Al-Kahtani, A.A.; Ubaidullah, M.; Alam, M.A.; Keong, L.K.; Hellgardt, K.; Abdullah, B. Syngas Production from Greenhouse Gases Using Ni–W Bimetallic Catalyst via Dry Methane Reforming: Effect of W Addition. *Int. J. Hydrogen Energy* **2021**, *46*, 27044–27061. [CrossRef]
71. Deng, J.; Liu, Z.; Qin, T.; Chen, X.; Li, K.; Meng, L.; Zhao, Y.; Zhou, Y.; Yuan, S. Development of Ni-Doped Fe/Ca Catalyst to Be Used for Hydrogen-Rich Syngas Production during Medicine Residue Pyrolysis. *Energy* **2022**, *254*, 124205. [CrossRef]
72. Raberg, L.; Jensen, M.; Olsbye, U.; Daniel, C.; Haag, S.; Mirodatos, C.; Sjastad, A. Propane Dry Reforming to Synthesis Gas over Ni-Based Catalysts: Influence of Support and Operating Parameters on Catalyst Activity and Stability. *J. Catal.* **2007**, *249*, 250–260. [CrossRef]
73. Tsodikov, M.V.; Ellert, O.G.; Nikolaev, S.A.; Arapova, O.V.; Konstantinov, G.I.; Bukhtenko, O.V.; Vasil'kov, A.Y. The Role of Nanosized Nickel Particles in Microwave-Assisted Dry Reforming of Lignin. *Chem. Eng. J.* **2017**, *309*, 628–637. [CrossRef]
74. Gai, C.; Zhu, N.; Hoekman, S.K.; Liu, Z.; Jiao, W.; Peng, N. Highly Dispersed Nickel Nanoparticles Supported on Hydrochar for Hydrogen-Rich Syngas Production from Catalytic Reforming of Biomass. *Energy Convers. Manag.* **2019**, *183*, 474–484. [CrossRef]
75. Li, Z.; Kathiraser, Y.; Ashok, J.; Oemar, U.; Kawi, S. Simultaneous Tuning Porosity and Basicity of Nickel@Nickel–Magnesium Phyllosilicate Core–Shell Catalysts for CO_2 Reforming of CH_4. *Langmuir* **2014**, *30*, 14694–14705. [CrossRef]
76. Wang, H.; Mao, D.; Qi, J.; Zhang, Q.; Ma, X.; Song, S.; Gu, L.; Yu, R.; Wang, D. Hollow Multishelled Structure of Heterogeneous Co_3O_4–CeO_{2-x} Nanocomposite for CO Catalytic Oxidation. *Adv. Funct. Mater.* **2019**, *29*, 1806588. [CrossRef]

77. Prieto, G.; Beijer, S.; Smith, M.L.; He, M.; Au, Y.; Wang, Z.; Bruce, D.A.; de Jong, K.P.; Spivey, J.J.; de Jongh, P.E. Design and Synthesis of Copper–Cobalt Catalysts for the Selective Conversion of Synthesis Gas to Ethanol and Higher Alcohols. *Angew. Chem. Int. Ed.* **2014**, *53*, 6397–6401. [CrossRef]
78. Zhang, R.; Liu, F.; Wang, B. Co-Decorated Cu Alloy Catalyst for C_2 Oxygenate and Ethanol Formation from Syngas on Cu-Based Catalyst: Insight into the Role of Co and Cu as Well as the Improved Selectivity. *Catal. Sci. Technol.* **2016**, *6*, 8036–8054. [CrossRef]
79. Nguyen, H.M.; Phan, C.M.; Liu, S.; Pham-Huu, C.; Nguyen-Dinh, L. Radio-Frequency Induction Heating Powered Low-Temperature Catalytic CO_2 Conversion via Bi-Reforming of Methane. *Chem. Eng. J.* **2022**, *430*, 132934. [CrossRef]
80. Chen, L.; Zhu, Q.; Hao, Z.; Zhang, T.; Xie, Z. Development of a Co–Ni Bimetallic Aerogel Catalyst for Hydrogen Production via Methane Oxidative CO_2 Reforming in a Magnetic Assisted Fluidized Bed. *Int. J. Hydrogen Energy* **2010**, *35*, 8494–8502. [CrossRef]
81. Liang, D.; Wang, Y.; Chen, M.; Xie, X.; Li, C.; Wang, J.; Yuan, L. Dry Reforming of Methane for Syngas Production over Attapulgite-Derived MFI Zeolite Encapsulated Bimetallic Ni-Co Catalysts. *Appl. Catal. B Environ.* **2023**, *322*, 122088. [CrossRef]
82. Daiyan, R.; Chen, R.; Kumar, P.; Bedford, N.M.; Qu, J.; Cairney, J.M.; Lu, X.; Amal, R. Tunable Syngas Production through CO_2 Electroreduction on Cobalt–Carbon Composite Electrocatalyst. *ACS Appl. Mater. Interfaces* **2020**, *12*, 9307–9315. [CrossRef]
83. Alaei, S.; Haghighi, M.; Toghiani, J.; Rahmani Vahid, B. Magnetic and Reusable $MgO/MgFe_2O_4$ Nanocatalyst for Biodiesel Production from Sunflower Oil: Influence of Fuel Ratio in Combustion Synthesis on Catalytic Properties and Performance. *Ind. Crops Prod.* **2018**, *117*, 322–332. [CrossRef]
84. Lan, P.-W.; Wang, C.-C.; Chen, C.-Y. Effect of Ni/Fe Ratio in Ni–Fe Catalysts Prepared under External Magnetic Field on CO_2 Methanation. *J. Taiwan Inst. Chem. Eng.* **2021**, *127*, 166–174. [CrossRef]
85. Guo, F.; Li, X.; Liu, Y.; Peng, K.; Guo, C.; Rao, Z. Catalytic Cracking of Biomass Pyrolysis Tar over Char-Supported Catalysts. *Energy Convers. Manag.* **2018**, *167*, 81–90. [CrossRef]
86. Han, T.; Yang, W.; Jönsson, P.G. Pyrolysis and Subsequent Steam Gasification of Metal Dry Impregnated Lignin for the Production of H_2-Rich Syngas and Magnetic Activated Carbon. *Chem. Eng. J.* **2020**, *394*, 124902. [CrossRef]
87. Gao, Y.; Jiang, J.; Meng, Y.; Aihemaiti, A.; Ju, T.; Chen, X.; Yan, F. A Novel Nickel Catalyst Supported on Activated Coal Fly Ash for Syngas Production via Biogas Dry Reforming. *Renew. Energy* **2020**, *149*, 786–793. [CrossRef]
88. Zhu, M.; Song, Y.; Chen, S.; Li, M.; Zhang, L.; Xiang, W. Chemical Looping Dry Reforming of Methane with Hydrogen Generation on Fe_2O_3/Al_2O_3 Oxygen Carrier. *Chem. Eng. J.* **2019**, *368*, 812–823. [CrossRef]
89. Hsieh, P.-H.; Lai, Y.-C.; Chen, K.-Y.; Hung, C.-H. Explore the Possible Effect of TiO2 and Magnetic Hematite Nanoparticle Addition on Biohydrogen Production by Clostridium Pasteurianum Based on Gene Expression Measurements. *Int. J. Hydrogen Energy* **2016**, *41*, 21685–21691. [CrossRef]
90. Zhu, L.-D.; Hiltunen, E.; Li, Z. Using Magnetic Materials to Harvest Microalgal Biomass: Evaluation of Harvesting and Detachment Efficiency. *Environ. Technol.* **2019**, *40*, 1006–1012. [CrossRef]
91. Ghasemi, H.; Aghabarari, B.; Alizadeh, M.; Khanlarkhani, A.; Abu-Zahra, N. High Efficiency Decolorization of Wastewater by Fenton Catalyst: Magnetic Iron-Copper Hybrid Oxides. *J. Water Process Eng.* **2020**, *37*, 101540. [CrossRef]
92. Yang, S.; Chen, L.; Sun, L.; Xie, X.; Zhao, B.; Si, H.; Zhang, X.; Hua, D. Novel Ni–Al Nanosheet Catalyst with Homogeneously Embedded Nickel Nanoparticles for Hydrogen-Rich Syngas Production from Biomass Pyrolysis. *Int. J. Hydrogen Energy* **2021**, *46*, 1762–1776. [CrossRef]
93. Guo, F.; Liang, S.; Zhao, X.; Jia, X.; Peng, K.; Jiang, X.; Qian, L. Catalytic Reforming of Biomass Pyrolysis Tar Using the Low-Cost Steel Slag as Catalyst. *Energy* **2019**, *189*, 116161. [CrossRef]
94. Zhang, X.; Yang, S.; Xie, X.; Chen, L.; Sun, L.; Zhao, B.; Si, H. Stoichiometric Synthesis of Fe/Ca_xO Catalysts from Tailored Layered Double Hydroxide Precursors for Syngas Production and Tar Removal in Biomass Gasification. *J. Anal. Appl. Pyrolysis* **2016**, *120*, 371–378. [CrossRef]
95. Yan, M.; Liu, J.; Hantoko, D.; Kanchanatip, E.; Grisdanurak, N.; Cai, Y.; Gao, Z. Hydrogen-Rich Syngas Production by Catalytic Cracking of Tar in Wastewater under Supercritical Condition. *Int. J. Hydrogen Energy* **2019**, *44*, 19908–19919. [CrossRef]
96. Khzouz, M.; Gkanas, E.I.; Du, S.; Wood, J. Catalytic Performance of $Ni-Cu/Al_2O_3$ for Effective Syngas Production by Methanol Steam Reforming. *Fuel* **2018**, *232*, 672–683. [CrossRef]
97. Liang, S.; Guo, F.; Du, S.; Tian, B.; Dong, Y.; Jia, X.; Qian, L. Synthesis of Sargassum Char-Supported Ni-Fe Nanoparticles and Its Application in Tar Cracking during Biomass Pyrolysis. *Fuel* **2020**, *275*, 117923. [CrossRef]
98. Zhang, S.; Zhu, S.; Zhang, H.; Liu, X.; Xiong, Y. High Quality H2-Rich Syngas Production from Pyrolysis-Gasification of Biomass and Plastic Wastes by Ni–Fe@Nanofibers/Porous Carbon Catalyst. *Int. J. Hydrogen Energy* **2019**, *44*, 26193–26203. [CrossRef]
99. Park, S.Y.; Oh, G.; Kim, K.; Seo, M.W.; Ra, H.W.; Mun, T.Y.; Lee, J.G.; Yoon, S.J. Deactivation Characteristics of Ni and Ru Catalysts in Tar Steam Reforming. *Renew. Energy* **2017**, *105*, 76–83. [CrossRef]
100. Khavryuchenko, O.V.; Khavryuchenko, V.D.; Su, D. Spin Catalysts: A Quantum Trigger for Chemical Reactions. *Chin. J. Catal.* **2015**, *36*, 1656–1661. [CrossRef]
101. Bhattacharyya, K.; Chowdhury, M. Environmental and Magnetic Field Effects on Exciplex and Twisted Charge Transfer Emission. *Chem. Rev.* **1993**, *93*, 507–535. [CrossRef]
102. Jiang, X.; Chen, Y.; Zhang, X.; You, F.; Yao, J.; Yang, H.; Xia, B.Y. Magnetic Field-Assisted Construction and Enhancement of Electrocatalysts. *ChemSusChem* **2022**, *15*, e202201551. [CrossRef]

103. Jiang, M.-P.; Huang, K.-K.; Liu, J.-H.; Wang, D.; Wang, Y.; Wang, X.; Li, Z.-D.; Wang, X.-Y.; Geng, Z.-B.; Hou, X.-Y.; et al. Magnetic-Field-Regulated TiO_2 {100} Facets: A Strategy for C-C Coupling in CO_2 Photocatalytic Conversion. *Chem* **2020**, *6*, 2335–2346. [CrossRef]
104. Kodaimati, M.S.; Gao, R.; Root, S.E.; Whitesides, G.M. Magnetic Fields Enhance Mass Transport during Electrocatalytic Reduction of CO_2. *Chem Catal.* **2022**, *2*, 797–815. [CrossRef]
105. Zhao, X.; Liang, H.; Chen, Y.; Chen, X.; Zhang, W.; Wang, J.; Zhang, G.; Belotelov, V.I.; Song, Y. Magnetic Field Coupling Microfluidic Synthesis of Diluted Magnetic Semiconductor Quantum Dots: The Case of Co Doping ZnSe Quantum Dots. *J. Mater. Chem. C* **2021**, *9*, 4619–4627. [CrossRef]
106. Wang, J.; Song, Y. Magnetic Field Coupling Microfluidic Synthesis of Fe_2Pt/C Nanocatalysts for Enhanced Electrochemical Catalytic Oxidation of Alcohol. *Int. J. Hydrogen Energy* **2020**, *45*, 13035–13044. [CrossRef]
107. He, Z.; Yu, S.-H.; Zhou, X.; Li, X.; Qu, J. Magnetic-Field-Induced Phase-Selective Synthesis of Ferrosulfide Microrods by a Hydrothermal Process: Microstructure Control and Magnetic Properties. *Adv. Funct. Mater.* **2006**, *16*, 1105–1111. [CrossRef]
108. Garcés-Pineda, F.A.; Blasco-Ahicart, M.; Nieto-Castro, D.; López, N.; Galán-Mascarós, J.R. Direct Magnetic Enhancement of Electrocatalytic Water Oxidation in Alkaline Media. *Nat. Energy* **2019**, *4*, 519–525. [CrossRef]
109. Kim, Y.-K.; Lee, H. Use of Magnetic Nanoparticles to Enhance Bioethanol Production in Syngas Fermentation. *Bioresour. Technol.* **2016**, *204*, 139–144. [CrossRef]
110. Xie, Y.; Wang, M.; Wang, X.; Wang, L.; Ning, P.; Ma, Y.; Lu, J.; Cao, R.; Xue, Y. Magnetic-Field-Assisted Catalytic Oxidation of Arsine over Fe/HZSM-5 Catalyst: Synergistic Effect of Fe Species and Activated Surface Oxygen. *J. Clean. Prod.* **2022**, *337*, 130549. [CrossRef]
111. Zhao, B.; Yang, H.; Zhang, H.; Zhong, C.; Wang, J.; Zhu, D.; Guan, H.; Sun, L.; Yang, S.; Chen, L.; et al. Study on Hydrogen-Rich Gas Production by Biomass Catalytic Pyrolysis Assisted with Magnetic Field. *J. Anal. Appl. Pyrolysis* **2021**, *157*, 105227. [CrossRef]
112. Sriakkarin, C.; Umchoo, W.; Donphai, W.; Poo-arporn, Y.; Chareonpanich, M. Sustainable Production of Methanol from CO_2 over 10Cu-10Fe/ZSM-5 Catalyst in a Magnetic Field-Assisted Packed Bed Reactor. *Catal. Today* **2018**, *314*, 114–121. [CrossRef]
113. Donphai, W.; Piriyawate, N.; Witoon, T.; Jantaratana, P.; Varabuntoonvit, V.; Chareonpanich, M. Effect of Magnetic Field on CO_2 Conversion over Cu-ZnO/ZrO_2 Catalyst in Hydrogenation Reaction. *J. CO2 Util.* **2016**, *16*, 204–211. [CrossRef]
114. Pan, H.; Jiang, X.; Wang, X.; Wang, Q.; Wang, M.; Shen, Y. Effective Magnetic Field Regulation of the Radical Pair Spin States in Electrocatalytic CO_2 Reduction. *J. Phys. Chem. Lett.* **2020**, *11*, 48–53. [CrossRef]
115. Li, J.; Li, J.; Zhu, Q.; Li, H. Magnetic Field Acceleration of CO_2 Reforming of Methane over Novel Hierarchical Co/MgO Catalyst in Fluidized Bed Reactor. *Chem. Eng. J.* **2018**, *350*, 496–506. [CrossRef]
116. Li, J.; Zhou, L.; Zhu, Q.; Li, H. Enhanced Methanation over Aerogel NiCo/Al_2O_3 Catalyst in a Magnetic Fluidized Bed. *Ind. Eng. Chem. Res.* **2013**, *52*, 6647–6654. [CrossRef]

catalysts

MDPI

Article

Performance of Particulate and Structured Pt/TiO_2-Based Catalysts for the WGS Reaction under Realistic High- and Low-Temperature Shift Conditions

Andreas Kouroumlidis [1], Georgios Bampos [1,*], Paraskevi Panagiotopoulou [2] and Dimitris I. Kondarides [1]

[1] Department of Chemical Engineering, University of Patras, 6504 Patras, Greece
[2] School of Chemical and Environmental Engineering, Technical University of Crete, 73100 Chania, Greece
* Correspondence: geoba@chemeng.upatras.gr; Tel.: +30-2610997223

Abstract: The water–gas shift (WGS) activity of Pt/TiO_2-based powdered and structured catalysts was investigated using realistic feed compositions that are relevant to the high-temperature shift (HTS) and low-temperature shift (LTS) reaction conditions. The promotion of the TiO_2 support with small amounts of alkali- or alkaline earth-metals resulted in the enhancement of the WGS activity of 0.5%Pt/TiO_2(X) catalysts (X = Na, Cs, Ca, Sr). The use of bimetallic (Pt–M)/TiO_2 catalysts (M = Ru, Cr, Fe, Cu) can also shift the CO conversion curve toward lower temperatures, but this is accompanied by the production of relatively large amounts of unwanted CH_4 at temperatures above ca. 300 °C. Among the powdered catalysts investigated, Pt/TiO_2(Ca) exhibited the best performance under both HTS and LTS conditions. Therefore, this material was selected for the preparation of structured catalysts in the form of pellets as well as ceramic and metallic catalyst monoliths. The 0.5%Pt/TiO_2(Ca) pellet catalyst exhibited comparable activity with that of a commercial WGS pellet catalyst, and its performance was further improved when the Pt loading was increased to 1.0 wt.%. Among the structured catalysts investigated, the best results were obtained for the sample coated on the metallic monolith, which exhibited excellent WGS performance in the 300–350 °C temperature range. In conclusion, proper selection of the catalyst structure and reaction parameters can shift the CO conversion curves toward sufficiently low temperatures, rendering the Pt/TiO_2(Ca) catalyst suitable for practical applications.

Keywords: water gas shift; platinum; TiO_2; promoter; structured catalysts; pellets; monoliths

Citation: Kouroumlidis, A.; Bampos, G.; Panagiotopoulou, P.; Kondarides, D.I. Performance of Particulate and Structured Pt/TiO_2-Based Catalysts for the WGS Reaction under Realistic High- and Low-Temperature Shift Conditions. *Catalysts* **2023**, *13*, 372. https://doi.org/10.3390/catal13020372

Academic Editor: Rufino Navarro Yerga

Received: 14 January 2023
Revised: 31 January 2023
Accepted: 6 February 2023
Published: 8 February 2023

1. Introduction

The water–gas shift (WGS) reaction (Equation (1)) is a well-established industrial process that is commonly used to reduce the CO concentration and increase the hydrogen content in the synthesis gas produced via reforming/partial oxidation of carbon-containing fuels [1–4].

$$CO + H_2O \rightleftharpoons CO_2 + H_2 \ \Delta H^o_{298\ K} = -41.1\ kJ\ mol^{-1} \quad (1)$$

The reaction is equilibrium-controlled and moderately exothermic and is therefore thermodynamically favored at low temperatures, and kinetically favored at high temperatures. As a result, the reaction is typically performed in industrial processes using two adiabatic reactors in sequence, with intermediate cooling. The first reactor operates at 350–450 °C using a Fe–Cr based oxide catalyst (high-temperature shift, HTS), and the second at 200–250 °C using Cu–Zn based catalysts (low-temperature shift, LTS) [1–4]. The inherent disadvantages of the classic commercial HTS catalysts (e.g., toxicity of hexavalent Cr and demanding reduction process to avoid sintering) and LTS catalysts (e.g., relatively low activity, pyrophoricity, thermal sintering, and leaching of active component) render them unsuitable for mobile and small-to-medium scale applications [2,3,5]. Therefore, efforts are currently focused on the development of alternative WGS catalysts that fulfill the requirements for use in compact fuel processors. In this respect, metal oxide-supported noble metals have been extensively studied due to their superior kinetic performance and stability under frequent start-up/shut-down and varied duty cycle operations [1,4–6]. Among the noble metal catalysts investigated, platinum supported on partially reducible oxides such as CeO_2 and TiO_2 exhibit the highest WGS activity [1,7–9]. However, most of the above studies reported results with powder catalyst samples, which are not directly relevant to practical applications where the catalyst is in the form of pellets or monoliths. Pellets provide a high surface area of catalyst per unit reactor volume and are easy to handle and replace, whereas monolithic catalysts offer several advantages including lower pressure drop, good thermal shock resistance, robustness, and fast response to transient operation [10,11]. The competitive performance of structured noble metal-containing WGS catalysts has been investigated in several studies [6,10–17], but only a few of them have been performed under realistic reaction conditions. In our previous work, we investigated in detail the performance for the WGS reaction of noble metal catalysts (Pt, Rh, Ru, Pd) dispersed on reducible (TiO_2, CeO_2, La_2O_3, YSZ) and irreducible (Al_2O_3, MgO, SiO_2) metal oxide supports [7,8]. Among the metal–support combinations investigated, Pt/TiO_2 was found to exhibit the highest WGS activity, which could be further enhanced following the promotion of TiO_2 with small amounts of alkali (Li, Na, K, Cs) [18] or alkaline earth metals (Mg, Ca, Sr, Ba) [19]. This has been attributed to the promoter-induced creation of dual function $Pt - \square_s - Ti^{3+}$ sites at the metal–support interface, where $\square_s$ denotes an oxygen defect, which are required for the WGS reaction to proceed. The adsorption strength of these sites toward hydrogen was found to depend on the nature of the promoter and decreased with an increase in its content [18–20].

The objective of the present work was to investigate the performance of the previously developed Pt/TiO_2-based catalysts under realistic reaction conditions using feed compositions that are relevant to the HTS and LTS reactions. The effects of the dispersion of a second metal (Fe, Cu, Cr, Ru) on the support were also studied. The performance of the optimized catalysts in the form of pellets or wash-coating on ceramic and metallic monoliths was further investigated under realistic conditions and was compared with that of a commercial catalyst obtained from Johnson Matthey. To the best of our knowledge, this is one of the few reports to deal with the development of structured Pt/TiO_2-based catalysts and testing under HT- and LT-WGS conditions that are relevant to practical applications and can be regarded as an essential step toward the commercialization of these materials.

2. Results and Discussion

The powdered and structured catalysts synthesized are listed in Table 1. The composition of the $Pt/TiO_2(X)$ catalysts (X = Na, Cs, Ca, Sr) was selected based on the results of our previous studies, which showed that the WGS activity of platinum dispersed on alkali- and alkaline earth metal-promoted TiO_2 exhibited a volcano-type dependence on the promoter content and was maximized for samples containing 0.06 wt.% Na, 0.34 wt.% Cs, 2.0 wt.% CaO, or 1.0 wt.% SrO [18,19]. Regarding the bimetallic $(Pt–M)/TiO_2$ catalysts (M = Ru, Fe, Cr, Cu), the second metal (M) was chosen based on the well-known activity of Fe/Cr and Cu for the HTS and LTS reactions, respectively, [1–4] and the high WGS activity of Ru supported on reducible metal oxides [7,8].

Table 1. Physicochemical characteristics of the synthesized catalysts.

Catalyst Notation	Nominal Composition (wt.%)	Specific Surface Area (SSA) [1] (m^2/g)	Anatase Content [2] (%)	Pt Dispersion [3] (%)	Pt Crystallite Size [3] (nm)
Pt/TiO_2	$0.5\%Pt/TiO_2$	49	85	85	1.2
$Pt/TiO_2(Na)$	$0.5\%Pt/TiO_2(0.06\%Na)$	30	47	93	1.1
$Pt/RiO_2(Cs)$	$0.5\%Pt/TiO_2(0.34\%Cs)$	31	58	95	1.1
$Pt/TiO_2(Ca)$	$0.5\%Pt/TiO_2(2\%CaO)$	40	65	78	1.3
$Pt/TiO_2(Sr)$	$0.5\%Pt/TiO_2(1\%SrO)$	38	62	94	1.1
$(Pt–Ru)/TiO_2$	$(0.5\%Pt–0.1\%Ru)/TiO_2$	46	68	n.m. [4]	n.m.

Table 1. *Cont.*

Catalyst Notation	Nominal Composition (wt.%)	Specific Surface Area (SSA) [1] (m^2/g)	Anatase Content [2] (%)	Pt Dispersion [3] (%)	Pt Crystallite Size [3] (nm)
(Pt–Fe)/TiO_2	(0.5%Pt–0.5%Fe)/TiO_2	49	67	n.m.	n.m.
(Pt–Cr)/TiO_2	(0.5%Pt–0.5%Cr)/TiO_2	50	62	n.m.	n.m.
(Pt–Cu)/TiO_2	(0.5%Pt–0.5%Cu)/TiO_2	45	66	n.m.	n.m.
TiO_2-P	TiO_2 pellets	42	n.m.	n.m.	n.m.
0.5Pt/TiO_2(Ca)-P	0.5%Pt/TiO_2(2%CaO) pellets	37	n.m.	90	1.1
1.0Pt/TiO_2(Ca)-P	1.0%Pt/TiO_2(2%CaO) pellets	38	n.m.	92	1.1
Pt/TiO_2(Ca)-MM	0.5%Pt/TiO_2(2%CaO) coated on metallic monolith				
Pt/TiO_2(Ca)-CM	0.5%Pt/TiO_2(2%CaO) coated on ceramic monolith				

[1] Estimated with the BET method; [2] Anatase content of TiO_2 calculated from the XRD patterns; [3] Estimated from the H_2 chemisorption measurements; [4] Not measured.

2.1. Physicochemical Characteristics of the Synthesized Catalysts

The X-ray diffraction (XRD) patterns obtained for the representative powdered catalysts are shown in Figure 1. It was observed that, in all cases, the diffractograms consisted of peaks attributable to the anatase (JCPDS Card No. 4-477) and rutile (JCPDS Card No. 21-1276) phases of TiO_2. Qualitatively similar results were obtained for all of the catalyst samples investigated (not shown for brevity). The specific surface area (SSA) and anatase content of TiO_2 (x_A) were considerably lower for the samples promoted with Na or Cs compared to the unpromoted Pt/TiO_2 (Table 1). This can be attributed to the synthesis method employed, which involved the heat treatment of the corresponding supports at 600 °C. This treatment is known to result in the partial transformation of the anatase phase of TiO_2 to rutile and is accompanied by particle growth and a decrease in the SSA [19]. In contrast, the SSA and anatase content were partially retained for the alkaline earth (Ca, Sr)-promoted catalysts (Table 1). This is in agreement with the results of our previous studies, which showed that the incorporation of foreign cations in the crystal matrix of TiO_2 resulted in an increase in the anatase-to-rutile phase transition temperature and in the stabilization of the specific surface area of the host material [19,21]. Regarding the bimetallic (Pt–M)/TiO_2 (M = Ru, Fe, Cr, Cu) catalysts, it was observed that their SSA and anatase content were not significantly different compared to the unpromoted Pt/TiO_2 sample (Table 1).

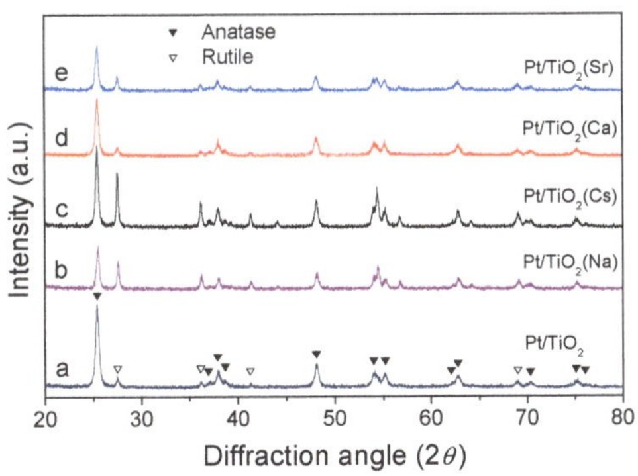

Figure 1. X-ray diffraction patterns of the indicated catalysts. (**a**) Pt/TiO_2; (**b**) Pt/TiO_2(Na); (**c**) Pt/TiO_2(Cs); (**d**) Pt/TiO_2(Ca); (**e**) Pt/TiO_2(Sr).

The average crystallite size of the TiO_2 particles, estimated using the Scherrer equation, did not vary appreciably from one catalyst to another, taking values in the ranges of 20–26 nm for anatase and 25–35 nm for rutile for all of the samples investigated. Therefore, the observed differences in the performance of these samples should not be attributed to this parameter, which is known to affect the WGS activity of TiO_2-supported Pt catalysts [7].

The Pt dispersion and average crystallite size were estimated for the monometallic catalysts from the H_2 chemisorption measurements and the results obtained are listed in Table 1. It should be noted that this method could not be applied for the bimetallic (Pt–M) samples because hydrogen interacts with both metals. It was observed that Pt was very well-dispersed on all supports studied, forming crystallites with an average size in the range 1.1–1.3 nm.

Representative TEM images obtained for the Pt/TiO_2 and Pt/TiO_2(Ca) catalysts are shown in Figure 2. It was observed that platinum was very well-dispersed on the surface of both the TiO_2 (Figure 2a) and TiO_2(Ca) (Figure 2b) supports. The size of the Pt crystallites was in the range of 1–2 nm, in very good agreement with the results of the H_2-chemisorption measurements (Table 1). The SEM images with EDS mapping of the above two catalysts are shown in the Supplementary Materials (Figures S1 and S3). It was observed that Pt was homogeneously distributed on the surface of both catalysts (Figures S1 and S3), and the same was true for the Ca present in the TiO_2(Ca) support (Figure S2).

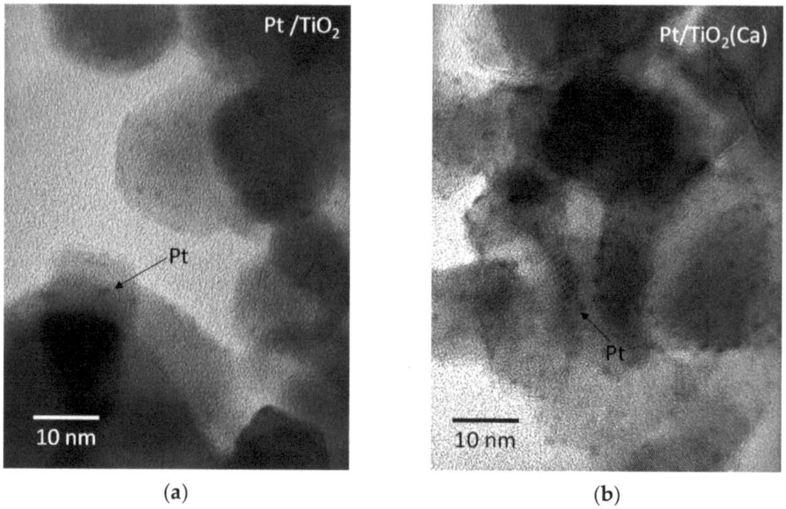

Figure 2. The TEM images of the (**a**) Pt/TiO_2 and (**b**) Pt/TiO_2(Ca) powdered catalysts.

2.2. *Performance of Particulate Catalysts*

2.2.1. Alkali- and Alkaline Earth Metal-Promoted Catalysts

The WGS activity of the synthesized alkali- and alkaline earth-promoted Pt/TiO_2 catalysts was investigated using two different feed compositions, which are representative of the high temperature shift (HTS) and the low temperature shift (LTS) reaction conditions. Figure 3a shows the CO conversion curves obtained under HTS conditions. It was observed that the unpromoted Pt/TiO_2 catalyst became active at temperatures around 250 °C. Conversion of CO (X_{CO}) progressively increased with the increase in the temperature and reached equilibrium at ca. 450 °C. Promotion of the TiO_2 support with small amounts of alkali metals resulted in a significant enhancement in the WGS activity, which was evidenced by the shift of the CO conversion curves obtained for the Pt/TiO_2(Na) and Pt/TiO_2(Cs) catalysts toward lower temperatures by ca. 40 °C and 50 °C, respectively, compared to the unpromoted sample. The performance of the two alkaline earth metal-promoted catalysts was even better than that of the alkali-promoted samples. Optimal results were obtained for Pt/TiO_2(Ca), which achieved an equilibrium CO conversion at ca. 390 °C (Figure 3a).

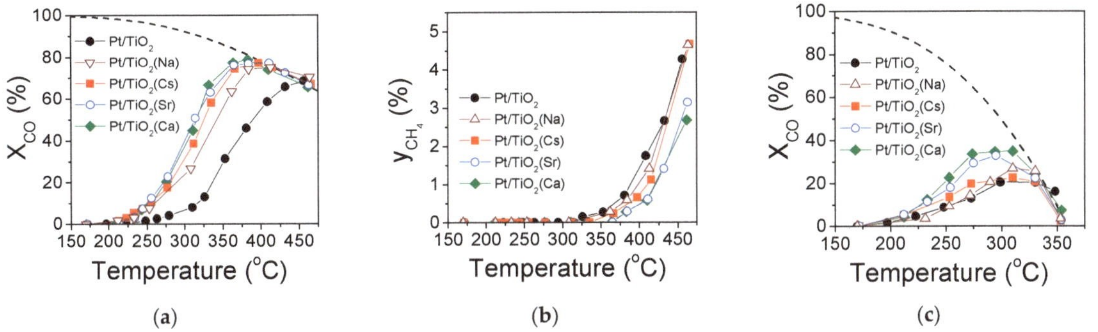

Figure 3. Performance of the alkali- and alkaline earth metal-promoted catalysts: Effect of reaction temperature on (**a**) the conversion of CO and (**b**) the yield of CH_4 under HTS conditions (feed composition: 9.7%CO, 38.7%H_2O, 44.8%H_2, 6.8%CO_2). (**c**) Effect of reaction temperature on the conversion of CO under LTS conditions (feed composition: 1.6%CO, 29.9%H_2O, 52.2%H_2, 16.3%CO_2). Mass of the catalyst: 750 mg; total flow rate: 220 cm^3 min^{-1}. The equilibrium CO conversion curves are shown with dashed lines.

Under the HTS conditions employed in Figure 3a, all samples catalyzed the CO/CO_2 methanation reactions at elevated temperatures:

$$CO + 3H_2 \rightarrow CH_4 + H_2O \ \Delta H^{o}_{298\ K} = -206 \text{ kJ mol}^{-1} \tag{2}$$

$$CO_2 + 4H_2 \rightarrow CH_4 + 2H_2O \ \Delta H^{o}_{298\ K} = -165 \text{ kJ mol}^{-1} \tag{3}$$

These reactions are unwanted because they reduce the yield of H_2 and may result in large exotherms. As shown in Figure 3b, methane formation started at ca. 320 °C and the yield of methane (y_{CH4}) increased considerably with the increase in the reaction temperature. Interestingly, the best performing HTS catalysts, namely, Pt/TiO_2(Ca) and Pt/TiO_2(Sr), produced lower amounts of methane compared to the unmodified Pt/TiO_2 and the alkali-promoted samples. It should be noted that the occurrence of the CO methanation reaction in parallel to the WGS reaction can explain the observation that X_{CO} slightly exceeded the equilibrium CO conversion at high temperatures under these conditions (Figure 3a).

Results of the catalytic performance tests obtained under LTS conditions are shown in Figure 3c. It was observed that catalyst ranking follows the same trend with that of the HTS experiments, with Pt/TiO_2(Ca) exhibiting the highest activity. It should be noted that under the LTS conditions, only negligible amounts of CH_4 were detected at the reactor effluent in the temperature range investigated.

In our previous studies [18–20], the enhanced WGS activity of the alkali/alkaline earth-promoted Pt/TiO_2 catalysts was attributed to the creation of $Pt - \square_s - Ti^{3+}$ active sites located at the metal–support interface. The population of this type of site is expected to increase under conditions that favor the decoration of the metal crystallites by TiO_x suboxides [22], in accordance with the SMSI effect [8,18,20,23–26]. It should be noted, however, that the SMSI effect is not expected to be significant under the conditions employed here. This is because the migration of suboxide species on the surface of dispersed metal crystallites is known to become important following reduction at temperatures typically higher than 500 °C and is not operable in the presence of water [23,24]. The promotion of TiO_2 by alkali/alkaline earth metals induces the same effect as SMSI as it also leads to the creation of $\square_s - Ti^{3+}$ defects. Moreover, this is undertaken in a controlled manner, depending on the dopant content, and is permanent. Therefore, it can be viewed as a "permanent" SMSI effect [18,20].

2.2.2. Bimetallic Catalysts

An attempt was made to enhance the WGS activity of the 0.5%Pt/TiO_2 catalyst by dispersing a second metal (M = Ru, Fe, Cr, Cu) on the support, in addition to Pt. Results obtained for the HTS reaction over the (0.5%Pt–0.5%M)/TiO_2 catalysts are shown in Figure 4a. It was observed that the addition of Cu, which is a component of commercial LTS catalysts [3], resulted in a significant shift in the CO conversion curve toward higher temperatures compared with the monometallic Pt/TiO_2 sample. On the other hand, the addition of Fe or Cr, which are components of commercial HTS catalysts [3], has a favorable effect as the X_{CO} curves shift toward lower temperatures by ca. 10–20 °C. A similar and more pronounced beneficial effect was observed following the addition of Ru. However, the (Pt–Ru)/TiO_2 catalyst promotes the unwanted CO/CO_2 methanation reactions at elevated temperatures (Figure 4b), rendering it unsuitable for practical applications. This observation is in agreement with the results of our previous study, which showed that the CO/CO_2 methanation activity of Ru was higher than that of Pt, and was not significantly affected by the presence of water in the feed [27]. Regarding the activity of (Pt–M)/TiO_2 catalysts under LTS conditions, all samples were less active compared to the monometallic Pt/TiO_2 (Figure 4c).

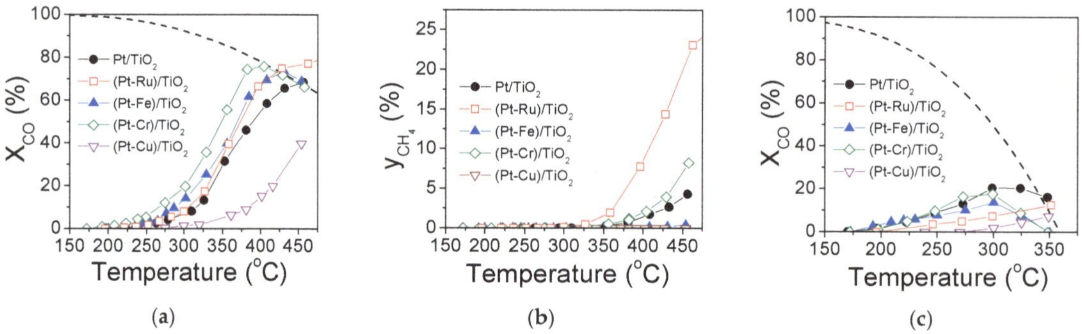

Figure 4. Performance of the bimetallic catalysts: Effect of reaction temperature on (**a**) the conversion of CO and (**b**) the yield of CH_4 under HTS conditions. (**c**) Effect of reaction temperature on the conversion of CO under LTS conditions. Experimental conditions are the same as in Figure 3.

Increasing the Fe or Cr loading of the bimetallic catalysts resulted in decreased performances under both the HTS and LTS conditions. Typical results obtained over the (0.5%Pt–M)/TiO_2 catalysts are shown in the Supplementary Materials (Figure S4). It was observed that increasing the Fe or Cr content from 0.5% to 5.0% resulted in a shift in the CO conversion curves toward higher temperatures, which was more pronounced for (Pt–Cr)/TiO_2. The yield of methane did not vary appreciably with increasing the Fe or Cr content from 0.5 to 5.0 wt.%. Furthermore, it was found that the performance of bimetallic Pt–M catalysts dispersed on the optimal TiO_2(Ca) support does not result in materials with improved WGS activity, compared to the monometallic Pt/TiO_2(Ca) sample (results not shown for brevity).

2.3. Performance of Structured Catalysts

Results presented in Figures 3 and 4 show that, among the alkali/alkaline earth metal-promoted Pt/TiO_2(X) catalysts and the bimetallic (Pt–M)/TiO_2 catalysts investigated, 0.5%Pt/TiO_2(Ca) exhibited the best performance under both the HTS and LTS reaction conditions. Therefore, this material was selected to prepare structured catalysts in the form of pellets or coatings on ceramic and metallic monoliths to investigate its performance for the WGS reaction under conditions that are more relevant to practical applications. The performance of the above structured samples was compared with that of a commercial catalyst in pellet form obtained from Johnson Matthey (JM) (Product name: W21, 10876, UK).

Results obtained over the structured catalysts with the use of the feed composition corresponding to the HTS conditions are presented in Figure 5a, where the conversion of CO is plotted as a function of the reaction temperature (measured at the exit of the catalyst bed). Regarding the 0.5%Pt/TiO_2(Ca)-P catalyst (i.e., in pellet form), it was observed that X_{CO} became measurable at temperatures around 200 °C and reached equilibrium at ca. 400 °C. Comparison with the corresponding results obtained over the powdered Pt/TiO_2(Ca) sample (Figure 3a) showed that the present CO conversion curve shifted toward lower temperatures (by ca. 50 °C). This was due to the larger amount of catalyst loaded in the reactor (1.8 g vs. 0.75 g) under the same experimental conditions. The performance of the Pt/TiO_2(Ca)-P catalyst was considerably improved with an increase in the Pt loading from 0.5% to 1.0% (Figure 5a). This is in agreement with the results of our previous studies that showed that the CO conversion curve over TiO_2-supported noble metal catalysts shifted monotonically toward lower temperatures with an increase in the metal loading from 0.1 to 5.0 wt.% [7,8]. It should be noted, however, that the specific activity (TOF) did not vary appreciably with metal loading, indicating that the WGS is a structure insensitive reaction [7,8]. It is of interest to note that the 1.0%Pt/TiO_2(Ca) pellet catalyst achieved equilibrium CO conversion at a temperature around 330 °C and was more active than the commercial JM catalyst across the whole temperature range investigated (Figure 5a).

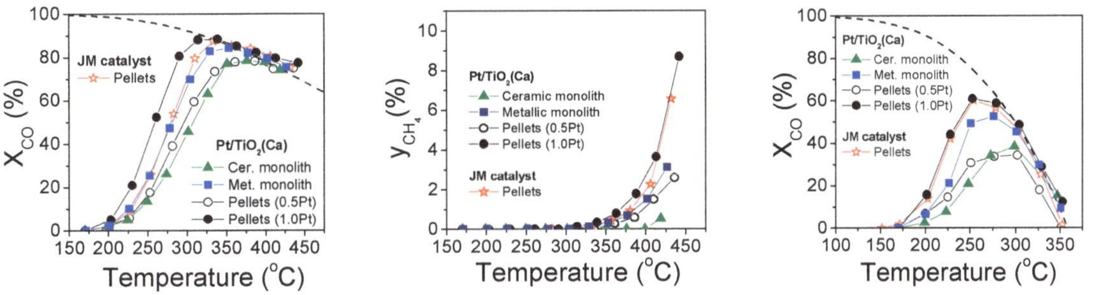

Figure 5. Performance of the structured Pt/TiO_2(Ca) catalysts in the form of pellets or coatings in the channels of ceramic and metallic monoliths in comparison with that of a commercial catalyst (Johnson Matthey, JM): Effect of reaction temperature on (**a**) the conversion of CO and (**b**) the yield of CH_4 under HTS conditions. (**c**) Effect of the reaction temperature on the conversion of CO under LTS conditions. Mass of the synthesized and JM catalyst pellets and the monolith coatings: 1.8 g. Other experimental conditions were the same as in Figure 3.

The CO conversion curve obtained for the ceramic monolith coated with 1.8 g of Pt/TiO_2(Ca) shifted toward slightly higher temperatures (ca. 10 °C) compared to the catalyst in the form of pellets (Figure 5a). On the other hand, the metallic monolith catalyst exhibited improved performance compared to the previous two samples, as the CO conversion reached equilibrium at ca. 350 °C. It is of interest to note that the performance of the latter sample was comparable to that of the commercial JM pellet catalyst. Under the experimental conditions in Figure 5a, all of the structured catalysts investigated produced methane at temperatures above ca. 320 °C (Figure 5b). The yield of methane followed the same trend as that of the WGS activity (i.e., it was lowest for Pt/TiO_2(Ca)-CM and highest for the 1.0Pt/TiO_2(Ca)-P and JM samples in the whole temperature range investigated) (Figure 5b).

The performance of the structured catalysts under the LTS conditions is shown in Figure 5c. It was observed that the best performing sample was 1.0%Pt/TiO_2(Ca)-P, which exhibited a similar performance with the commercial JM catalyst. Regarding the structured catalysts of the same Pt content (0.5 wt.%), the most active one was Pt/TiO_2(Ca)-MM. It should be noted that the blank experiments performed over the bare ceramic and metallic monoliths did not show measurable CO conversion under the HTS and LTS conditions employed here.

The effects of the total flow rate (space velocity) on the WGS performance of the best-performing Pt/TiO_2(Ca)-MM catalyst are shown in Figure 6. Results of similar experiments performed over the cordierite monolithic catalyst are presented in the Supplementary Materials (Figure S5). As expected, an increase in the flow rate from 220 to 400 cm^3 min^{-1} resulted in a shift in the CO conversion curves toward higher temperatures under both the HTS and LTS conditions. On the other hand, the X_{CO} curves shifted toward lower temperatures upon decreasing the total flow rate to 100 cm^3 min^{-1}. It is of interest to note that under these conditions and the use of the HTS feed, the Pt/TiO_2(Ca)-MM catalyst achieved equilibrium CO conversion at around 320 °C (Figure 6a), where the production of CH_4 is negligible (Figure 6b). The catalyst also performed well under LTS conditions, where equilibrium CO conversion was achieved at ca. 270 °C (Figure 6c).

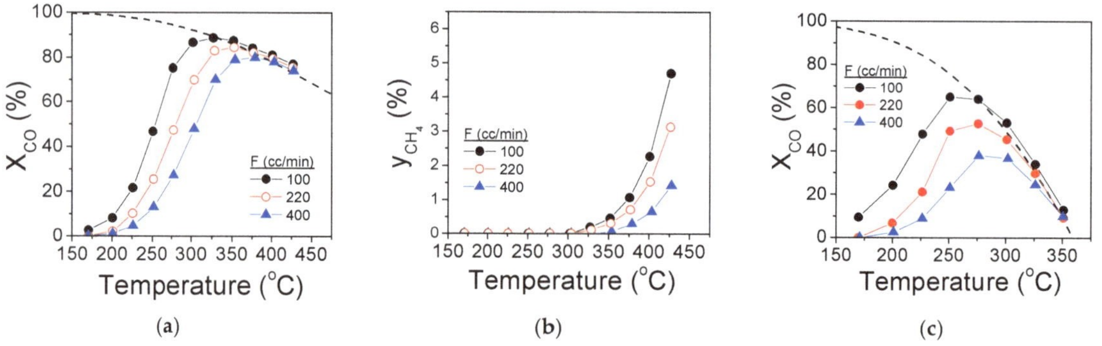

Figure 6. Effects of the total flow rate (*F*) on the performance of the Pt/TiO_2(Ca) catalyst coated on the channels of the metallic monolith: (**a**) Conversion of CO and (**b**) yield of CH_4 under HTS conditions. (**c**) Conversion of CO under LTS conditions. Other experimental conditions are the same as in Figure 5.

It should be noted that the temperature profile along the Pt/TiO_2(Ca)-MM catalyst was practically uniform under both the HTS and the LTS conditions employed here (see Supplementary Materials, Figure S6). In contrast, the temperature along the ceramic monolithic catalyst was found to increase toward the exit of the reactor (Figure S7). This was more pronounced under HTS conditions, elevated temperatures, and higher flow rates, where the CO conversion was also higher. This can be explained considering that metallic monoliths present high thermal conductivities and improved heat transfer characteristics compared to, for example, ceramic monoliths [15,28,29]. This advantage, combined with the easy fabrication and superior mechanical shock resistance of metallic monoliths, makes them suitable for practical applications.

The results presented in Figure 6a show that the Pt/TiO_2(Ca)-MM catalyst has the potential to be used in a single-step, medium temperature shift reactor operating in the 300–350 °C range to reduce the CO content in the reformatted gas to ca. 1%, which can be tolerated by preferential oxidation (PROX) catalysts [1]. Replacing the two HTS and LTS reactors by a single medium-temperature reactor is highly desirable, especially for small-scale and mobile applications where volume/weight and ruggedness become more important [30,31]. The WGS performance of this catalyst may be further improved by optimizing the catalyst loading, washcoat thickness, and cell density of the monolithic substrate as well as by using novel multi-shelled and core-shell nano/microstructured catalyst formulations [32,33], the study of which was beyond the scope of the present work and may be the subject of our future investigation.

3. Materials and Methods

3.1. Catalyst Preparation

3.1.1. Powdered Catalysts

Titanium dioxide supports promoted with small amounts of alkali or alkaline earth metals, denoted as TiO_2(X) (X = Na, Cs, Ca, Sr), were prepared following the procedures described in detail elsewhere [18–20]. In a typical synthesis, a pre-weighed amount of commercial TiO_2 powder (Degussa P25) is added under continuous stirring in an aqueous solution containing the appropriate amount of the corresponding promoter salt, namely, $NaNO_3$, $CsNO_3$, $Sr(NO_3)_2$, or CaO (Alfa Aesar), the latter diluted in HNO_3 at pH = 1. The resulting dispersion was heated at 70 °C and maintained at that temperature until nearly all the water had evaporated. The solid residue was dried overnight at 110 °C and then calcined in stagnant air at 600 °C for 5 h. The nominal contents (wt.%) of the promoters in the samples thus prepared were 0.06% for Na, 0.34% for Cs, 2.0% for CaO, and 1.0% for SrO (Table 1) and were selected based on their optimal concentrations derived in our previous studies [18–20].

Platinum catalysts dispersed on the pristine TiO_2 and the promoted TiO_2(X) supports were prepared via the wet impregnation method [7]. Briefly, a pre-weighed amount of the respective support was added under continuous stirring into an aqueous solution of $(NH_3)_2Pt(NO_2)_2$ (Alfa Aesar) followed by heating at 70 °C to evaporate water, drying at 110 °C overnight, and subsequent reduction under flowing H_2 at 300 °C for 2 h. The nominal Pt loading of the catalysts thus prepared was 0.5 wt.%.

The bimetallic (Pt–M)/TiO_2 catalysts (M = Fe, Cu, Cr, Ru) were synthesized by employing a two-step procedure. In the first step, the metal M (0.5 or 5.0 wt.%) was first deposited on the support following the impregnation of TiO_2 (P25) in an aqueous solution of the corresponding metal precursor salt, namely, ($Fe(NO_3)_3 \cdot 9H_2O$, $Cu(NO_3)_2 \cdot 3H_2O$, $Cr(NO_3)_3 \cdot 9H_2O$, or $Ru(NO)(NO_3)_3$ (Alfa Aesar). This was followed by the evaporation of water at 70 °C, drying overnight at 110 °C, and subsequent calcination in stagnant air at 300 °C for 3 h. Platinum (0.5 wt.%) was then added onto the resulting materials following the procedure described above.

3.1.2. Structured Catalysts

The best performing Pt/TiO_2(Ca) catalyst was prepared in the form of pellets as well as in the form of coating on ceramic and metallic monoliths (Figure 7). Two pellet catalyst samples with different Pt loadings (0.5 and 1.0 wt.%) were synthesized as follows: An appropriate amount of CaO was diluted in an aqueous solution of HNO_3 (pH = 1) in a BUCHI beaker followed by the addition of TiO_2 pellets (Aerolyst 7711, 1/16"). The beaker was adjusted in a Rotavapor to remove water under vacuum at 40 °C. The pellets were then dried at 110 °C overnight and subsequently calcined at 600 °C for 5 h in stagnant air. The CaO-promoted TiO_2 pellets thus obtained were added in an aqueous $(NH_3)_2Pt(NO_2)_2$ solution contained in a BUCHI beaker adjusted to a Rotavapor. After the removal of water under vacuum at 40 °C, the pellets were dried overnight at 110 °C and finally reduced under flowing H_2 at 300 °C for 2 h.

The commercial catalyst used as the reference was supplied by Johnson Matthey and was in the form of 1/16" pellets (Product name: 205 W21, 10876, UK). No detailed information is available regarding the composition of this catalyst except that it contains a mixture of Al_2O_3 ($\geq$90%) and $CeZrO_4$ ($\leq$10%).

The ceramic monolithic catalyst was prepared by coating the Pt/TiO_2(Ca) powder (see Section 3.1.1) in the channels of a cordierite monolith (400 cell per square inch, cpsi) of a cylindrical shape with a diameter of 2.4 cm and length of 3.6 cm. A hole was drilled at the center of the monolith (Figure 7b) to place a 1/16" quartz tube with a K-type thermocouple running throughout it to allow the temperature profile along the monolith to be monitored during the catalytic performance tests. The catalyst sample was first pulverized in a mortar and the resulting powder (d < 63 μm) was suspended under continuous stirring in triple-distilled water at a concentration of 0.33 g/mL. The monolith was then immersed in the catalyst suspension, blown with compressed air to remove the excess slurry left in the channels, dried at 110 °C overnight, and subsequently calcined at 300 °C for 3 h in stagnant air. The amount of catalyst loaded was determined by weighing the monolith before and after the catalyst deposition. This procedure (immersion, drying, calcination, weighing) was repeated until the desired amount of catalyst (1.8 g) was deposited in the channels of the monolith.

(a) (b) (c)

Figure 7. Photographs of the structured catalysts: (**a**) 1/16″ pellets; (**b**) ceramic monolith; (**c**) metallic monolith.

A metallic monolith (custom made by rolling a corrugated Fe-Cr-Al alloy film, 300 cpsi) with the same dimensions as the ceramic monolith (diameter: 2.4 cm; length: 3.6 cm) was coated with the $Pt/TiO_2(Ca)$ catalyst using the same technique. The total amount of catalyst deposited was 1.8 g.

3.2. Physicochemical Characterization

The specific surface areas (SSAs) of the synthesized catalysts were determined by employing the N_2 physisorption method at the temperature of liquid nitrogen (−196 °C) (BET method) on a Micromeritics (Gemini III 2375) instrument. Powder X-ray diffraction patterns were obtained in the 2θ range between 20° and 80° (scan rate 0.01° s^{-1}) on a Brucker D8 Advance (Cu Kα) apparatus operated at 40 kV and 40 mA, and were analyzed using JCPDS data files. The anatase content of TiO_2 in the catalyst samples was estimated from the following equation [7].

$$x_A = [1 + 1.26(I_R/I_A)]^{-1} \quad (4)$$

where I_A and I_R are the integral intensities of the anatase (101) and rutile (110) reflections, respectively.

Hydrogen chemisorption measurements were performed at 25 °C on a modified Fisons Instruments (Sorptomatic 1900) apparatus and the results were used to estimate the Pt dispersion and average crystallite size. Details of the experimental procedure and methods employed can be found elsewhere [7].

Transmission electron microscopy (TEM) images were recorded on a JEOL JEM-2100 apparatus (JEOL, Tokyo, Japan) operated at 200 kV (point resolution 0.23 nm) by means of an Erlangshen CCD Camera (Gatan Model 782 ES500W, Pleasanton, CA 94588, United States). SEM images were obtained using a JEOL 6300 scanning electron microscope (Akishima, Tokyo, Japan) equipped with an energy dispersive spectrometer (EDS, ISIS Link 300, Oxford Instruments, UK).

3.3. Catalytic Performance Tests

The experimental setup employed for the investigation of the WGS activity of the synthesized catalysts consisted of a flow system, a down-flow reactor, and an analysis

system. The flow system was equipped with mass flow controllers to adjust the flow of gases (He, CO, CO_2, H_2) and an HPLC pump (Marathon II) for feeding H_2O. Water was pumped into an evaporator operating at 170 °C and then mixed with the gas coming from the mass flow controllers. The resulting mixture was fed into the reactor through insulated stainless-steel tubing heated at 170 °C. The reactor consisted of a quartz tube (6 mm OD) with an expansion in its upper part (12 mm in the case of powdered samples; 20 mm in the case of pellets and 32 mm in the case of monoliths) where the catalyst sample was placed. The monoliths were wrapped with quartz wool to avoid by-pass flow. The oven temperature was controlled using a K-type thermocouple located between the oven walls and the reactor. The temperature profile at the catalyst bed was measured with a mobile K-type thermocouple placed in a thin quartz tube (1/16") running through the reactor. The reactor effluent passed through a condenser to trap unreacted H_2O and was then introduced to the analysis system, which consisted of a gas chromatograph (Shimadzu GC 14B) operating with He as the carrier gas, equipped with two columns (Porapak Q and Carbosieve) and two detectors (TCD and FID). Porapak Q was used for the separation of CH_4, C_2H_4, C_2H_6, C_3H_6, and C_3H_8 whereas Carbosieve was used for the separation of CO, CO_2, and CH_4.

Catalytic performance tests were performed at near atmospheric pressure in the temperature range of 170–450 °C using two different feed compositions: one corresponding to the HTS conditions (9.7%CO, 38.7%H_2O, 44.8%H_2, 6.8%CO_2) and one corresponding to the LTS conditions (1.6%CO, 29.9%H_2O, 52.2%H_2, 16.3%CO_2), which are relevant to practical applications [1]. Unless otherwise indicated, the total flow rate was 220 cm^3 min^{-1}. The amount of catalyst used for the powdered samples (0.18 < *d* < 0.25 mm) was 750 mg, which corresponds to a space velocity of ca. 10,000 h^{-1}. For the structured samples, the amount of catalyst in the form of pellets or coating on the monoliths was 1.8 g (W/F= 0.5 g s cm^{-3}). In a typical experiment, the catalyst was heated under He flow at 300 °C where the sample was in situ reduced under H_2 flow (60 cm^3 min^{-1}) for 1 h. The temperature was then increased to 450 °C under He flow and the feed was switched to the reaction mixture. Measurements were recorded after conditioning of the catalyst for 1 h-on-stream. The reaction temperature was then stepwise decreased under the flowing reaction mixture and similar measurements were obtained. The CO conversion (X_{CO}) was calculated by employing the following equation:

$$X_{CO} = \frac{[CO]^{in} - [CO]^{out}}{[CO]^{in}} \times 100 \tag{5}$$

where $[CO]^{in}$ and $[CO]^{out}$ are the inlet and outlet concentrations of CO, respectively. The yield of methane (y_{CH4}), which was formed at elevated temperatures via the methanation of CO (Equation (2)) and/or CO_2 (Equation (3)), was estimated using the equation:

$$y_{CH4} = \frac{[CH_4]^{out}}{[CO]^{in} + [CO_2]^{in}} \times 100 \tag{6}$$

where $[CH_4]^{out}$ is the outlet concentration of CH_4, and $[CO]^{in}$ and $[CO_2]^{in}$ are the inlet concentrations of CO and CO_2, respectively. All values reported are the averages of at least three measurements and correspond to the temperature measured at the outlet of the catalyst bed.

4. Conclusions

The WGS performance of 0.5%Pt/TiO_2(X) (X = Na, Cs, Ca, Sr) and (0.5%Pt–M)/TiO_2 (M = Fe, Cu, Cr, Ru) catalysts in powder form was investigated under realistic reaction conditions. The best results were obtained for the 0.5%Pt/TiO_2(Ca) catalyst, which exhibited high activity and selectivity toward H_2 under both HTS and LTS reaction conditions. The catalytic performance of the optimized catalyst in pellet form was comparable to that of a commercial WGS pellet catalyst provided by JM and was further enhanced by increasing the Pt content from 0.5 to 1.0 wt.%. The results obtained using a ceramic and a metallic monolithic catalyst coated with 0.5%Pt/TiO_2(Ca) showed that the latter exhibited superior performance and operated under nearly isothermal conditions with the minimal production of undesired methane. The CO conversion curve could be shifted toward lower temperatures by decreasing the space velocity, demonstrating the feasibility of the metallic monolithic catalyst for small-scale and mobile applications and its potential use in a single-step, medium temperature shift reactor operating in the 300–350 °C range.

Supplementary Materials: The following supporting information can be downloaded at: https://www.mdpi.com/article/10.3390/catal13020372/s1, Figure S1: (a) SEM image of the as prepared Pt/TiO_2 catalyst, and EDS mapping results showing the distribution of (b) Pt, (c) O and (d) Ti elements; Figure S2: (a) SEM image of the as prepared TiO_2(Ca) support, and EDS mapping results showing the distribution of (b) Ca, (c) O and (d) Ti elements; Figure S3: (a) SEM image of the Pt/TiO_2(Ca) catalyst and (b) EDS mapping results showing the distribution of Pt element; Figure S4: Effects of the metal content on the performance of bimetallic (0.5%Pt-x%M)/TiO_2 catalysts (x = 0.5 or 5.0; M = Fe or Cr) under (A) HTS and (B) LTS conditions. Experimental conditions same as in Figure 3; Figure S5: Effects of total flow rate on the performance of Pt/TiO_2(Ca) catalysts coated on the channels of the ceramic monolith: (A) Conversion of CO and (B) yield of CH4 under HTS conditions. (C) Conversion of CO under LTS conditions. Other experimental conditions same as in Figure 5; Figure S6: Temperature profile along the Pt/TiO_2(Ca)-coated metallic monolith (inlet at L = 0) at the indicated total flow rates under the (a) HTS and (b) LTS reaction conditions shown in Figure 6; Figure S7: Temperature profile along the Pt/TiO_2(Ca)-coated ceramic monolith (inlet at L = 0) at the indicated total flow rates under the (a) HTS and (b) LTS reaction conditions shown in Figure 6.

Author Contributions: Conceptualization, A.K., G.B., P.P. and D.I.K.; Methodology, A.K., G.B. and D.I.K.; Software, A.K. and G.B.; Validation, A.K., G.B., P.P. and D.I.K.; Investigation, A.K. and G.B.; Writing—original draft preparation, A.K., G.B., P.P. and D.I.K.; Writing—review and editing, G.B. and D.I.K.; Supervision, D.I.K. All authors have read and agreed to the published version of the manuscript.

Funding: This research was funded by the European Union and Greek national funds through the Operational Program Competitiveness, Entrepreneurship and Innovation, under the call RESEARCH-CREATE-INNOVATE (Project acronym: "Eco-Bio-H2-FCs", project code: MIS 5074538/T2EΔK 00955).

Data Availability Statement: Not applicable.

Conflicts of Interest: The authors declare no conflict of interest.

References

1. Ratnasamy, C.; Wagner, J.P. Water Gas Shift Catalysis. *Catal. Rev.* **2009**, *51*, 325–440. [CrossRef]
2. Smith, R.J.B.; Loganathan, M.; Shantha, M.S. A Review of the Water Gas Shift Reaction Kinetics. *Int. J. Chem. React. Eng.* **2010**, *8*, Review R4. [CrossRef]
3. Pal, D.; Chand, R.; Upadhyay, S.; Mishra, P. Performance of water gas shift reaction catalysts: A review. *Renew. Sustain. Energy Rev.* **2018**, *93*, 549–565. [CrossRef]
4. Chen, W.-H.; Chen, C.-Y. Water gas shift reaction for hydrogen production and carbon dioxide capture: A review. *Appl. Energy* **2020**, *258*, 114078. [CrossRef]
5. Baraj, E.; Ciahotný, K.; Hlinčík, T. The water gas shift reaction: Catalysts and reaction mechanism. *Fuel* **2021**, *288*, 119817. [CrossRef]
6. Farrauto, R.J.; Liu, Y.; Ruettinger, W.; Ilinich, O.; Shore, L.; Giroux, T. Precious Metal Catalysts Supported on Ceramic and Metal Monolithic Structures for the Hydrogen Economy. *Catal. Rev.* **2007**, *49*, 141–196. [CrossRef]
7. Panagiotopoulou, P. Effect of morphological characteristics of TiO_2-supported noble metal catalysts on their activity for the water-gas shift reaction. *J. Catal.* **2004**, *225*, 327–336. [CrossRef]
8. Panagiotopoulou, P.; Kondarides, D.I. Effect of the nature of the support on the catalytic performance of noble metal catalysts for the water–gas shift reaction. *Catal. Today* **2006**, *112*, 49–52. [CrossRef]
9. Ebrahimi, P.; Kumar, A.; Khraisheh, M. A review of recent advances in water-gas shift catalysis for hydrogen production. *Emergent Mater.* **2020**, *3*, 881–917. [CrossRef]
10. Giroux, T.; Hwang, S.; Liu, Y.; Ruettinger, W.; Shore, L. Monolithic structures as alternatives to particulate catalysts for the reforming of hydrocarbons for hydrogen generation. *Appl. Catal. B Environ.* **2005**, *56*, 95–110. [CrossRef]

11. Du, X.; Gao, D.; Yuan, Z.; Liu, N.; Zhang, C.; Wang, S. Monolithic $Pt/Ce_{0.8}Zr_{0.2}O_2$/cordierite catalysts for low temperature water gas shift reaction in the real reformate. *Int. J. Hydrogen Energy* **2008**, *33*, 3710–3718. [CrossRef]
12. Marín, P.; Ordóñez, S.; Díez, F.V. Performance of reverse flow monolithic reactor for water–gas shift reaction. *Catal. Today* **2009**, *147*, S185–S190. [CrossRef]
13. Özyönüm, G.N.; Yildirim, R. Water gas shift activity of Au–Re catalyst over microstructured cordierite monolith wash-coated by ceria. *Int. J. Hydrogen Energy* **2016**, *41*, 5513–5521. [CrossRef]
14. González-Castaño, M.; Ivanova, S.; Laguna, O.H.; Martínez, T.L.M.; Centeno, M.; Odriozola, J. Structuring $Pt/CeO_2/Al_2O_3$ WGS catalyst: Introduction of buffer layer. *Appl. Catal. B Environ.* **2017**, *200*, 420–427. [CrossRef]
15. García-Moncada, N.; Groppi, G.; Beretta, A.; Romero-Sarria, F.; Odriozola, J.A. Metal Micro-Monoliths for the Kinetic Study and the Intensification of the Water Gas Shift Reaction. *Catalysts* **2018**, *8*, 594. [CrossRef]
16. Palma, V.; Ruocco, C.; Cortese, M.; Martino, M. Recent Advances in Structured Catalysts Preparation and Use in Water-Gas Shift Reaction. *Catalysts* **2019**, *9*, 991. [CrossRef]
17. Portela, R.; Wolf, P.; Marinkovic, J.M.; Serrano-Lotina, A.; Riisager, A.; Haumann, M. Tailored monolith supports for improved ultra-low temperature water-gas shift reaction. *React. Chem. Eng.* **2021**, *6*, 2114–2124. [CrossRef]
18. Panagiotopoulou, P.; Kondarides, D.I. Effects of alkali promotion of TiO_2 on the chemisorptive properties and water–gas shift activity of supported noble metal catalysts. *J. Catal.* **2009**, *267*, 57–66. [CrossRef]
19. Panagiotopoulou, P.; Kondarides, D.I. Effects of promotion of TiO_2 with alkaline earth metals on the chemisorptive properties and water–gas shift activity of supported platinum catalysts. *Appl. Catal. B Environ.* **2011**, *101*, 738–746. [CrossRef]
20. Panagiotopoulou, P.; Kondarides, D.I. Effects of alkali additives on the physicochemical characteristics and chemisorptive properties of Pt/TiO_2 catalysts. *J. Catal.* **2008**, *260*, 141–149. [CrossRef]
21. I Halkides, T.; I Kondarides, D.; E Verykios, X. Catalytic reduction of NO by C_3H_6 over Rh/TiO_2 catalysts: Effect of W6+-cation doping of TiO_2 on morphological characteristics and catalytic performance. *Appl. Catal. B Environ.* **2003**, *41*, 415–426. [CrossRef]
22. Alexeev, O.S.; Chin, S.Y.; Engelhard, M.H.; Ortiz-Soto, L.; Amiridis, M.D. Effects of Reduction Temperature and Metal—Support Interactions on the Catalytic Activity of Pt/γ-Al_2O_3 and Pt/TiO_2 for the Oxidation of CO in the Presence and Absence of H_2. *J. Phys. Chem. B* **2005**, *109*, 23430–23443. [CrossRef] [PubMed]
23. Tauster, S.J.; Fung, S.C.; Garten, R.L. Strong metal-support interactions. Group 8 noble metals supported on titanium dioxide. *J. Am. Chem. Soc.* **1978**, *100*, 170–175. [CrossRef]
24. Haller, G.L.; Resasco, D.E. Metal–Support Interaction: Group VIII Metals and Reducible Oxides. *Adv. Catal.* **1989**, *36*, 173–235. [CrossRef]
25. Huang, R.; Kwon, O.; Lin, C.; Gorte, R.J. The effects of SMSI on m-Cresol hydrodeoxygenation over Pt/Nb_2O_5 and Pt/TiO_2. *J. Catal.* **2021**, *398*, 102–108. [CrossRef]
26. He, K.; Wang, Q. Activation of Pt Nanoclusters on TiO_2 via Tuning the Metallic Sites to Promote Low-Temperature CO Oxidation. *Catalysts* **2021**, *11*, 1280. [CrossRef]
27. Panagiotopoulou, P.; Kondarides, D.; Verykios, X.E. Selective methanation of CO over supported noble metal catalysts: Effects of the nature of the metallic phase on catalytic performance. *Appl. Catal. A Gen.* **2008**, *344*, 45–54. [CrossRef]
28. Cybulski, A.; Moulijn, J.A. Modelling of heat transfer in metallic monoliths consisting of sinusoidal cells. *Chem. Eng. Sci.* **1994**, *49*, 19–27. [CrossRef]
29. Tronconi, E.; Groppi, G. A study on the thermal behavior of structured plate-type catalysts with metallic supports for gas/solid exothermic reactions. *Chem. Eng. Sci.* **2000**, *55*, 6021–6036. [CrossRef]
30. Van Dijk, H.; Boon, J.; Nyqvist, R.; Brink, R.V.D. Development of a single stage heat integrated water-gas shift reactor for fuel processing. *Chem. Eng. J.* **2010**, *159*, 182–189. [CrossRef]
31. Palma, V.; Pisano, D.; Martino, M. Comparative Study Between Aluminum Monolith and Foam as Carriers for The Intensification of The CO Water Gas Shift Process. *Catalysts* **2018**, *8*, 489. [CrossRef]
32. Wang, H.; Mao, D.; Qi, J.; Zhang, Q.; Ma, X.; Song, S.; Gu, L.; Yu, R.; Wang, D. Hollow Multishelled Structure of Heterogeneous Co_3O_4–CeO_{2-x} Nanocomposite for CO Catalytic Oxidation. *Adv. Funct. Mater.* **2019**, *29*, 1806588. [CrossRef]
33. Wang, Z.; Qi, J.; Yang, N.; Yu, R.; Wang, D. Core–shell nano/microstructures for heterogeneous tandem catalysis. *Mater. Chem. Front.* **2021**, *5*, 1126–1139. [CrossRef]

MDPI AG
Grosspeteranlage 5
4052 Basel
Switzerland
Tel.: +41 61 683 77 34

Catalysts Editorial Office
E-mail: catalysts@mdpi.com
www.mdpi.com/journal/catalysts